T0212779

Lecture Notes in Computer Science 9259

Commenced Publication in 1973
Founding and Former Series Editors:
Gerhard Goos, Juris Hartmanis, and Jan van Leeuwen

More information about this series at http://www.springer.com/series/7407

Javier Campos · Boudewijn R. Haverkort (Eds.)

Quantitative Evaluation of Systems

12th International Conference, QEST 2015
Madrid, Spain, September 1–3, 2015
Proceedings

Editors
Javier Campos
Universidad de Zaragoza
Zaragoza
Spain

Boudewijn R. Haverkort
University of Twente
Enschede
The Netherlands

ISSN 0302-9743   ISSN 1611-3349   (electronic)
Lecture Notes in Computer Science
ISBN 978-3-319-22263-9   ISBN 978-3-319-22264-6   (eBook)
DOI 10.1007/978-3-319-22264-6

Library of Congress Control Number: 2015944718

LNCS Sublibrary: SL1 – Theoretical Computer Science and General Issues

Springer Cham Heidelberg New York Dordrecht London

Printed on acid-free paper

Springer International Publishing AG Switzerland is part of Springer Science+Business Media
(www.springer.com)

Preface

Welcome to the proceedings of QEST 2015, the 12th International Conference on Quantitative Evaluation of Systems. QEST is a leading forum on quantitative evaluation and verification of computer systems and networks, through stochastic models and measurements. QEST was first held in Enschede, The Netherlands (2004), followed by meetings in Turin, Italy (2005), Riverside, USA (2006), Edinburgh, UK (2007), St. Malo, France (2008), Budapest, Hungary (2009), Williamsburg, USA (2010), Aachen, Germany (2011), London, UK (2012), Buenos Aires, Argentina (2013), and, most recently, Florence, Italy (2014).

This year's QEST was held in Madrid, Spain, and collocated with the 26th Conference on Concurrency Theory (CONCUR 2015), the 13th International Conference on Formal Modeling and Analysis of Timed Systems (FORMATS 2015), the 12th European Workshop on Performance Engineering (EPEW 2015), the 10th International Symposium on Trustworthy Global Computing (TGC 2015), the International Symposium on Web Services, Formal Methods and Behavioural Types (WS-FM/BEAT 2015), the Combined 22nd International Workshop on Expressiveness in Concurrency and 12th Workshop on Structured Operational Semantics (EXPRESS/SOS 2015), the Second International Workshop on Parameterized Verification (PV 2015), the 14th International Workshop on Foundations of Coordination Languages and Self-Adaptation (FOCLASA 2015), the 4th IFIP WG 1.8 Workshop on Trends in Concurrency Theory (TRENDS 2015), and the 4th International Workshop of Hybrid Systems and Biology (HSB 2015). Together these conferences and workshops formed MADRID MEET 2015, a one-week scientific event in the areas of formal and quantitative analysis of systems, performance engineering, computer safety, and industrial critical applications.

As one of the premier fora for research on quantitative system evaluation and verification of computer systems and networks, QEST covers topics including classic measures involving performance and reliability, as well as quantification of properties that are classically qualitative, such as safety, correctness, and security. QEST welcomes measurement-based studies as well as analytic studies, diversity in the model formalisms and methodologies employed, as well as development of new formalisms and methodologies. QEST also has a tradition in presenting case studies, highlighting the role of quantitative evaluation in the design of systems, where the notion of system is broad. Systems of interest include computer hardware and software architectures, communication systems, embedded systems, infrastructure systems, and biological systems. Moreover, tools for supporting the practical application of research results in all of the aforementioned areas are also of interest to QEST. In short, QEST aims to encourage all aspects of work centered around creating a sound methodological basis for assessing and designing systems using quantitative means.

The Program Committee (PC) consisted of 35 experts and we received a total of 42 submissions. Each submission was reviewed by four reviewers, either PC members or external reviewers. The review process included a rebuttal phase after the first notification to authors and before the PC discussion. In the end, 17 full papers and two tool demonstration papers were selected for the conference program. The program was greatly enriched with the QEST keynote talk of Boris Köpf (IMDEA Software Institute, Spain) and the joint keynote talk with FORMATS 2015 of Jozef Hooman (RU Nijmegen and TNO-ESI, The Netherlands). We believe the overall result is a high-quality conference program of interest to QEST 2015 attendees and other researchers in the field.

We would like to thank a number of people. Firstly, all the authors who submitted papers, as without them there simply would not be a conference. In addition, we would like to thank the PC members and the additional reviewers for their hard work and for sharing their valued expertise with the rest of the community as well as EasyChair for supporting the electronic submission and reviewing process. We are also indebted to Alfred Hofmann and Anna Kramer for their help in the preparation of this volume. Also thanks to the MADRID MEET 2015 Web Manager Gustavo Santos, Local Organization Chair Ismael Rodríguez, and General Chair David de Frutos-Escrig for their dedication and excellent work. In addition, we thank the Facultad de Ciencias Matemáticas of the Universidad Complutense de Madrid for providing the venue location. Furthermore, we gratefully acknowledge the financial support of the Spanish Ministerio de Economía y Competitividad. Finally, we would like to thank Joost-Pieter Katoen, chair of the QEST Steering Committee, for his guidance throughout the past year.

We hope that you find the conference proceedings rewarding and will consider submitting papers to QEST 2016 in Quebec, Canada.

September 2015   Javier Campos
                                                                    Boudewijn R. Haverkort

Organization

General Chair

David de Frutos-Escrig   Universidad Complutense de Madrid, Spain

Program Committee Co-chairs

Javier Campos   University of Zaragoza, Spain
Boudewijn Haverkort   University of Twente, The Netherlands

Steering Committee

Alessandro Abate   University of Oxford, UK
Nathalie Bertrand   Inria Rennes, France
Luca Bortolussi   University of Trieste, Italy
Pedro R. D'Argenio   Universidad Nacional de Córdoba, Argentina
Holger Hermanns   Saarland University, Germany
Jane Hillston   University of Edinburgh, UK
Joost-Pieter Katoen   RWTH Aachen University, Germany
Peter Kemper   College of William and Mary, USA
William Knottenbelt   Imperial College London, UK
Anne Remke   University of Münster, Germany
Enrico Vicario   University of Florence, Italy

Program Committee

Alessandro Abate   University of Oxford, UK
Erika Abraham   RWTH Aachen University, Germany
Gul Agha   University of Illinois, USA
Simona Bernardi   Centro Universitario de la Defensa, Spain
Nathalie Bertrand   Inria Rennes, France
Luca Bortolussi   University of Trieste, Italy
Javier Campos   University of Zaragoza, Spain
Josée Desharnais   Laval University, Canada
Paulo Fernandes   PUCRS, Brazil
Martin Fränzle   University of Oldenburg, Germany
Boudewijn Haverkort   University of Twente, The Netherlands
Jane Hillston   University of Edinburgh, UK
Mohamed Kaâniche   LAAS-CNRS, France
William Knottenbelt   Imperial College London, UK
Jan Krcál   Saarland University, Germany

Boris Köpf	IMDEA Software Institute, Spain
Jan Madsen	Technical University of Denmark, Denmark
Annabelle McIver	Macquarie University, Australia
Raffaela Mirandola	Politecnico di Milano, Italy
Sayan Mitra	University of Illinois, USA
Gethin Norman	University of Glasgow, UK
Catuscia Palamidessi	Inria Saclay and LIX, France
David Parker	University of Birmingham, UK
Anne Remke	University of Münster, Germany
Eric Rozier	University of Cincinnati, USA
Gerardo Rubino	Inria Rennes, France
Markus Siegle	Bundeswehr University Munich, Germany
Evgenia Smirni	College of William and Mary, USA
Jeremy Sproston	University of Turin, Italy
Miklós Telek	Technical University of Budapest, Hungary
Mirco Tribastone	University of Southampton, UK
Valentín Valero	University of Castilla-La Mancha, Spain
Benny Van Houdt	University of Antwerp, The Netherlands
Enrico Vicario	University of Florence, Italy
Lijun Zhang	ISCAS Beijing, China

Additional Reviewers

Dieky Adzkiya	Alexander Gouberman	Andras Meszaros
Luedtke Andreas	Marco Gribaudo	Rasha Osman
Alexander Andreychenko	Dennis Guck	Marco Paolieri
Marco Beccuti	Ernst Moritz Hahn	Diego Perez-Palacin
Sergiy Bogomolov	Illes Horvath	Juan F. Perez
Yuliya Butkova	Gábor Horváth	Ajitha Rajan
Eckard Böde	Zhenqi Huang	Roberta Sirovich
Laura Carnevali	Nils Jansen	Lei Song
Fernando Cuartero	Joachim Klein	Petr Sosik
Frits Dannenberg	Lubos Korenciak	Bharath Siva Kumar Tati
Yuxin Deng	Jan Kretinsky	Andrea Turrini
Gregorio Díaz	Youngmin Kwon	Rafael Brundo Uriarte
Desmond Elliott	Michele Loreti	Sergey Verlan
Luis María Ferrer-Fioriti	Hermenegilda Macia	Steen Vester
Sibylle Froeschle	Istvan Majzik	Ralf Wimmer
Sicun Gao	Andrea Marin	Paolo Zuliani

Abstracts of Keynote Talks

Reasoning About the Trade-Off Between Security and Performance

Boris Köpf

IMDEA Software Institute

Introduction

Side-channel attacks break the security of systems by exploiting signals that are unwittingly emitted by the implementation. Examples of such signals are the consumption of power [6] and execution time [5]. Execution time is a particularly daunting signal because it can be measured and exploited from a long distance [2].

In theory, one can get rid of timing side channels by avoiding the use of secret-dependent control ow and by avoiding performance-enhancing features of the hardware architecture, such as caches and branch prediction units. However, this defensive approach comes at the price of a performance penalty, which can be unacceptably large. In practice, one is hence faced with the problem of identifying a sweet spot between performance and security against timing attacks.

Identifying this sweet spot poses a number of challenges: The first is to determine the degree of performance of a specific implementation; the second is to determine the degree of security of this implementation; the third is to come to a decision based on this information. Our work focusses on the second and third challenges; we rely on existing work for the first challenge.

Quantifying Security

We address the challenge of quantifying the security of a system by static analysis of the implementation. To this end, we are developing CacheAudit [4], a modular platform for the security analysis of programs, based on abstract interpretation. The security guarantees delivered by CacheAudit are quantitative upper bounds on the amount of information that an adversary can extract by measuring execution time, for different kinds of observational capabilities. The distinguishing feature of CacheAudit is that it bases the analysis on executable code and realistic models of CPU caches, which is crucial for reliably capturing timing leaks. We have applied CacheAudit to library implementations of symmetric cryptosystems and have formally analyzed the effect on side-channel leakage of different software countermeasures, cache configurations, and concurrency [1, 4].

Decision-Making

We address the challenge of decision-making by casting optimal defenses against timing attacks as equilibria in a two-stage (Stackelberg) game between an adversary and a defender. The adversary strives to maximize the probability of recovering a secret (such as a cryptographic key) by distributing bounded resources between timing measurements and brute-force search. The defender strives to minimize the cost of protection, while maintaining a certain security level. The defender's means to achieve this are choosing the size of the secret and the configuration of the countermeasure.

The core novelty in this approach is that we instantiate the bounds for the probability of the adversary recovering the secret with those obtained from the quantitative security analysis of the implementation, i.e. we derive the parameters of the game from the program semantics [3]. We have put our techniques to work for a library implementation of ElGamal. A highlight of our results is that we can formally justify the use of an aggressively tuned but (slightly) leaky implementation over a defensive constant-time implementation, for some parameter ranges.

The Talk

The talk will present the core ideas and most recent practical advances of this line of research, and will conclude with a number of open challenges.

References

1. Barthe, G., Köpf, B., Mauborgne, L., Ochoa, M.: Leakage resilience against concurrent cache attacks. In: Abadi, M., Kremer, S. (eds.) POST 2014. LNCS, vol. 8414, pp. 140–158. Springer, Heidelberg (2014)
2. Brumley, D., Boneh, D.: Remote timing attacks are practical. Comput. Netw. 48(5), 701–716 (2005)
3. Doychev, G., Köpf, B.: Rational protection against timing attacks. In: Proceedings of 28th IEEE Computer Security Foundations Symposium (CSF 2015). IEEE (2015)
4. Doychev, G., Köpf, B., Mauborgne, L., Reineke, J.: CacheAudit: a tool for the static analysis of cache side channels. ACM Trans. Inf. Syst. Secur. 18(1) (2015)
5. Kocher, P.C.: Timing attacks on implementations of Diffie-Hellman, RSA, DSS, and other systems. In: Koblitz, N. (ed.) Advances in Cryptology - CRYPTO 1996. LNCS, vol. 1109, pp. 104–113. Springer, Heidelberg (1996)
6. Kocher, P., Jaffe, J., Jun, B.: Differential power analysis. In: Wiener, M. (ed.) CRYPTO 1999. LNCS, vol. 1666, pp. 388–397. Springer, Heidelberg (1999)

Uniting Academic Achievements on Performance Analysis with Industrial Needs

Bart Theelen[1] and Jozef Hooman[1,2]

[1] Embedded Systems Innovation by TNO, Eindhoven, The Netherlands
{bart.theelen, jozef.hooman}@tno.nl
[2] Radboud University, Nijmegen, The Netherlands

Abstract. In our mission to advance innovation by industrial adoption of academic results, we perform many projects with high-tech industries. Favoring formal methods, we observe a gap between industrial needs in performance modeling and the analysis capabilities of formal methods for this goal. After clarifying this gap, we highlight some relevant deficiencies for state-of-the-art quantitative analysis techniques (focusing on model checking and simulation). As an ingredient to bridging the gap, we propose to unite domain-specific industrial contexts with academic performance approaches through Domain Specific Languages (DSLs). We illustrate our vision with examples from different high-tech industries and discuss lessons learned from the migration process of adopting it.

This work was supported by the ARTEMIS Joint Undertaking through the Crystal project on Critical System Engineering Acceleration and the Dutch program COMMIT through the Allegio project on Composable Embedded Systems for Healthcare.

Contents

Applications

Queueing Systems and Hybrid Systems

Keynote Presentation

Uniting Academic Achievements
on Performance Analysis with Industrial Needs

Bart Theelen[1] and Jozef Hooman[1,2]([✉])

[1] Embedded Systems Innovation by TNO, Eindhoven, The Netherlands
{bart.theelen,jozef.hooman}@tno.nl
[2] Radboud University, Nijmegen, The Netherlands

Abstract. In our mission to advance innovation by industrial adoption of academic results, we perform many projects with high-tech industries. Favoring formal methods, we observe a gap between industrial needs in performance modeling and the analysis capabilities of formal methods for this goal. After clarifying this gap, we highlight some relevant deficiencies for state-of-the-art quantitative analysis techniques (focusing on model checking and simulation). As an ingredient to bridging the gap, we propose to unite domain-specific industrial contexts with academic performance approaches through Domain Specific Languages (DSLs). We illustrate our vision with examples from different high-tech industries and discuss lessons learned from the migration process of adopting it.

Keywords: Performance modeling · Performance analysis · Quantitative analysis · Domain specific languages · Model checking · Simulation

1 Introduction

Quantitative qualities of high-tech (embedded or cyber-physical) systems are often key selling factors second to the ability to perform certain functionality. As opposed to functionality, quantitative qualities can often be balanced against each other, possibly across multiple technologies or engineering disciplines. One may for example realize a better performance by either changing the software algorithms that implement the functionality, by changing the configuration of the resources executing the algorithms or by some combination of both, where the trade-off can have an effect on other quantitative qualities such as physical size of the system. The diversity in possible design trade-offs that can be made for quantitative qualities has resulted in a plethora of academic results on how to design high-tech systems effectively and efficiently. Despite these achievements, which are commonly based on the use of abstract models to predict the qualities of a design proposal or even construct an appropriate design, high-tech industry

This work was supported by the ARTEMIS Joint Undertaking through the Crystal project on Critical System Engineering Acceleration and the Dutch program COM-MIT through the Allegio project on Composable Embedded Systems for Healthcare.

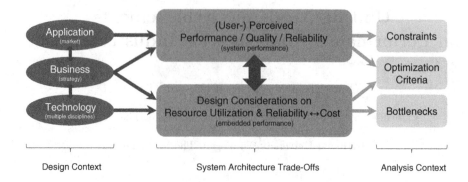

Fig. 1. Performance in an industrial context.

still relies mostly on traditional approaches leading to a reactive way of working in resolving unexpected issues during integration and test or even later.

Figure 1 illustrates that achieving appropriate quantitative qualities during industrial design processes requires system architects to make trade-offs subject to a complex design context. This design context is characterized by the application functionality demanded by (multiple) markets and the business strategy towards serving the diverse and sometimes conflicting market demands. The design context also covers combining technologies from multiple engineering disciplines that may be subject to business strategy-depend restrictions on availability or that may be developed as part of the business' innovation processes. A related aspect is that most designs in industry increment from a long-standing legacy for which the rationale behind certain trade-offs may no longer be known or valid.

In trading off design alternatives, the center part of Fig. 1 expresses that system architects have to consider the impact on the user-perceived system qualities (i.e., the qualities listed in a product catalogue for which users pay) and the qualities that are relevant from a business and design perspective (like cost and physical size). Examples of user-perceived qualities include the throughput of a printer system in terms of pages printed per minute or the quality of a picture on a TV display. Notice that such qualities are sometimes difficult to express in measurable objective metrics. An important difficulty to deal with is the unclear or complex relation between user-perceived system qualities and the measurable objective metrics used for evaluating design alternatives. How does the load of a processor for example impact the quality of the picture on the TV display?

Evaluating alternative designs involves exploiting analysis techniques for different kinds of quantitative qualities. The right-hand side of Fig. 1 categorizes qualities as constraints, optimization criteria or bottlenecks. The difference between constraints and optimization criteria is visible from whether the requirement specification includes a concrete target quantitative value that must be satisfied. For example, the printer throughput must be at least 80 pages per minute (constraint) at a minimal cost (optimization criterion). Bottlenecks relate to design considerations causing a difficulty in satisfying constraints under specific, often unforeseen, usage conditions. Bottlenecks often arise due to legacy.

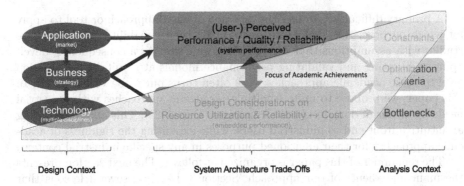

Fig. 2. Focus of academic achievements in relation to industrial needs.

In this paper, we investigate the difficulty of getting academic approaches adopted by industry. We start with highlighting a mismatch in focus of the academic achievements in relation to the industrial needs as shown in Fig. 2.

It is natural for academic work to disregard most business and market aspects. Trade-off analysis is commonly limited to a single engineering discipline or just to the technologies of embedded systems (i.e., computer electronics and software). The industrial practice may however also cover for instance mechanical or optical technologies combined with choices for different materials. A well-known example from the past is the CD-player tray where the original choice was for expensive stiff materials to give support to CDs in order to realize accurate reading by laser (requiring simple control software) as opposed to the cheaper plastic trace developed later which is too flexible from a mechanical viewpoint and therefore requires more complex control software to read the CD accurately.

Due to the diversity of applications, academic approaches rarely cover the relation between embedded performance and system performance unless the relation is reasonably straightforward. Although for example the quality of a picture on a TV display and the accuracy of overlaying ink by a printer may depend on the same embedded performance metrics, the relation between them is very different and therefore hampers the typical generalization pursued by academia.

With respect to the analysis context, academia concentrate on constraints and optimization criteria. A mismatch we observe here is that industry often only cares about getting approximate results for a large variety of quantitative qualities very fast whereas academia promote the use of formal methods because they give high quality results, while suffering from limited scalability. An additional issue with formal methods is that they are commonly perceived as not being very intuitive. Moreover, industry struggles mostly with identifying bottlenecks. This shows from the need to evaluate the concrete value for a wide range of quantitative qualities under certain conditions instead of/next to pursuing satisfaction of a certain bound or an optimization. Many academic approaches (in particular formal methods) do not support such analysis (straightforwardly).

A primary difficulty is in fact the question of what approach or tool to apply for which purpose. Academic approaches claim their capabilities on small case studies with assumptions that rarely hold in practice. As a consequence, developing, calibrating and validating performance models at a suitable abstraction level to make academic approaches work often turns out to be very difficult. In this paper, we propose to unite (multiple) academic approaches with industrial needs by means of domain specific languages (DSLs) [29]. This allows providing an intuitive front-end to industrial users, while exploiting the rigidness of academic approaches for their envisioned purposes in any specific industrial context.

The remainder of this paper is organized as follows. The next section presents the main ingredients of our approach. Section 3 describes several lessons that we learned on getting our approach adopted by industry. Section 4 illustrates application of our approach in three different industrial domains. After reflecting on related work in Sect. 5, we summarize our conclusions in Sect. 6.

2   Approach

Industrial design processes are commonly *reactive* when it comes to performance qualities, i.e., performance is realized/optimized as an 'afterthought' when the design is already fixed. This is often due to the (business) needs of exploiting certain legacy or reducing costs while extending the functionality to support (market). As a consequence, performance is only considered when substantial bottlenecks arise during test and integration (i.e., the right-hand side of the V-model [8]) or even later. To counter this, we promote taking performance aspects into account right from the start of the design and system architecting phases (i.e., the left-hand side of the V-model). Academia proposed many predictive and some constructive approaches to support this vision. Following primarily an analysis strategy, *predictive* approaches exploit performance models to allow iterative evaluation of design alternatives devised by system architects before selecting the most appropriate one that is eventually realized by certain implementation technologies. A step further are *constructive* approaches, which follow an automated synthesis strategy in realizing 'performance-by-construction'.

As ingredient to introducing 'performance-by-construction' into industry, we propose the approach in Fig. 3. Crux is exploitation of a *reference architecture* or *architecture template* that describes what components a system may consist of and how these components can be combined into systems[1]. Such a reference architecture is available as design knowledge from the legacy in existing products or product families. Reconstructing the design knowledge and rationale on how components can be combined and configured *exactly* and how different concepts in the reference architecture interrelate is a very difficult process since this information is commonly ill-documented. We propose the use of DSLs to formalize this knowledge such that it can be processed automatically for the purposes

---

[1] Notice that the terms *component* and *system* are relative to the level of detail at which the design takes place where the highest level denotes complete products and the lowest level comprises automatically synthesizable or bought off-the-shelf parts.

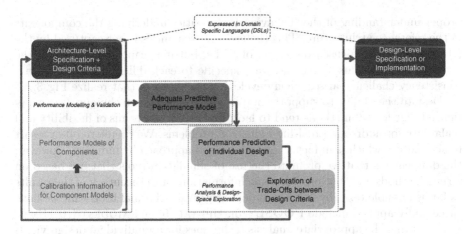

Fig. 3. An approach towards achieving 'performance-by-construction'.

of both (system & embedded) performance analysis and synthesis. Exploiting DSLs allows specifying designs and any qualitative quantity of interest in a way that is close to the intuition of the system architects by using their domain specific terminology and way of describing things [16], while hiding the complicated details of applying the (formal) analysis techniques from them. We experienced that deducting the knowledge underlying a reference architecture by formalizing it in DSLs usually reveals already many ambiguities and inconsistencies in what different system architects (often of different departments) understand about their designs, particularly at the level of detail necessary to automate validation of such possible ambiguities and inconsistencies in design specifications and to automate performance analysis and synthesis from valid design specifications.

The approach of Fig. 3 exploits the architecture-level specification of a design in terms of a DSL instance of the reference architecture to automatically generate (an) adequate performance model(s) for the quantitative qualities of interest. This is enabled by a collection of appropriately calibrated performance models for the individual components that conform to the reference architecture. Calibration of such model components is commonly a very difficult aspect in industrial practice. Although techniques like static code analysis can aid obtaining the required calibration data, one often has to rely on (extrapolation of) measurements of existing products (if measurements are possible at all). Nevertheless, isolating the contribution of individual components to specific measurement results, which is often needed for appropriate calibration, is in many cases infeasible and we observe a lack of academic approaches to improve on this.

Given the diversity of quantitative qualities an industry may be interested in, our approach supports the use of different analysis technologies. From the very same DSL instance, we may for example generate complementary analysis models ranging from basic formula-based computations to simulation models and models for model checking tools depending, amongst others, on the metric of interest, desired accuracy and analysis speed. Notice that this requires appropriately calibrated component models for each of these analysis technologies and a proper understanding of the (behavioral) semantics underlying the components in the reference architecture. This (behavioral) semantics is not captured by the (static) semantics focused definition of DSLs, but it is captured as part of the generation algorithms. Since these are specific to each different target, a clear consistency challenge arises when developing design tools that realize Fig. 3.

Despite the ability to support complementary analysis techniques, we experienced that formal methods tend to lack applicability in terms of flexibility and scalability for addressing real-life industrial problems. We therefore often resort to simulation, which is in fact a commonly used approach by industry. However, the difference is that we promote the use of simulation tools that are based on formal methods and that provide some information on the accuracy of the results as far as possible (e.g., by using confidence intervals). Formal tools that we have successfully applied include POOSL [25], UPPAAL [5] and MODEST [10, 11].

Executing the appropriate analysis techniques for an individual design yields a performance prediction that serves as input for an overall trade-off analysis when comparing alternative designs. The analysis results for individual designs and the comparison of alternatives are again to be presented to system architects in an intuitive form. Hence, some post-processing is usually required in the design tooling to represent the raw quantitative results produced by an exploited analysis tool as they are not made available in the domain specific terminology. For this purpose, we created the TRACE visualization tool [13] that can be configured to show quantitative results conform the domain specific terminology.

When a design alternative is selected for realizing the system, the DSL instance of the corresponding architecture-level specification is used as starting point for generating a design-level specification DSL instance or synthesizing an implementation. If no suitable design alternative is identified, the exploration process iterates. Notice that such iteration may request for (re-)designing components for which calibrated performance models must be provided in order to complete the process. We therefore stimulate developing and calibrating performance models as an integral part of the design process (see also Sect. 3).

A final ingredient to the approach in Fig. 3 is to structure the DSLs conform the organization of an industry in terms of groups or departments. Hence, we have not one but a coherent collection of inter-dependent DSLs to capture a single design specification. The dependencies cover in fact the cross-department relations between their individual responsibilities. By the ability of DSL technology to automate validation of possible ambiguities and inconsistencies in design specifications, communication between system architects from different departments (often covering different engineering disciplines) improves substantially.

Starting from a performance modeling perspective, we have observed that the Y-chart modeling paradigm [14] shown in Fig. 4 fits very well to most industrial organizations. The Y-chart modeling paradigm follows the separation of concerns from [20] by isolating modeling the *application* functionality from modeling the *platform* resources that execute this functionality. The platform model covers the raw resources (time, space, bandwidth and energy) and the scheduling and arbitration mechanisms used to allow sharing them by multiple parts of the application. The *mapping* part of the Y-chart modeling paradigm covers

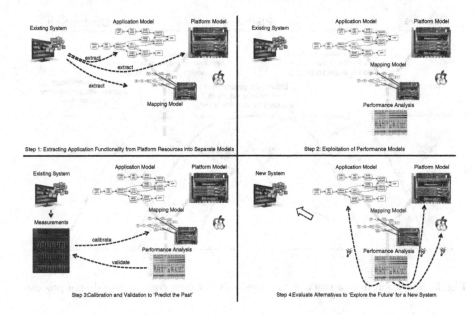

Fig. 4. Introducing the Y-chart modeling paradigm into industry.

the deployment of application functionality onto platform resources and the corresponding configuration of the scheduling and arbitration mechanisms in the platform. The crux of the Y-chart is to ease design-space exploration without the need to change the complete design specification when evaluating alternative mappings or platform configurations or when changing the application functionality for product variants. Moreover, the ingredients of application, platform and mapping are often addressed in industry by different departments as we exemplify in Sect. 4, while system architects (often organized in yet another department) are responsible for the quantitative qualities of their combination.

3   Lessons Learned

Industrial adoption of model-based engineering in general and formal methods in particular does not happen over night. We experienced that such adoption entails a migration process covering a cultural change in the way of working (and possibly also an organizational change), which takes about 2–4 years. Figure 5 identifies the phases we observed in this migration process, which we will classify in this section onto 'performance-by-construction' maturity levels 0–5.

Starting from a traditional design process (maturity level 0) where performance modeling and design-space exploration are processes during integration and test, the introduction of the Y-chart modeling paradigm in Fig. 4 provides some important added values to industry. The first is that the performance models extracted from an existing system are often a first-time formalization of design specifications, which thereby provides an aid in resolving inconsistencies in the design. The second (and possibly more important) value is the fact

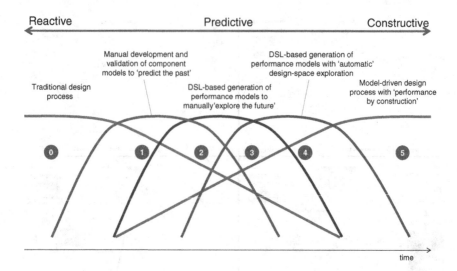

Fig. 5. Migration from a reactive to a predictive or even constructive design process.

that the developed performance models give appropriate insight in existing bottlenecks as an aid to suggest better designs. A third added value comes from
executing the last step in Fig. 4, which allows to predict the performance of any
such design proposal. We classify this situation, where the exploration of design
alternatives is performed based on manually constructed performance models
(i.e., not generated from DSLs), as maturity level 1. The performance models
are in this phase made by modeling experts who will however need to learn the
industrial domain. To accelerate this and to allow the industry observing the
added values, it is of utmost importance to perform real-life case studies without compromises and which are reasonably 'hot' in the sense that the observed
issues may actually imply a loss in profit. These case studies must be performed
on-site to really understand the design context (see Fig. 1) and to have close
contact with the system architects and their managers responsible for solving
the 'hot' issue. It also accelerates obtaining information to calibrate and validate performance models. We therefore advocate the 'industry-as-laboratory'
approach [19].

When model-based engineering has shown (substantial) measurable successes
by resolving performance issues effectively and efficiently, system designers may
either become eager to learn how it works or show resistance due to feeling
threatened in no longer being the one expert capable of solving those issues (with
traditional approaches). Although most academic work stops after successfully
performing a case study, this is actually the moment where the cultural change
in the way of working starts. Supported by their management, system architects
should be able to adopt the new way of working in terms of applying the academic approaches themselves: the domain experts also need to become experts on
performance modeling *their own* systems. Next to some minimal form of educating system architects on the new techniques and tools, introducing DSLs is a

crucial aspect of this phase. The ability to automatically generate performance models from reverse-engineered DSLs as described in Sect. 2, where design alternatives are mostly still explored manually defines maturity level 2. Maturity level 3 supersedes level 2 when proper embedding of performance prediction in the industrial design process is realized such that performance are indeed taken into account right from the start of the design and system architecting phases.

Maturity level 3 marks a predictive attitude towards performance instead of a reactive one. It may also be the highest achievable maturity level when state-of-the-art academic approaches do not (yet) provide solutions to further automate the design process. In some cases, especially when the intrinsic complexity[2] of the domain (most prominently determined by the amount of dynamism in systems) is relatively low, academic work does provide some approaches suitable for adoption by industry. An example is the static-order scheduling of loop-control functionality on (multi-core) processors in the lithography systems of ASML (see Sect. 4). Maturity level 4 is achieved when the exploration of alternative designs is supported with (semi-)automatic tooling. Maturity level 5 classifies the situation where the creative insights of system architects have been completely captured in a fully automated design process that realizes 'performance-by-construction'.

As mentioned above, some form of education is required in order to achieve adoption by industry. We experienced that it is important to distinguish the roles of a user of the new approach and of those who will maintain the corresponding design tooling with any relevant support. Such maintenance and support is crucial in an industrial setting. Where users shift to applying performance modeling and analysis, educating the developers of design tools in understanding the exploitation of the formal methods hidden behind the user-friendly DSLs and the DSL-based infrastructure is an even bigger challenge. We observe that different industries address this challenge differently, ranging from outsourcing to (small-scale) in-house education programs that focus on how to make performance models for the specific domain and to maintain the design tooling (DSLs with their generation algorithms) instead of 'coding implementations'.

4 Industrial Applications

We briefly report on three different application domains, where the approach described in Fig. 3 has been or is being introduced into industry.

4.1 Lithography Systems (ASML)

ASML develops lithography systems for the semiconductor industry. To achieve the required nanometer precision in combination with high throughput, these systems consist of hundreds of sensors and actuators which are controlled by thousands of control tasks. The development of such a system requires multidisciplinary trade-offs involving mechatronic engineers, electronic engineers and

---

[2] The sheer size of systems (in the number of parts) is a different kind of complexity.

software engineers. An early system-wide insight in timing bottlenecks is crucial to avoid costly redesigns and to meet time-to-market and quality constraints.

The loop-control systems are developed using a reference architecture that is captured by a coherent collection of inter-dependent DSLs [21]. These DSLs are used to specify the system according to the Y-chart modeling paradigm, where the origin of the application, platform and mapping lies in different departments.

- The application defines the mechatronic control logic, including networks of so-called servo groups and transducer groups.
- The platform level is captured by DSLs to describe electronic hardware and their physical architecture consisting of high performance multi-core processors, IO boards, and switched interconnection networks. Also physical limitations, such as the maximal frequency of an IO board, can be expressed.
- The mapping describes static deployment of application elements onto the platform resources. In addition, it also configures certain timing synchronization aspects of the platform.

The approach relies on a constructive scheduling algorithm to compute a static order schedule for each processor core such that the latency requirements of the loop-control systems are met [3]. To able to compute these schedules, a number of decisions have to be taken about the platform configuration, such as the number of processors, the topology of the switched interconnects and the setting of synchronization timers. To support these decisions, the DSLs are transformed into an executable POOSL model which is also structured according to the Y-chart paradigm. It formalizes the total system in terms of stochastically-timed communicating parallel processes. During simulation of the POOSL model, performance requirements of the applications are checked automatically. Moreover, the user obtains information about jitter, timing averages, and the load of processors and communication switches.

As reported in [30], the performance of the POOSL simulator easily deals with 4000 control tasks. The POOSL models were calibrated using an existing embedded control system. For a new system release, various significant performance improvements were identified. Important for the industrial adoption of the performance analysis techniques is the automatic generation of analysis models. Moreover, there is a strong coupling with synthesis, because the DSLs are also used for the generation of schedulers and code.

4.2   Multi-functional High-End Printer (Océ)

In projects with the company Océ the performance of high-end printers has been addressed. An important performance measurement is the throughput as perceived by the user, i.e., the number of pages printed per minute. Part of the throughput is determined by the data processing part which has to deal with different types of jobs, such as scan jobs and copy jobs. Each job has a number of characteristics which determine the processing steps to be taken. For instance, the paper size (A3, A4, ...), the number of images per paper sheet, and the export format for scan jobs (pdf, jpeg, ...). Since this involves large size bit maps, there are also compression and decompression steps.

Designers of the data processing part have to make a large number of decisions about the processing units to use (CPUs, FPGAs, ..) and the type and size of memory. Many questions have to be answered, e.g., whether certain jobs are allowed to execute in parallel or whether certain steps may share a certain resource. Since costs are an important aspect, designers have to investigate whether the required performance can be achieved with reduced hardware and which performance can be achieved on a certain hardware platform.

To address these performance questions, a DSL for data processing architectures has been devised [24]. To achieve a separation of concerns, the DSL is structured in the three levels of the Y-chart paradigm.

- The application level contains a number of use cases, such as scanning a stack of pages and printing it three times double-sided. A use case consists of a number of atomic steps and a partial order on these steps.
- The platform level uses three types of resources: memory (with a certain capacity), bus (with a maximum bandwidth), and executor (with a processing speed). Connections between these resources limit the data flow.
- The mapping level specifies which steps run on which executors and which data is stored where.

Having the data processing architecture represented in a DSL allows an easy translation to various analysis tools with different capabilities. For instance UPPAAL [5] has been used, because it allows exhaustive model checking of timed automata. However, for large models the time needed for the analysis became too large. Hence, a dedicated simulator, tuned to this particular domain, has been built to perform large scale simulations [12]. This allows numerous simulation runs over multiple design, for a fast identification of bottlenecks and promising designs.

4.3   Interventional X-Ray Systems (Philips)

Performance aspects of interventional X-ray systems are addressed in a project with Philips. These systems are used for minimally invasive treatments. The number of medical procedures is quickly increasing and includes cardiovascular applications, e.g., placing a stent via a catheter, neurology and oncology. In these procedures, the surgeon is guided by X-ray images which have been processed extensively to obtain a clear image of the essential medical structures using a minimal amount of X-ray.

For the eye-hand coordination of the surgeon, there are clear upper bounds on the latency and the jitter of the image processing chain. However, experiments with surgeons show that a small percentage of deadline misses is allowed.

For every new release, the designers have to make a careful trade off between the required X-ray dose, the hardware configuration (typically PFGAs and multi-core PCs), the algorithms to use, the order and the allocation of processing steps, the image quality, and the supported image resolutions. For instance, to reduce costs and maintenance there is a wish to reduce the number of PCs in the system, but it is very difficult to reason about the impact on the performance.

To support the designers of the image chain, the iDSL approach has been developed [27], consisting of a DSL for service systems, a mapping to the MODEST toolset [10,11], and visualization tools. The iDSL approach supports a high level description of the image processing chain, with the three levels of the Y-chart paradigm. Moreover, three additional aspects can be specified:

- A scenario which restricts the performance analysis to a particular set of service requests. E.g., for the image processing, we can specify assumptions about the arrival rate of raw images from the detector.
- A measure defines the required performance observations and how to obtain them. E.g., a cumulative distribution function (CDF) obtained via model checking or average response times by means of simulation.
- A study allows the definition of a number of design instances that have to be evaluated.

The iDSL tools translate the DSL specification to a number of models in MODEST. Next the iDSL tools call the MODEST tools iteratively to obtain a particular performance observation. For instance, a number of simulation runs or repeated calls to a model checker to construct a CDF. There is a high degree of automation which makes it easy to switch between model checking and simulation.

Performance also plays a crucial role in the movement control part of an interventional X-ray system. This concerns, for instance, movements of patient table and the C-arm on which generator and detector are mounted. In our project we have addressed the redesign of the collision prevention component [17]. A DSL has been developed to express the rules of collision prevention at a high level of abstraction. Next a number of transformations have been defined to models for simulation, formal verification and performance analysis. The performance analysis of collision prevention concentrates on the computations needed to compute distances between objects [28]. It uses a generated POOSL model to perform simulations and to obtain statistics about expected execution times. The model uses performance profiles of the basic computation steps. Moreover, it has been calibrated using performance measurements on an existing legacy component. Finally, code has been generated from the same DSL, which was an important factor for the industrial adoption of the approach.

5   Related Work

An extensive overview of research that aims at the integration of performance analysis in the software development process can be found in [4]. This survey discusses methodologies based on queueing networks, process algebra, Petri nets, simulation methods, and stochastic processes. To achieve integration, many methods are connected to existing notations which describe software designs, often based on UML. More recently, [7] reported on the design of an ultra-modern satellite platform using the AADL notation in combination with several formal modeling and analysis techniques, including performance analysis. In our work, we do not start from these general purpose modeling languages, because usually such models are not available at the right level of abstraction. Instead, we use DSLs and transformations to suitable performance models.

Traditionally, DSLs have been developed for the construction and maintenance of software systems [16,29]. The construction of DSLs has been facilitated by new technology such as the Eclipse Modeling Framework (EMF) [23] which is supported by a large collection of open source tools, including EMFText [22], Xtext [2], and Acceleo [1]. With such tools, it becomes relatively easy to generate different artefacts from a DSL. Relevant for our approach is the possibility to generate analysis models and code from the same DSL instance [6]. This turns out to be an important factor for industrial adoption. Note, however, that our focus is not on software only, but we consider a broader multi-disciplinary system scope.

Methodologies based on the Y-chart paradigm [14] can be found in [15] which compares a number of models of computation and three methodologies: Metropolis, Modular Performance Analysis (MPA), and Software/Hardware Engineering (SHE) supported by POOSL. An application of MPA with the Real-Time Calculus [26] to a distributed in-car radio navigation system has been described in [31]. Related tool support for this approach is provided by SYMTA/S [9]. A comparison of these techniques with, e.g., analysis based on timed automata using UPPAAL [5], can be found in [18].

6   Concluding Remarks

Projects with industry on performance analysis revealed a number of needs:

- Support for system-wide trade-offs that involve multiple disciplines and takes business and market aspects into account.
- Tools that allow the fast generation of a large number of approximate results for large-scale systems. These results should correspond to industrially relevant questions.
- Convenient calibration of analysis results based on measurements of existing systems.
- Light-weight modeling where it is easy to change software design, hardware, and deployment.
- Guidelines about which analysis method to use for which purpose. It should be easy to switch between methods.

Our approach is based on DSLs to capture the essential domain knowledge in a concise and readable way. By defining appropriate transformations, formal analysis models can be generated automatically. Together with the use of the Y-chart paradigm this leads to a clear separation of concerns and a framework which allows changes easily and supports design space exploration. Adding a new analysis method is relatively easy by defining an additional generator. To get fast results for large-scale systems, we often use simulation tools that are based on formal methods. Visualization tools present the results in terms of the domain.

This approach is most effective when the performance models are generated from DSLs which are also used for the functional design. A close connection with the reference architecture and models from which code is generated stimulates industrial acceptance and the incorporation into the industrial work flow. Nevertheless, reaching a high level of maturity (see Sect. 3), such as achieved at ASML, requires many years of close collaboration.

References

1. Acceleo (2015). http://www.eclipse.org/acceleo/
2. Xtext (2015). http://www.eclipse.org/Xtext/
3. Adyanthaya, S., Geilen, M., Basten, T., Schiffelers, R., Theelen, B., Voeten, J.: Fast multiprocessor scheduling with fixed task binding of large scale industrial cyber physical systems. In: 2013 Euromicro Conference on Digital System Design, DSD 2013, Los Alamitos, CA, USA, September 4–6, 2013, pp. 979–988 (2013)
4. Balsamo, S., di Marco, A., Inverardi, P., Simeoni, M.: Model-based performance prediction in software development: a survey. IEEE Trans. Softw. Eng. **30**(5), 295–310 (2004)
5. Behrmann, G., David, A., Larsen, K.G.: A tutorial on UPPAAL. In: Bernardo, M., Corradini, F. (eds.) SFM-RT 2004. LNCS, vol. 3185, pp. 200–236. Springer, Heidelberg (2004)
6. Edwards, G., Brun, Y., Medvidovic, N.: Automated analysis and code generation for domain-specific models. In: 2012 Joint Working IEEE/IFIP Conference on Software Architecture (WICSA) and European Conference on Software Architecture (ECSA), pp. 161–170 (2012)
7. Esteve, M.-A., Katoen, J.-P., Nguyen, V.Y., Postma, B., Yushtein, Y.: Formal correctness, safety, dependability, and performance analysis of a satellite. In: Proceedings of the 34th International Conference on Software Engineering, ICSE 2012, pp. 1022–1031. IEEE Press (2012)
8. Forsberg, K., Mooz, H.: The relationship of system engineering to the project cycle. INCOSE Int. Symp. **1**(1), 57–65 (1991)
9. Hamann, A., Henia, R., Racu, R., Jersak, M., Richter, K., Ernst, R.: SymTA/S - symbolic timing analysis for systems. In: Work In Progress session - Euromicro Workshop on Real-time Systems (2004)
10. Hartmanns, A.: Modest - a unified language for quantitative models. In: The 2012 Forum on Specification and Design Languages (FDL), pp. 44–51. IEEE (2012)
11. Hartmanns, A., Hermanns, H.: The modest toolset: an integrated environment for quantitative modelling and verification. In: Ábrahám, E., Havelund, K. (eds.) TACAS 2014 (ETAPS). LNCS, vol. 8413, pp. 593–598. Springer, Heidelberg (2014)
12. Hendriks, M., Basten, T., Verriet, J., Brassé, M., Somers, L.: A blueprint for system-level performance modeling of software-intensive embedded systems. Int. J. Softw. Tools Technol. Transf. 1–20 (2014)
13. Hendriks, M., Verriet, J., Basten, T., Theelen, B., Brassé, M., Somers, L.: Analyzing execution traces - critical path analysis and distance analysis. Submitted to: Software Tools for Technology Transfer (2015)
14. Kienhuis, B., Deprettere, E., Vissers, K., van der Wolf, P.: An approach for quantitative analysis of application-specific dataflow architectures. In: ASAP 1997: Proceedings of the IEEE International Conference on Application-Specific Systems, Architectures and Processors, p. 338. IEEE Computer Society (1997)

15. Lapalme, J., Theelen, B., Stoimenov, N., Voeten, J., Thiele, L., Aboulhamid, E.M.: Y-chart based system design: a discussion on approaches. In: Nouvelles approches pour la conception d'outils CAO pour le domaine des systems embarqu'es, pp. 23–56. Universite de Montreal (2009)
16. Mernik, M., Heering, J., Sloane, A.M.: When and how to develop domain-specific languages. ACM Comput. Surv. **37**(4), 316–344 (2005)
17. Mooij, A.J., Hooman, J., Albers, R.: Early fault detection using design models for collision prevention in medical equipment. In: Gibbons, J., MacCaull, W. (eds.) FHIES 2013. LNCS, vol. 8315, pp. 170–187. Springer, Heidelberg (2014)
18. Perathoner, S., Wandeler, E., Thiele, L.: Evaluation and comparison of performance analysis methods for distributed embedded systems. Technical report, ETH Zurich, Switzerland (2006)
19. Potts, C.: Software-engineering research revisited. IEEE Softw. **10**(5), 19–28 (1993)
20. Sangiovanni-Vincentelli, A., Martin, G.: Platform-based design and software design methodology for embedded systems. IEEE Des. Test **18**(6), 23–33 (2001)
21. Schiffelers, R., Alberts, W., Voeten, J.: Model-based specification, analysis and synthesis of servo controllers for lithoscanners. In: Proceedings of the 6th International Workshop on Multi-Paradigm Modeling, MPM 2012, pp. 55–60. ACM, New York (2012)
22. Software Technology Group, TU Dresden. EMFText (2015). http://www.emftext.org/
23. Steinberg, D., Budinsky, F., Paternostro, M., Merks, E.: Eclipse Modeling Framework. Pearson Education, London (2008)
24. Teeselink, E., Somers, L., Basten, T., Trcka, N., Hendriks, M.: A visual language for modeling and analyzing printer data path architectures. In: Proceedings of the ITSLE, p. 20 (2011)
25. Theelen, B., Florescu, O., Geilen, M., Huang, J., van der Putten, P., Voeten, J.: Software/hardware engineering with the parallel object-oriented specification language. In: Proceedings of the 5th IEEE/ACM International Conference on Formal Methods and Models for Codesign, MEMOCODE 2007, pp. 139–148. IEEE Computer Society. Washington, DC (2007)
26. Thiele, L., Chakraborty, S., Naedele, M.: Real-time calculus for scheduling hard real-time systems. In: Proceedings of the IEEE International Symposium on Circuits and Systems, vol. 4, pp. 101–104 (2000)
27. van den Berg, F., Remke, A., Haverkort, B.: A domain specific language for performance evaluation of medical imaging systems. In: Proceedings of the 5th Workshop on Medical Cyber-Physical Systems, MCPS 2014, Berlin, Germany. OpenAccess Series in Informatics (OASIcs), vol. 36, pp. 80–93, Schloss Dagstuhl-Leibniz-Zentrum fuer Informatik, Dagstuhl (2014)
28. van den Berg, F., Remke, A., Mooij, A., Haverkort, B.: Performance evaluation for collision prevention based on a domain specific language. In: Balsamo, M.S., Knottenbelt, W.J., Marin, A. (eds.) EPEW 2013. LNCS, vol. 8168, pp. 276–287. Springer, Heidelberg (2013)
29. van Deursen, A., Klint, P., Visser, J.: Domain-specific languages: an annotated bibliography. SIGPLAN Not. **35**(6), 26–36 (2000)

30. Voeten, J., Hendriks, T., Theelen, B., Schuddemat, J., Suermondt, W.T., Gemei, J., Kotterink, K., van Huët, C.: Predicting timing performance of advanced mechatronics control systems. In: 2011 IEEE 35th Annual Computer Software and Applications Conference Workshops (COMPSACW), pp. 206–210 (2011)
31. Wandeler, E., Thiele, L., Verhoef, M., Lieverse, P.: System architecture evaluation using modular performance analysis: a case study. Int. J. Softw. Tools Technol. Transf. (STTT) 8(6), 649–667 (2006)

Modelling and Applications

Stochastic Modeling for Performance Evaluation of Database Replication Protocols

Peter Popov, Kizito Salako, and Vladimir Stankovic[✉]

Centre for Software Reliability, City University London, London, UK
{P.T.Popov,K.O.Salako,V.Stankovic}@city.ac.uk

Abstract. Performance is often the most important non-functional property for database systems and associated replication solutions. This is true at least in industrial contexts. Evaluating performance using real systems, however, is computationally demanding and costly. In many cases, choosing between several competing replication protocols poses a difficulty in ranking these protocols meaningfully: the ranking is determined not so much by the quality of the competing protocols but, instead, by the quality of the available implementations. Addressing this difficulty requires a level of abstraction in which the impact on the comparison of the implementations is reduced, or entirely eliminated. We propose a stochastic model for performance evaluation of database replication protocols, paying particular attention to: (i) empirical validation of a number of assumptions used in the stochastic model, and (ii) empirical validation of model accuracy for a chosen replication protocol. For the empirical validations we used the TPC-C benchmark. Our implementation of the model is based on Stochastic Activity Networks (SAN), extended by bespoke code. The model may reduce the cost of performance evaluation in comparison with empirical measurements, while keeping the accuracy of the assessment to an acceptable level.

Keywords: Stochastic modeling · Database replication protocols · Performance evaluation · Diverse redundancy

1   Introduction

Performance evaluation of database systems is important and usually affects the decision about which among many competing products to use. Performance evaluation via measurements with real database systems is a complex and expensive process. If comparison of replication protocols is sought, the above problems are compounded by the fact that the ranking among the compared protocols might be affected by both the quality of the available implementations of the compared protocols and the precision/accuracy of the methods employed in the evaluation process itself.

Performance evaluation via stochastic modelling has some potential advantages in comparison with performance measurements using real systems:

- Saving on both the ICT infrastructure costs and the time taken for evaluation. In our own experience, performance measurement takes a lot of time, especially if

© Springer International Publishing Switzerland 2015
J. Campos and B.R. Haverkort (Eds.): QEST 2015, LNCS 9259, pp. 21–37, 2015.
DOI: 10.1007/978-3-319-22264-6_2

performance needs to be measured in a wide range of conditions (varying work-load, size of the data processed by the tested system, hardware constraints, etc.). The cost will further escalate if getting high confidence in the performance evaluation is necessary. Performance measures, e.g. average transaction response time or throughput, may vary significantly between individual measurements. Stochastic models may offer dramatic cost-savings in these circumstances;

- Replication protocols – which one to choose? Being able to compare replication protocols at a high level of abstraction (eliminating the limitations of a particular implementation) is useful before one commits to a particular solution. Typically, new protocols are compared with the performance of a single solution, but rarely can two competing protocols be measured without the comparison being affected by the quality of the particular implementations. The ranking of the products may reflect the quality of the implementations, rather than the quality, or lack thereof, of the compared replication solutions. Even if the protocol implementations have been suitably optimised, the ranking might still depend significantly on precisely how, and the conditions under which, the evaluation was conducted (e.g. performance over-heads due to network delays or inaccurate measurement routines).

Model-based performance evaluation of database replication protocols has been studied by others in the past. For instance, in [1] the authors opt to explicitly model disk, cache, CPU, network and concurrency control. Similarly, the research in [2] is based on modeling CPU, disks, database log disks and network with the aim of evaluating their proposed replication protocol – Database State Machine (DBSM). In [3] the authors scrutinize various assumptions in modelling the performance of single DBMS concurrency control mechanisms and conclude that different assumptions may lead to "contradictory results". Also, a variety of analytical models for distributed and replicated databases have been proposed, e.g. as in the survey [4]. Likewise, a thorough survey and classification of queuing network models for database systems performance is given in [5]. Many models referred to above do not validate the modelling assumptions seriously while, contrastingly, the main contribution of [3] is in demonstrating how building a credible model starts with validating the modelling assumptions; this is the view that we take in this paper. The building of stochastic models that give trustworthy performance evaluation results requires significant effort in model validation. Only then can any benefits from model-based evaluation (e.g. cost savings) be taken seriously.

The foregoing suggests that we need a trustworthy, implementation-agnostic, model-based performance evaluation of database replication protocols. We, therefore, put forward and rigorously validate a stochastic model for performance evaluation. This model operates at a relatively high level of abstraction, e.g. hardware resources (HW) are not explicitly modelled; we demonstrate how this level of fidelity, which implicitly takes HW resources into account, is appropriate for our aims.

The stochastic model was implemented in the Mobius [6] modelling environment, enhanced with our own code: we created suitably expressive representations of a database replication protocol, clients and diverse database servers using the *Stochastic Activity Network* (SAN) [7] formalism. The model and its assumptions were validated (Sect. 5) for a diverse database replication protocol [8] using our implementation of the TPC-C, an industry benchmark for performance evaluation of database servers.

To investigate the efficacy of our proposed approach we proceeded as follows. The aforementioned server models were calibrated and validated using statistical distributions obtained from experiments with real systems, each experiment consisting of a single client and a single server. In particular, for each server model, the validation consisted of a detailed statistical analysis to check/refute the model assumptions, as well as comparing the distribution of transaction durations observed in experiments with those obtained via simulation. Upon gaining sufficient agreement between experiment and simulation, we used the validated server models in simulating the behavior of the protocol under 1 client and 5 client loads. We examined how well the simulation results agree with reality, by comparing the transaction duration distributions obtained with those observed in real system experiments.

The remaining sections of the paper are as follows. In Sect. 2 we present related work. Section 3 describes a stochastic model of a database server operation applicable to a range of database replication protocols, while Sect. 4 explains the model we implemented for a particular replication protocol. Using data from real system experiments, in Sect. 5 we present the validation of both the modelling assumptions and the model's behavior when suitably calibrated. The main simulation results are detailed in Sect. 6. Section 7 is a discussion of the stochastic model approach and the results, and Sect. 8 concludes the paper, highlighting future work.

2 Related Work

In addition to the references of "related work" given in the Introduction, we would like to summarise some relevant research on database replication protocols. Database replication has proved a viable method for enhancing both dependability and performance of DBMSs. Performance is typically improved by balancing the load between deployed replicas, while fail-over mechanisms are normally used to redistribute the load of a failed replica among the operational ones to improve availability.

The common assumption in building database replication protocols is that *crashes* are the main type of DBMS failures. Under this assumption, using several *identical* replicas (e.g. MS SQL servers) provides appropriate protection. Under this assumption, which is used as the basis for all commercial solutions and most academic ones (e.g. [9]), various performance and scalability improvements are viable. This common belief is, nonetheless, hard to justify – recent research resulted in overwhelming evidence against crash failures being of primary concern [10, 11]. Using the log of known bugs reported for major DBMSs, it was observed that a majority lead to non-crash failures; these failures can be tolerated only by diverse replication. This is why we chose a diverse replication protocol – one of the few we know of – to illustrate our approach to model-based performance evaluation. Our research focuses entirely on performance evaluation and defers dependability modelling, and its impact on performance, for future work. In so doing, the paper studies the best performance achievable by a replication protocol.

3   Stochastic Model of Database Server Operation

3.1   Model of a Database Server

Modern relational database servers implement a client-server architecture: a client sends requests to a server in the form of SQL statements for the server to execute, and the server returns the result of each execution to the client. In this way, a server may serve multiple clients, concurrently. Each client establishes a connection (or several may share the same one), via which a series of transactions is executed. Each transaction is a set of serially executed operations (SQL statements). A transaction is completed according to the ACID properties, e.g., it is either committed (all the changes are made permanent) or aborted (all the changes are discarded). In general, transactions are not guaranteed to be committed; there is uncertainty here, from both the client-server system and its environment, which ultimately determines the transaction's fate.

According to the TPC-C benchmark, each transaction made by a client belongs to one of 5 types – *Delivery, New Order, Order Status, Payment* or *Stock level*. The proportions in which each of these types occur are specified and define a probability distribution (transaction profile); during simulation, each client determines its next transaction by choosing it according to this distribution. In this way, a random sequence of transactions is generated by the clients. These TPC-C transaction types are defined by 34 SQL statement types, each of which can be a SELECT, DELETE, INSERT or UPDATE.

Executing a transaction takes time. Typically, how long a transaction will take is not known beforehand; uncertainty lurks here, as well. The duration is dependent on the nature of the transaction and its constituent statements, the relevant data for the transaction, the conditions under which the transaction is executed and the server (including the database state and load). For most applications, a transaction's duration is well-approximated by the sum of its constituent statements' durations. For a given server executing statements of a given type, the statement durations follow some statistical distribution – one that is adequately approximated by collecting a suitably large sample of such durations from a server deployed on a real system testbed. As a first pass, we made direct use of these distributions *as is*, foregoing any attempts to fit them to some member of a *yet to be determined* theoretical probability distribution family. These distributions are vital for our model. When a server executes a statement during simulation, in order to determine how long this simulated execution will take, a random sampling of an amount of time is made according to the statement's related statistical distribution (which, implicitly, captures HW configuration effects).

So, the time taken for server i to execute a given statement type is a random variable, X_i, with distribution function $F_i(x_i)$. With n replicas, each statement is modelled by a random vector, $X_1, X_2, \ldots, X_n$. And, a continuous joint distribution function, $F_{1,2,\ldots,n}(x_1, x_2, \ldots, x_n)$, governs the stochastic process of executing the statement on the replicas. What is the precise form of this function? One might suspect that the random variables, X_i, are not independently distributed. Indeed, these random variables model the time taken by each server to process a statement. So, if the statement leads to an execution with a small amount of data being processed, then the time taken by the replicas is likely to be short. Contrastingly, if the statement leads to a complex analysis of a large amount of data, then a long time might be necessary for all replicas. Admittedly, the picture is more complex than this simplistic intuitive speculation: the time taken by the server to process a statement involves processing some data, data exchange between disk and RAM, etc. However, despite our detailed investigation of the correlations between TPC-C statement durations across the two servers we chose, we uncovered no statistically significant evidence to reject an assumption that the durations are independent (Sect. 5.3). Consequently, in our model, the joint distribution factors into the product of the individual replicas related distributions for sampling the statement's duration:

$$F_{1,2,\ldots,n}\left(x_1, x_2, \ldots, x_n\right) = F_1\left(x_1\right) \ldots F_n\left(x_n\right).$$

On a related note, the duration, T, of a transaction on server i is the sum of the durations, $X_i, Y_i, \ldots, Z_i$, of its constituent statements and transaction edge operations (*begin*, *commit* and *abort*). That is, the relationship between these random variables is

$$T = X_i + Y_i + \ldots + Z_i.$$

We sought evidence from experiment to suggest plausible relationships between the random variables $T, X_i, Y_i, \ldots, Z_i$, and found evidence for two relationships: one indicating how, with an appropriate definition of "fast" and "slow", fast/slow transactions tend to be comprised of a significant number of fast/slow statements, and another relationship which suggests that those statements in the transaction with significantly larger average durations almost completely determine the transaction speed. As a first pass, we chose to model a gross approximation to both of these two effects in the following way. When a transaction begins, we uniformly sample a number, α, between 0 and 1, and treat this number as defining the ($\alpha \times 100$)th-quantile duration for each statement that will eventually form a part of the transaction. Then, at the point in simulated time when a statement duration is sampled, the sampling will either default to using this prespecified quantile with probability ρ, or it won't with probability $1 - \rho$. This determines the sampled durations for the statements and, thus, the distribution of T. Here, ρ is a model calibration parameter – with a unique value per server – which gives us two-degrees of freedom to achieve the best fit between measurements and simulation results when calibrating the model.

3.2 Concurrency Control

When serving multiple clients, database servers use various concurrency control schemes to guarantee adequate data consistency, while also striving for the best performance possible. They handle conflicts between simultaneously executing transactions, which can result in aborting some. Thus, these schemes provide different levels of isolation between concurrent transactions. We have chosen to model Snapshot Isolation (SI) [12] - although not a standardized transaction isolation level, it is offered by many database servers (Oracle, MSSQL, PostgreSQL, etc.) as it improves performance by eliminating conflicts between concurrent readers and writers of the same data items. Modeling the isolation level is also an important characteristic of our approach.

Under SI, the detection and resolution of *write-write* conflicts – where two or more concurrent transactions attempt to write to the same database item (stored at a logical location) – is of primary importance. *Write-write* conflicts arise naturally, and at random, in multi-client applications. Consider that, for modelling these conflicts, an explicit notion of "database logical location" is absolutely necessary, but an explicit notion of data at these locations is not! And, even these database locations are only relevant for conflict detection as long as there are concurrent transactions attempting to write to that location. This is very convenient: it allowed us to use a level of abstraction for our model which consisted of choosing (in accordance with TPC-C) a database location to accompany each simulated *write* statement, and recording this location (for conflict detection on each server) when executing such statements.

In many situations, a priori, one may know neither where the next conflict will arise in the database, nor precisely when it will arise. "Where" depends on the transaction profile. For a given database location, if each client independently chooses to execute a transaction that attempts to write to that location, then the probability of a conflict at that location is the product of the probabilities describing each client's concurrent choice. "When" depends on the order in which the server executes the statements.

3.3 Measures of Interest

Performance measurements are dictated by the specific context. A number of performance benchmark standards exist for databases (e.g., www.tpc.org). These standards commonly recommend steady-state measures, such as number of transactions per time unit or average transaction duration. TPC-C mandates using the number of New-Order transactions (one of the 5 transaction types defined by the benchmark) per minute. This metric implicitly takes into account all transaction types in the mix, as their individual throughput is controlled by the minimum percentages defined by the standard. Such measures are not constants – they vary over time even for steady-state mode of operation. This variation can be captured by establishing the distribution of the measure. We note here that a TPC-C requirement is to report the 90th percentile of frequency distribution of response times for each transaction type. In our approach, we analysed the distributions of transaction durations.

4   DivRep Replication Protocol and SAN-Based Implementation

We demonstrate our approach for stochastic modeling of database replication protocols using DivRep replication protocol [8, 13] as an example.

4.1 The Chosen Database Replication Protocol - DivRep

DivRep is an eager, multi-master replication protocol, implemented as a middleware on top of diverse database servers. It assumes the database servers are configured with SI. Full data replication is performed. Replication is performed at SQL statement level. While replicas execute SQL statements asynchronously inside a transaction, detection of incorrect results failures proceeds in parallel by comparing the results of SQL statements produced by diverse database servers. Replica consistency is achieved by executing transaction edge operations "atomically" – the same order of commits and begins is guaranteed on both replicas. This atomicity is achieved using a variant of 2-Phase Commit (2PC). DivRep operates with two diverse database replicas configured in a Fault-tolerant node (FT-node).

An integral part of DivRep is the use of the NOWAIT feature of a DBMS, which raises an exception as soon as the DBMS detects that two concurrent transactions attempt to modify the same data item (e.g. to modify the same row of a database table). This feature is typically implemented as part of a locking protocol: the transaction, which finds the exclusive lock on a data item being taken by another concurrent transaction will be interrupted by a NOWAIT exception, and the modifications of the particular data item by the interrupted transaction will be discarded by the DBMS. Many off-the-shelf products (e.g. Oracle, MSSQL, PostgreSQL, etc.) offer NOWAIT functionality. An important feature of DivRep is that only one replica has NOWAIT enabled. This asymmetric configuration is important: write-write conflicts are *typically* reported by a single replica – the one on which the NOWAIT is enabled – while, on the other replica, transaction blocking will take place in case of write-write conflicts.

4.2 SAN-Based Model of Client, Servers and DivRep

Our performance evaluation is carried out via Monte-Carlo simulation. In creating an application capable of running our simulation campaigns, faithful representations of a TPC-C client, diverse servers configured with SI, and the DivRep protocol, were all realized as atomic SAN models in a Mobius project (Fig. 1.); for further details see http://openaccess.city.ac.uk/4744/.

The client SAN interacts with any of the server SANs (referred to as TransactionSANs in Fig. 1.), either directly or through the DivRep SAN. The client generates transactions (statements and database location data) according to the TPC-C specification; this is the only source of uncertainty in the client. The DivRep SAN operates deterministically, coordinating the receipt and forwarding of statements and the results of statement executions between the client and a pair of replicas. It does

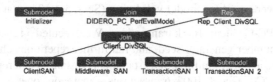

Fig. 1. Composed model of Client, Server and DivRep, in Mobius modelling environment.

this while ensuring the same order of transaction edges across the replicas and synchronizing the concurrency control actions across the replicas. And, the server SANs themselves are capable of executing and coordinating concurrent transaction statements, using NOWAIT and transaction blocking. The primary sources of uncertainty for these server models are the SQL statement durations.

Each server model takes as inputs empirical distributions of statement durations; obtained from real system experiments involving a client and the server, and representing the duration distribution for each TPC-C statement type (34 in total). Together with duration distributions for the transaction edges (*begin, commit, abort*) and the fact that there are two servers, the number of input distributions to the model is 74. While this is a large number of distributions the intention here is that the same distributions will be used for different scenarios – we may vary the number of clients or change the protocol altogether. In this sense, *the work of collecting the data to parameterize the model is done only once.*

This SAN-based model is meant to be reusable; e.g. evaluating the performance of a protocol under a new transaction profile could require simply changing input parameters; evaluating a different protocol involves simply replacing the DivRep *atomic* model with an adequate *atomic* model of the chosen protocol, etc.

5   Statistical Analysis and Validation of Modeling Assumptions

5.1   Test Harness for Real System Experimentation

We built a test harness to evaluate the performance of DivRep and inform our model-based approach. It was deployed in a virtualized environment: 3 physical servers (HP ProLiant DL165 G5p) run VMware ESXi v4 hypervisor. Each physical machine deploys a set of virtual machines (VMs), which ran either Windows 2008 Server (64 bit), or Linux Fedora Core 11 (64 bit). The client application is our own Java implementation of the TPC-C benchmark. We performed the evaluation using FT-node consisting of 2 open source servers: Firebird (FB), v2.1 and PostgreSQL (PG), v8.4. Each server and the client application run in a separate VM, deployed on a separate physical machine. The servers and the client application run on the Windows OS. Our implementation of DivRep has been executed on a separate VM, which runs FC 11.

We collected detailed logs: transaction durations, SQL statement durations, abort counts etc. The precision for transaction and statement durations is nanoseconds (nsec). We ran several types of experiments. Firstly, we conducted single server experiments without DivRep. The SQL statements' durations obtained from these experiments were used as input to the model (Sect. 3.1); the distributions of transaction durations were used for model validation (Sect. 5.4). They are as follows:

- Single FB, 1 TPC-C Client, 100 k transactions. We executed 5 repetitions using the same random number generator seed, and ran further repetitions changing the seed value. There are about 2.8 million SQL statements instances in a repetition.
- Single PG, 1 TPC-C Client, 100 k transactions – analogous ones as for 1FB above.

We also ran two types of DivRep experiments, configured with an FT node (1FB, 1PG): (i) 1 Client, 100 k transactions, and (ii) 5 Clients, each executing 20 k transactions (100 k in total). There were 5 repetitions of each type. The same seeds as for the single server experiments were used. The resulting transaction duration distributions were used for model validation.

5.2 Model Assumptions Validation

As indicated in Sect. 3.1, the time a server takes to execute a given SQL statement is modeled as being sampled from a distribution of statement times derived from measurements on a real system testbed with the respective server. As is standard practice, we estimated and excluded an initial transient period from the measurements. Out of 100 k recorded transactions, the length of the transient period for each server was established to be about the first 30 k transactions for FB, and about the first 25 k transactions for PG. To determine these lengths, we divided the duration of the experiment into 5 min. periods. For each of these "bins", we identified the response times of those transactions that complete at times which fall in the same bin, from which we computed the mean transaction response time for each bin (Fig. 2). We observed consistent transients across repetitions of the same experiment; the graphs shown here are based on only one of 5 repetitions of each experiment type (Sect. 5.1).We also produced exponential moving averages. By considering a range of lags over which to perform the moving averages, a lag of 10 k observations manifested the transient trend component most accurately. The "moving averages" results were consistent with the "binned averages" results.

Additional verification of the truncated data representing a steady-state process can be seen from the Autocorrelation function (ACF) graphs of the transaction duration time-series. The maximum lag is roughly half of all the transaction instances observed in our proposed steady state. Figure 3 shows the ACF graphs of transaction durations for the 1FB and 1PG experiments. The graphs are clearly consistent with that of a purely stationary random process. And, while not strictly applicable, we also ran Augmented

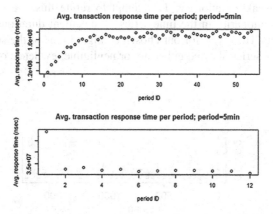

Fig. 2. "Binned averages" graphs for determining Transients period for the experiments: "1FB, 1 Client, 100 k transactions" (top) and "1PG, 1 Client, 100 k transactions" (bottom)

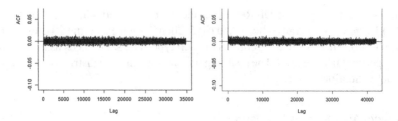

Fig. 3. Autocorrelation function (ACF) of the steady-state transaction durations, for the experiments: "1FB, 1 Client, 100 k transactions" (left) and "1PG, 1 Client, 100 k transactions" (right).

Dickey-Fuller (ADF) [14] statistical tests to check if the transaction durations form a stationary process. For both servers, at the 1 % significance-level, there was statistically significant evidence to reject the null-hypothesis that the observed processes are non-stationary. The ADF is from a family of so-called *unit-root* tests; it assumes a certain functional form (a shifted *integrated autoregressive process*) as a suitable model for the time-series under the null hypothesis. We do not have any reason to believe such a functional form might be applicable here.

After discarding the transactions in the transient periods, we used the remaining SQL statement logs to construct populations of times representing the durations for each statement of a particular statement type. Recall, from Sect. 3.1, that each of the 5 transaction types consists of a series of statements. For instance, the number of statements in New-Order transactions ranges from 26 to 66 (there is a loop and a "conditional" in the implementation). For each of these statement types and each server – such as *New-Order* 1 (NO1) on FB – we constructed a population of all instances of the statement type seen within those transactions belonging to the steady state. This defines the duration distribution, for the statement type on a given server, to be used during simulation.

For most statement types during simulation, a statement's duration on a given server is sampled, independently, according to the statement's duration distribution. Consequently, for the given statement type, the time-series of observed durations for the statement is trivially weakly stationary. To attempt to refute this, we sought statistical evidence of non-stationarity in the time-series of statement durations from the real-system experiments. However, for our chosen configurations of the servers, the ACF graphs for the time-series showed either no, or negligible, evidence of non-stationary behavior (Fig. 4).

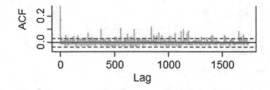

Fig. 4. ACF of Payment 9 statement steady-state durations, "1PG, 1Client, 100 k transactions".

5.3   Correlation of SQL Statements Durations

Are the SQL statement durations correlated across the servers; for instance, does a statement with a long duration on FB imply that the statement will also have a long duration on PG? The same RNG seed value was used when executing either of the single server (1FB or 1PG) experiments under the "1 Client, 100 k transactions" load. Since these are single client experiments, the same sequence of database transactions and SQL statements was executed in both. For each of the 34 TPC-C statement types and the transaction edge operations, we calculated the Pearson Correlation Coefficient (CC) between the two servers. The calculation, based on steady-state data, showed no significant correlation: the CC values for SQL statements types were in the range [−0.01, 0.05], while the values for the transaction edges in the range [−0.001, 0.041]; indeed, a surprising result. Although not a proof of statistical independence, it provides evidence against linearly correlated statement/edges durations, and is consistent with our choice of modelling the statement/transaction edges durations as statistically independent across two servers (see Sect. 3.1).

5.4   Client-Server System Model Validation

The validation of our model proceeds by simulating the single-client/single-server experiments (Sect. 5.1) and comparing the results thereof with experimental observations from the real system. But first, we studied the variability in the real system experiments to determine the extent to which these observations are reproducible.

The necessary step here was to test if the samples of transactions durations obtained from different repetitions of the same real system experiment come from the same distribution. If true, we could then compare transaction duration distributions from our Mobius-based model with *any* of the respective distributions obtained from real system experiments, using an appropriate statistical test to determine if the distributions are suitably similar. So, with this goal in mind, we constructed the Empirical Cumulative Distribution Functions (ECDFs) of transaction durations from two 1FB experiment repetitions (Fig. 5). Visual inspection of these ECDFs revealed no significant difference.

We, however, performed a two-sample, two-sided Kolmogorov-Smirnov (KS) test (see the text box in Fig. 5 for its results) – one of the most general non-parametric tests for checking if two samples are drawn from the same continuous distribution. It is sensitive to differences in both location and shape of the ECDFs of the two samples. The test indicated that, at a 1 % significance-level, there is statistically significant evidence to reject the hypothesis that the two samples of transaction durations come from the same distribution. Similar surprising results were obtained for other pairs of the 1FB experiment with the same or different seed values; this is true for 1PG experiment too. The point to note here is that our sample sizes are atypically large, ranging between 65–70 K transaction durations in each experiment – this is a veritable "embarrassment of riches". With so much sample data comes a lot of "tiny deviations between the distributions" evidence, amounting to significant evidence as far as the discriminatory power of the KS test is concerned [15]. To illustrate the unhelpful level of sensitivity at play here, consider that the maximum

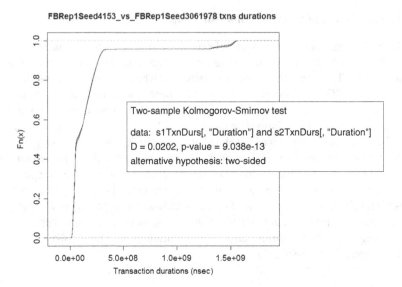

Fig. 5. ECDFs of transaction durations for "1FB, 1C, 100 k transactions" experiment. The dark line represents the repetition with the seed 4153; the light-colored line the seed 3061978.

vertical distance between the ECDFs being tested – the D value from the KS test – is 2 %; a very small distance indeed. And, the distributions "fared no better" when other statistical tests were applied (e.g. Chi-Square test).

What, therefore, would be the basis of a useful comparison between simulation results and experimental observation, if even repetitions of the same experiment are not guaranteed to pass a KS test? We chose to compare the distributions using: (1) ECDFs plots, (2) QQ-plots, (3) average transaction durations and (4) the sample standard deviations, all of which give acceptable agreement across experiment repetitions.

In our simulation campaigns, each simulated run simulates 1 min 40 s of (steady state) server operation, with a total of 1000 runs per campaign. Our criteria for a suitable number of runs was that the number of statement requests made by a client and the number of statements executed by each server were required to converge with a relative standard error of less than 10 %. We simulated a single client interacting with a single FB server, from which we obtained a distribution of transaction durations. This distribution and the corresponding distribution from experiment are plotted in Fig. 6 (left hand side). The close agreement between the distributions, manifested in the plot, is further evidenced by the corresponding QQ-plot (Fig. 6, right hand side); the only noteworthy deviations between the distributions occur in the upper 2.5 % of the distributions. Furthermore, the average transaction durations from simulation and experiment are 163 ms and 166 ms, respectively, with related standard deviations of 1.65 s and 1.75 s. A similar exercise was carried out for the PG server. Again, the simulation and experiment were in close agreement: the average transaction durations are 46 ms and 44 ms for the simulation and experiments, respectively, with the standard deviations of 1.34 s and 0.99 s.

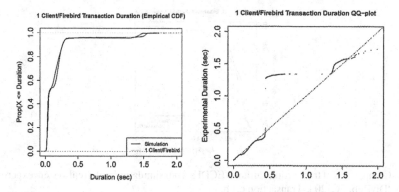

Fig. 6. Comparison of transaction-duration ECDFs (left), and QQ-plot for transaction durations (right), from simulation and a real system experiment of type "1FB, 1C, 100 k Transactions". Similar graphs were obtained for the "1PG, 1C, 100 k Transactions" experiment.

6 Results

Having gained sufficient confidence in the model, we simulated a single client interacting with 1FB and 1PG server via DivRep. The distributions deviated significantly in this case, with the simulated transaction duration distribution having faster durations than its experimental counterpart. The shapes of the distributions, however, were quite similar. This suggested that there might exist a systematic overhead in our experiment. To illustrate the extent of this overhead, we introduced a log-normally distributed overhead in our simulations,[1] resulting in much better agreement between the simulated and experimental distributions (Fig. 7). The average transaction durations from simulation and experiment are 225 ms and 236 ms, respectively, with related standard deviations of 1.21 s and 1.10 s.

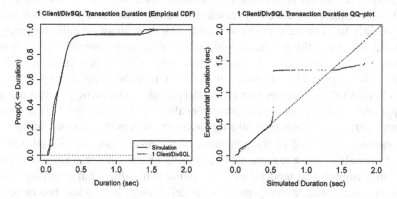

Fig. 7. Comparison of transaction-duration ECDFs (left), and QQ-plot for transaction durations (right), from simulation and a real system experiment of type "DivRep, 1C, 100 k Transactions"

---

[1] The use of the lognormal distribution is purely illustrative, and not based on this distribution family being optimal, in some sense, for modelling overheads.

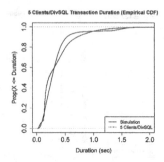

Fig. 8. Comparison of transaction-duration ECDFs from simulation and a real system experiment of type "DivRep, 5C, 20 k Transaction each".

The conjecture about the overhead is, indeed, plausible: the middleware implementing the replication protocol relies on Java RMI (Remote Method Invocation) technology, which is known to introduce significant transport latencies.

Using this lognormal overhead, we proceeded with a final comparison: simulating 5 clients interacting with the servers via DivRep. The ECDF and QQ-plots illustrate how the deviations in the shapes of the distributions have become more noticeable (Fig. 8). Despite these deviations, the average transaction durations from simulation and experiment are 380 ms and 338 ms, respectively, with related standard deviations of 0.836 s and 0.880 s. Given the size of the standard deviation (more than twice the mean), this indicates that the two averages are remarkably close.

7   Discussion

Arguably, model-based performance evaluation of replication protocols is a rather ambitious premise; there are many reasons to doubt whether it is practical or, indeed, even possible to evaluate a protocol's performance without actually building and deploying it. For one thing, what would such a performance measure actually mean for a real system? And, certainly, in the process of building an accurate model, one should not be surprised by: (i) having to make choices about the level of abstraction (model fidelity) at which to operate; (ii) model parameterization of, possibly unknowable, parameters. Even when performance evaluation is conducted on real systems, there are many considerations to take into account, such as the accuracy of the measuring methods employed or the generality of the results obtained.[2] And, yet, it is precisely for all of these reasons that the results of our studies are not only promising – they suggest that model-based evaluation is feasible – but surprisingly so.

For instance, consider the question of model fidelity. In building our models we abstracted away many seemingly important details. Our clients and servers have no

---

[2] We changed the precision of the measurements from msec to nsec, because a significant proportion of the statement durations were in the sub msec range, preventing us from simulating these durations accurately.

explicit notion of detailed data that is manipulated, stored to and retrieved from a database; only time-sensitive, sufficiently detailed database locations are generated for the purposes of conflict occurrence and detection. The servers do not have explicit notions of executing the SQL statements. Despite these simplifications the model achieved the correspondence with experiment reported in Sect. 6. These modelling simplifications were made by, first, identifying those sources of uncertainty that are crucial for system performance – such as statement durations, conflict occurrence and the transaction profile – and then including only those aspects of the real world that we deemed necessary to adequately represent these uncertainties in our model.

Model parameterization was not daunting. As a first pass, the use of empirical CDFs as direct model inputs meant that we did not have to identify suitable theoretical distributions (and their attendant parameters) to approximate these ECDFs. At 74 distributions, this would have been a tedious exercise at best.

A possible advantage of simulation is the speed with which results are obtained. The 1 Client/1 FB experiment took 4 h 28 min to complete and about 10 min to simulate. The biggest difference was for the 1 Client/DivSQL experiment, which saw the experiment complete in 6 h 35 min, whereas the simulation took about 10 min.

While there is a good agreement between the simulation and experimental results (especially in terms of the averages and standard deviations), the disagreements are also worth noting. For, our current model represents a stylized system in which there are no overheads resulting from, say, the protocol itself or the network between the clients and servers. This means that our validated model gives a distribution which is an estimate of the best performance an implementer of the protocol can hope to achieve, *no matter how skilled the implementer is*. The usefulness of this bound is in pointing out that any significant discrepancy between the bound and the performance of a specific protocol implementation indicates that the implementation is sub-optimal.

8 Conclusions

We presented a stochastic model for performance evaluation of database replication protocols. An implementation of the model, using a combination of the Mobius modeling environment, its SAN formalism and our own codebase, is presented. The model is created to be reusable in a variety of scenarios. The model assumptions, and its accuracy with respect to the chosen database replication protocol's performance, are rigorously validated. The measurements obtained from experiments conducted on a real systems testbed were used to validate the model. For this validation, we used the TPC-C benchmark – executed in both the model and the real systems testbed.

The model enables performance evaluation of database replication protocol(s), and therefore their comparison. Model-based evaluation allows for eliminating the impact, on a replication protocol's performance, of overhead due to the protocol's specific implementation. If the ordering between several replication protocols is driven by implementation overheads, model-based evaluation gives a ranking of the replication protocols based on their optimal performance.

We also demonstrated that savings in the model-based evaluation time may be very significant in comparison with the measurements using real systems.

There are a number of ways in which the work can be furthered and the model can be improved. Currently, the model requires a more sophisticated mechanism to account for the effects of a significant increase in client load on server performance. Under a light load, the current server model is one of "infinite resource" [3]. However, under heavy client loads, a realistic mechanism of server resource is essential for modelling the chosen protocol accurately. We have begun experimenting with models of resource, and the validation of their accuracy – to be completed – follows the approach described above for validating the other modelling assumptions. This will include further investigation of how the increased load will affect the surprising result that SQL statement durations are uncorrelated across the servers.

We intend to evaluate other replication protocols, e.g. [11, 16, 17], to see how well the model behaves.

In Sect. 3.1, we highlighted the fact that the parameter ρ was a "blunt" approach for modeling the more nuanced relationships we observed between transactions and their statements. We expect that implementing algorithms which better approximate these relationships will result in even better agreement with experiment.

We also plan to explore the possibility of using the systematic discrepancies between simulated and observed distributions illustrated in the QQ-plots. We are aware that similar systematic differences between models and respective observations have been used in the past for model re-calibration to improve model accuracy [18].

The model has explanatory power which we would like to improve upon. When significant overheads are detected by comparing simulation results with experimental ones, a better explanatory distribution for the overheads might be a gamma distribution with parameters dependent on transaction length and transaction statement types.

The research presented in this paper considers a failure free environment. A natural extension would include evaluation of dependability attributes. For example, faults can be injected by an appropriate data corrupting daemon in the experimental testbed, or modeled in the model-based approach. Different likelihoods of crash, non-crash and Byzantine failures can be simulated to perform a number of "what-if" analyses and observe the effects they have on dependability, and performance, of the chosen system.

Acknowledgement. This work was supported in part by the UK's Engineering and Physical Sciences Research Council (EPSRC) through the DIDERO-PC project (EP/J022128/1). We would like to thank the anonymous reviewers and Bev Littlewood for useful comments about an earlier version of the paper.

References

1. Sousa, A., et al.: Testing the dependability and performance of group communication based database replication protocols. In: Proceedings of the International Conference on Dependable Systems and Networks (DSN 2005) (2005)

2. Pedone, F., Guerraoui, R., Schiper, A.: The database state machine approach. Distrib. Parallel Databases **14**(1), 71–98 (2003)
3. Agrawal, R., Carey, M.J., Livny, M.: Concurrency control performance modeling: alternatives and implications. ACM Trans. Database Syst. **12**(4), 609–654 (1987)
4. Nicola, M., Jarke, M.: Performance modeling of distributed and replicated databases. IEEE Trans. Knowl. Data Eng. **12**(4), 645–672 (2000)
5. Osman, R., Knottenbelt, W.J.: Database system performance evaluation models: a survey. Perform. Eval. **69**(10), 471–493 (2012)
6. Graham, C., et al.: The möbius modeling tool. In: Ninth International Workshop on Petri Nets and Performance Models (PNPM 2001), Aachen, Germany (2001)
7. Sanders, W.H., Meyer, J.F.: Stochastic activity networks: formal definitions and concepts. In: Brinksma, E., Hermanns, H., Katoen, J.-P. (eds.) EEF School 2000 and FMPA 2000. LNCS, vol. 2090, pp. 315–343. Springer, Heidelberg (2001)
8. Popov, P., Stankovic, V.: Improvements Relating to Database Replication, p. 60. City University London: EU (2013) (EPO, Editor)
9. Cecchet, E., Marguerite, J., Zwaenepoel, W.: C-JDBC: Flexible database clustering middleware. In: USENIX Annual Technical Conference, Freenix (2004)
10. Gashi, I., Popov, P., Strigini, L.: Fault tolerance via diversity for off-the-shelf products: a study with SQL database servers. IEEE Trans. Dependable Secure Comput. **4**(4), 280–294 (2007)
11. Vandiver, B., et al.: Tolerating byzantine faults in transaction processing systems using commit barrier scheduling. In: Proceedings of 21st ACM SIGOPS Symposium on Operating Systems Principles, pp. 59–72. ACM: Stevenson, Washington (2007)
12. Berenson, H., et al.: A Critique of ANSI SQL Isolation levels. In: SIGMOD International Conference on Management of Data, San Jose, California, United States. ACM Press, New York (1995)
13. Stankovic, V.: Performance implications of using diverse redundancy for database replication. In: Centre for Software Reliability, p. 169. City University London, London (2008)
14. Fuller, W.A.: Introduction to Statistical Time Series. Wiley, New York (1996)
15. Osborne, J.W.: Best Practices in Data Cleaning: A Complete Guide to Everything You Need to Do Before and After Collecting Your Data. Sage Publishing, Thousand Oaks (2012)
16. Garcia, R., Rodrigues, R., Preguica, N.: Efficient middleware for byzantine fault tolerant database replication. In: Proceedings of the Sixth Conference on Computer systems (EuroSys 2011), pp. 107–122. ACM, Salzburg (2011)
17. Vandiver, B.: Detecting and tolerating byzantine faults in database systems. In: Programming Methodology Group, p. 176. Massachusetts Institute of Technology, Boston (2008)
18. Brocklehurst, S., et al.: Recalibrating Software Reliability Models. IEEE Trans. Softw. Eng. **16**(4), 458–470 (1990)

A Continuous-Time Model-Based Approach to Activity Recognition for Ambient Assisted Living

Laura Carnevali[1]($\boxtimes$), Christopher Nugent[2], Fulvio Patara[1], and Enrico Vicario[1]

[1] Department of Information Engineering, University of Florence, Florence, Italy
{laura.carnevali,fulvio.patara,enrico.vicario}@unifi.it
[2] School of Computing and Mathematics, University of Ulster, Belfast, UK
cd.nugent@ulster.ac.uk

Abstract. In Ambient Assisted Living (AAL), Activity Recognition (AR) plays a crucial role in filling the semantic gap between sensor data and interpretation needed at the application level. We propose a quantitative model-based approach to on-line prediction of activities that takes into account not only the sequencing of events but also the continuous duration of their inter-occurrence times: given a stream of time-stamped and typed events, online transient analysis of a continuous-time stochastic model is used to derive a measure of likelihood for the currently performed activity and to predict its evolution until the next event; while the structure of the model is predefined, its actual topology and stochastic parameters are automatically derived from the statistics of observed events. The approach is validated with reference to a public data set widely used in applications of AAL, providing results that show comparable performance with state-of-the-art offline approaches, namely Hidden Markov Models (HMM) and Conditional Random Fields (CRF).

Keywords: Ambient Assisted Living (AAL) · Activity Recognition (AR) · Continuous-time stochastic models · Transient analysis · On-line prediction · Process enhancement

1 Introduction

Ambient Assisted Living (AAL) aims at providing assistance to people living within smart environments through the integration and exploitation of new sensing technologies, data processing techniques, and services [11]. To this end, Activity Recognition (AR) plays a crucial role in filling the gap between sensor data and high level semantics needed at the application level [18]. This comprises a major ground for the application of quantitative approaches to diagnosis, prediction, and optimization.

A large part of techniques applied for AR [30] rely on or compare with Hidden Markov Models (HMM) [27]: the current (hidden) activity is the state

© Springer International Publishing Switzerland 2015
J. Campos and B.R. Haverkort (Eds.): QEST 2015, LNCS 9259, pp. 38–53, 2015.
DOI: 10.1007/978-3-319-22264-6_3

of a Discrete Time Markov Chain (DTMC), and the observed event depends only on the current activity; stochastic parameters of the model can be determined through supervised learning based on some given statistics; and efficient algorithms are finally available to determine which *path* along hidden activities may have produced with maximum likelihood a given *trace* of observed events. While the ground truth is often based on data sets annotated w.r.t. predefined activities [19,27], more data driven and unsupervised approaches have been advocated where activities are identified through the clustering of emergent recurring patterns [3,20]. Various extensions were proposed to encode memory in HMM by representing sojourn times through discrete general or phase type distributions [16]. However, also in these cases, the discrete-time abstraction of the model prevents exploitation of continuous time observed between event occurrences.

To overcome this limitation, [8] proposes that the evolution of the hidden state be modeled as a non-Markovian stochastic Petri Net emitting randomized observable events at the firing of transitions. Approximate transient probabilities, derived through discretization of the state space, are then used as a measure of likelihood to infer the current hidden state from observed events. To avoid the complexity of age memory accumulated across subsequent states, the approach assumes that some observable event is emitted at every change of the hidden state. Moreover, the structure of the model and the distribution of transition durations are assumed to be given.

Automated construction of an unknown model that can accept sequences of observed events is formulated in [24] under the term of *process elicitation*, and solved by various algorithms [26] supporting the identification of an (untimed) Petri Net model. Good results are reported in the reconstruction of administrative workflows [25], while applicability appears to be more difficult for less structured workflows, such as healthcare pathways [15]. As a part of the process mining agenda, *process enhancement* techniques have been proposed to enrich an untimed model with stochastic parameters derived from the statistics of observed data [22].

In this paper, we propose a quantitative model-based approach to on-line prediction of activities that takes into account not only the type of events but also the continuous duration of activities and of inter-events time. Given a stream of time-stamped and typed events, transient probabilities of a continuous-time stochastic model are used to derive a measure of likelihood for on-line diagnosis and prediction of performed activities. Transient analysis based on transient stochastic state classes [12] maintains the continuous-time abstraction and keeps the complexity insensitive to the actual time between subsequent events. While the structure of the model is predefined, its actual topology and stochastic parameters are automatically derived from the statistics of observed events. Applicability to the context of AAL is validated by experimenting on a reference annotated data set [27], and results show comparable performance w.r.t. offline classification based on Hidden Markov Models (HMM) and Conditional Random Fields (CRF).

In the rest of the paper: the data set of [27] is introduced and commented in Sect. 2; assumed structure and stochastic enhancement of the model used for the evaluation of likelihood is introduced in Sect. 3; the experimental setup and results are reported in Sect. 4; limitations and further steps enabled by the results are discussed in Sect. 5.

2   Problem Formulation

2.1   Description of the Data Set Under Consideration

We base our experimentation on a well-known and publicly available annotated data set for AR [27] containing binary data generated by 14 state-change sensors installed in a 3-room apartment, deployed at different locations (e.g., kitchen, bathroom, bedroom) and placed on various objects (e.g., household appliances, cupboards, doors). Seven activity types derived from the Katz Activities of Daily Living (ADL) index [14], i.e., $\Gamma = \{$*Leaving house, Preparing a beverage, Preparing breakfast, Preparing dinner, Sleeping, Taking shower, Toileting*$\}$, were performed and annotated by a 26-year-old subject during a period of 28 days. The annotation process yielded a *ground truth* consisting of a stream of activities a_1, $a_2, \ldots, a_K$, each being a triple $a_k = \langle \gamma_k, t_k^{start}, t_k^{end} \rangle$ where $\gamma_k \in \Gamma$ is the activity type, t_k^{start} is the activity start time, and t_k^{end} is the activity end time. An additional activity (not directly annotated) named *Idling* is considered, consisting of the time during which the subject is not performing any tagged activity. The data set includes 245 activity instances, plus 219 occurrences of *Idling*. Activities are usually annotated in a *mutually exclusive* way (i.e., one activity at a time), with the only exception of some instances of *Toileting* which was annotated so as to be performed concurrently with *Sleeping* (21 times) or with *Preparing dinner* (1 time).

The data set includes 1319 sensor events, classified in 14 event types, and encoded in the so called *raw* representation, which holds a high signal in the interval during which the condition detected by a sensor is true, and low otherwise (see Fig. 1-left). In this case, each event is a triple $e_n = \langle \sigma_n, t_n^{start}, t_n^{end} \rangle$ where $\sigma_n \in \Sigma^{raw}$ is the event type, t_n^{start} is the event start time, and t_n^{end} is the event end time. As suggested in [2] for the handling of data sets with frequent object interaction, raw events where converted into a *dual change-point* representation, which emits a punctual signal when the condition goes true and when it goes back false (see Fig. 1-right). In this encoding, observations are a stream of punctual events $e_1, e_2, \ldots, e_N$ (doubled in number w.r.t. the raw representation, and sub-typed as *start_* and *end_*), each represented as a pair $e_n = \langle \sigma_n, t_n \rangle$, where $\sigma_n \in \Sigma$ is the event type, and t_n is the event occurrence time. In so doing, the number of events and event types has doubled, i.e., $N = 2638$, and $|\Sigma| = 28$.

Also for the limited accuracy of the tagging process (in [27], annotation was performed on-the-fly by tagging the start time t_k^{start} and the end time t_k^{end} of each performed activity a_k using a bluetooth headset combined with speech recognition software), the starting and ending points of activities are often delayed and anticipated, respectively. As a result, as shown in Fig. 2, the start (end) time of an activity does not necessarily coincide with the occurrence time of its first (last) event.

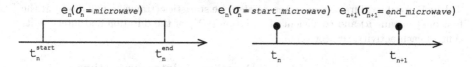

Fig. 1. Sensor representation: raw (left) and dual change-point (right).

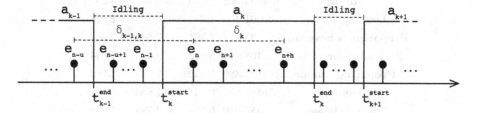

Fig. 2. A fragment of an events stream together with annotated activities.

2.2 Statistical Abstraction

We abstracted the data set content so as to capture four aspects of its statistics: the duration of each activity; the time elapsed between subsequent events within each activity; the type of events occurred in each activity; and, the type of events occurred as first events in each activity.

Let $\{e_n = \langle \sigma_n, t_n \rangle\}_{n=1}^N$ and $\{a_k = \langle \gamma_k, t_k^{start}, t_k^{end}\rangle\}_{k=1}^K$ be the streams of observed events and tagged activities, respectively. The duration δ_k of an activity instance a_k is computed as the time elapsed from the first to the last event observed during a_k, i.e., $\delta_k = \max_{n|t_k^{start} \leq t_n \leq t_k^{end}}\{t_n\} - \min_{n|t_k^{start} \leq t_n \leq t_k^{end}}\{t_n\}$ (e.g., in Fig. 2, $\delta_k = t_{n+h} - t_n$). The duration $\delta_{k-1,k}$ of an instance of *Idling* enclosed within two activities a_{k-1} and a_k is derived as the time elapsed from the last event observed during a_{k-1} to the first event observed during a_k, i.e., $\delta_{k-1,k} = \min_{n|t_k^{start} \leq t_n \leq t_k^{end}}\{t_n\} - \max_{n|t_{k-1}^{start} \leq t_n \leq t_{k-1}^{end}}\{t_n\}$ (e.g., in Fig. 2, $\delta_{k-1,k} = t_n - t_{n-u}$). The *duration statistic* provides the mean and variation coefficient (CV) of the duration of each activity type, as shown in Table 1.

The *inter-events time statistic* evaluates the time between consecutive events occurring within an activity. In so doing, we do not distinguish between event types, and we only consider times between events. The inter-events time of *Idling* is computed taking into account *orphan events*, i.e., events not belonging to any tagged activity (e.g., e_{n-1} in Fig. 2), and the first event of each activity (e.g., e_n in Fig. 2). Also the inter-events time statistic provides mean and CV for each activity type, as shown in Table 1. Most of measured time spans have a CV higher than 1, thus exhibiting a hyper-exponential trend, as expected in ADL where timings may follow different patterns from time to time. Only the duration of *Leaving house* has a CV nearly equal to 1.

Table 2 shows the *event type statistic*, which computes the frequency $\psi_{\sigma,\gamma}$ of an event of type σ within an activity of type γ, $\forall \sigma \in \Sigma$, $\forall \gamma \in \Gamma$, i.e., $\psi_{\sigma,\gamma} = Prob\{$the type of the next event is $\sigma \mid$ the type of the current activity is $\gamma\}$.

Table 1. Activity duration and inter-events time statistics (trained on all days but the first one): mean (μ) and coefficient of variation (CV) of the duration and inter-events time of each activity, respectively.

	Duration		Inter-events time	
	μ (s)	**CV**	μ (s)	**CV**
Idling	1793.102	2.039	3906.167	3.292
Leaving house	40261.455	1.042	9354.190	2.810
Preparing a beverage	35.667	1.361	7.643	2.613
Preparing breakfast	108.684	0.713	9.928	1.844
Preparing dinner	1801.889	0.640	77.966	2.589
Sleeping	26116.571	0.442	1871.836	3.090
Taking shower	485.910	0.298	102.788	1.969
Toileting	88.742	1.175	14.814	2.449

Table 3 shows the *starting event type statistic*, which evaluates: (i) the frequency θ_σ of an event type σ either as the first event of an activity (regardless of the activity type) or as an orphan event, $\forall\,\sigma \in \Sigma$, i.e., $\theta_\sigma = Prob\{$the type of an event e is $\sigma \mid e$ is either the first event observed during the current activity or an orphan event$\}$; and, (ii) the frequency $\phi_{\sigma,\gamma}$ of an event of type σ as the first event of an activity of type γ, $\forall\,\sigma \in \Sigma$, $\forall\,\gamma \in \Gamma$, i.e., $\phi_{\sigma,\gamma} = Prob\{$an activity of type γ is started $\mid$ an event of type σ is observed and the subject was idling before the observation$\}$. As a by-product, $\forall\,\sigma \in \Sigma$, the statistic also computes $Prob\{$the subject remains idling $\mid$ an event of type σ is observed and the subject was idling before the observation$\} = 1 - \sum_{\gamma \in \Gamma} \phi_{\sigma,\gamma}$.

3   Classification Technique

In the proposed approach, a continuous-time stochastic model is constructed so as to fit the statistical characterization of the data set. Activity recognition is then based on a measure of likelihood that depends on the probability that observed time-stamped events have in this model.

3.1   Model Syntax and Semantics

The stochastic model is specified as a stochastic Time Petri Net (sTPN) [28]. As in [23], the formalism is enriched with flush functions which permit the marking of a set of places be reset to zero upon firing of a transition. This improves modeling convenience without any substantial impact on the complexity for the analysis.

Syntax. An sTPN is a tuple $\langle P; T; A^-; A^+; A^{\cdot}; m_0; EFT; LFT; \mathcal{F}; \mathcal{C}; L \rangle$ where: P is the set of places; T is the set of transitions; $A^- \subseteq P \times T$, $A^+ \subseteq T \times P$, and $A^{\cdot} \subseteq P \times T$ are the sets of precondition, postcondition, and inhibitor arcs,

Table 2. Event type statistic (trained on all days but the first one): frequency $\psi_{\sigma,\gamma}$ of each event type $\sigma \in \Sigma$ (rows) within each activity type $\gamma \in \Gamma$ (columns).

	Leaving house	Preparing a beverage	Preparing breakfast	Preparing dinner	Sleeping	Taking shower	Toileting
start_front door	0.497	-	-	-	-	-	-
end_front door	0.503	-	-	-	-	-	-
start_hall-bathroom door	-	-	-	0.018	0.111	0.008	0.261
end_hall-bathroom door	-	-	-	0.023	0.115	-	0.261
start_hall-bedroom door	-	-	-	-	0.274	-	0.019
end_hall-bedroom door	-	-	-	-	0.280	-	0.013
start_hall-toilet door	-	-	0.004	-	0.041	0.540	0.057
end_hall-toilet door	-	-	0.004	-	0.045	0.452	0.063
start_cups cupboard	-	0.176	0.009	0.018	-	-	-
end_cups cupboard	-	0.176	0.009	0.018	-	-	-
start_groceries cupboard	-	-	0.119	0.055	-	-	-
end_groceries cupboard	-	-	0.123	0.055	-	-	-
start_pans cupboard	-	-	0.009	0.115	-	-	-
end_pans cupboard	-	-	0.009	0.111	-	-	-
start_plates cupboard	-	-	0.106	0.083	-	-	-
end_plates cupboard	-	-	0.106	0.083	-	-	-
start_dishwasher	-	0.010	0.004	0.005	-	-	-
end_dishwasher	-	0.010	0.004	0.005	-	-	-
start_freezer	-	0.020	0.049	0.070	-	-	-
end_freezer	-	0.020	0.049	0.070	-	-	-
start_fridge	-	0.294	0.167	0.106	-	-	-
end_fridge	-	0.294	0.167	0.106	-	-	-
start_microwave	-	-	0.031	0.023	-	-	-
end_microwave	-	-	0.031	0.018	-	-	-
start_toilet flush	-	-	-	0.009	0.067	-	0.163
end_toilet flush	-	-	-	0.009	0.067	-	0.163

respectively; $m_0 : P \rightarrow \mathbb{N}$ is the initial marking associating each place with a number of tokens; $EFT : T \rightarrow \mathbb{Q}_0^+$ and $LFT : T \rightarrow \mathbb{Q}_0^+ \cup \{\infty\}$ associate each transition with an *earliest* and a *latest firing time*, respectively, such that $EFT(t) \leq LFT(t) \ \forall \ t \in T$; $\mathcal{F} : T \rightarrow F_t^s$ associates each transition with a static Cumulative Distribution Function (CDF) with support $[EFT(t), LFT(t)]$; $\mathcal{C} : T \rightarrow \mathbb{R}^+$ associates each transition with a weight; $L : T \rightarrow \mathcal{P}(P)$ is a a *flush function* associating each transition with a subset of P. A place p is termed an *input*, an *output*, or an *inhibitor* place for a transition t if $\langle p, t \rangle \in A^-$, $\langle t, p \rangle \in A^+$, or $\langle p, t \rangle \in A^\cdot$, respectively. A transition t is called *immediate* (IMM) if $[EFT(t), LFT(t)] = [0, 0]$ and *timed* otherwise; a timed transition t is termed *exponential* (EXP) if $F_t(x) = 1 - e^{-\lambda x}$ over $[0, \infty]$ for some rate $\lambda \in \mathbb{R}_0^+$ and *general* (GEN) otherwise; a GEN transition t is called *deterministic* (DET) if

Table 3. Starting event type statistic (trained on all days but the first one): for each event type $\sigma \in \Sigma$, (*i*) frequency θ_σ (first column), (*ii*) frequency $\phi_{\sigma,\gamma}$ for each activity type $\gamma \in \Gamma$ (from the second to the second-to-last column), (*iii*) $1 - \sum_{\gamma \in \Gamma} \phi_{\sigma,\gamma}$ (the last column). Note that only the 16/28 event types that have non-null θ_σ are shown.

	frequency θ_σ	Leaving house	Preparing a beverage	Preparing breakfast	Preparing dinner	Sleeping	Taking shower	Toileting	Idling
start_front door	0.076	0.780	-	-	-	-	-	-	0.220
end_front door	0.017	0.111	-	-	-	-	-	-	0.889
start_hall-bathroom door	0.183	-	-	-	-	-	-	0.707	0.293
end_hall-bathroom door	0.063	-	-	-	-	-	-	0.118	0.882
start_hall-bedroom door	0.050	-	-	-	-	0.148	-	0.296	0.556
end_hall-bedroom door	0.046	-	-	-	-	0.360	-	0.080	0.560
start_hall-toilet door	0.117	-	-	-	-	0.095	0.222	0.016	0.667
end_hall-toilet door	0.128	-	-	0.015	-	0.029	0.116	0.145	0.695
start_cups cupboard	0.048	-	0.308	-	-	-	-	-	0.692
start_groceries cupboard	0.044	-	-	-	0.125	-	-	-	0.875
start_pans cupboard	0.031	-	-	0.118	-	-	-	-	0.882
start_plates cupboard	0.046	-	-	0.520	-	-	-	-	0.480
start_dishwasher	0.024	-	-	-	0.077	-	-	-	0.923
start_freezer	0.020	-	0.091	-	0.273	-	-	-	0.636
start_fridge	0.068	-	0.243	0.081	0.054	-	-	-	0.622
start_toilet flush	0.039	-	-	-	-	-	-	0.524	0.476

$EFT(t) = LFT(t)$ and *distributed* otherwise. For each distributed transition t, we assume that F_t is absolutely continuous over its support and thus that there exists a Probability Density Function (PDF) f_t such that $F_t(x) = \int_0^x f_t(y)dy$.

Semantics. The *state* of an sTPN is a pair $\langle m, \tau \rangle$, where $m : P \to \mathbb{N}$ is a marking and $\tau : T \to \mathbb{R}_0^+$ associates each transition with a time-to-fire. A transition is *enabled* if each of its input places contains at least one token and none of its inhibitor places contains any token; an enabled transition is *firable* if its time-to-fire is not higher than that of any other enabled transition. When multiple transitions are firable, one of them is selected to fire with probability $Prob\{t \text{ is selected}\} = \mathcal{C}(t)/\sum_{t_i \in T^f(s)} \mathcal{C}(t_i)$, where $T^f(s)$ is the set of firable transitions in s. When t fires, $s = \langle m, \tau \rangle$ is replaced by $s' = \langle m', \tau' \rangle$, where m' is derived from m by: (*i*) removing a token from each input place of t and assigning zero tokens to the places in $L(t) \subseteq P$, which yields an intermediate marking m_{tmp}, (*ii*) adding a token to each output place of t. Transitions enabled both by m_{tmp} and by m' are said *persistent*, while those enabled by m' but not by m_{tmp} or m are said *newly-enabled*; if t is still enabled after its own firing, it is regarded as newly enabled [4]. The time-to-fire of persistent transitions is reduced by the time elapsed in s, while the time-to-fire of newly-enabled transitions takes a random value sampled according to their CDF.

3.2   Model Structure and Enhancement

Model Structure. The model used to evaluate the likelihood of observed events is organized by composition of 7+1 submodels, which fit the observed behavior in the 7 activities classified in the Van Kasteren data set [27] and in the remaining *Idling* periods. Figure 3 shows a fragment focused on *Idling* and *Preparing a beverage*.

In the *Idling* submodel, places p_IDLE_durationStart and p_IDLE_eventWait receive a token each when the *Idling* starts, on completion of any activity. The token in p_IDLE_eventWait is removed whenever an event is observed (firing of the GEN transition t_IDLE_interEventsTime); in this case, different IMM transitions are fired depending on the type of the observed event (t_START_FRIDGE, t_START_FREEZER, ...), and then for each different type, a choice is made on whether the event is interpreted as a continuation of the *Idling* period (e.g., t_skip_START_FRIDGE) which will restore a token in p_IDLE_eventWait, or as the starting of each of the possible activities (e.g. t_START_FRIDGE_starts_GET_DRINK, t_START_FRIDGE_starts_PREPARE_DINNER,...). In parallel to all this, the token in p_IDLE_durationStart will be removed when the duration of *Idling* expires (at the firing of the GEN transition t_IDLE_duration) or when an observed event is interpreted as the beginning of any activity (e.g., at the firing of t_START_FRIDGE_starts_GET_DRINK); note that the latter case is not shown in the graphical representation and it is rather encoded in a flush function. Similarly, when the duration of *Idling* expires, the token in p_IDLE_eventWait will be removed by a flush function associated with transition t_GET_IDLE_duration.

In the *Preparing a beverage* submodel, places p_GET_DRINK_durationStart and p_GET_DRINK_eventWait receive a token each when an event observed during the *Idling* period is interpreted as the beginning of an instance of the *Preparing a beverage* activity. The token in p_GET_DRINK_eventWait is removed whenever an event is observed (firing of the GEN transition t_GET_DRINK_interEventsTime), and restored after the event is classified according to its type (IMM transitions t_GET_DRINK_END_FRIDGE, t_GET_DRINK_END_FREEZER, ...). In parallel to this, the token in p_GET_DRINK_durationStart will be removed when the duration of the *Preparing a beverage* activity expires (at the firing of the GEN transition t_GET_DRINK_duration). Note that, in this case, also the token in p_GET_DRINK_eventWait will be removed: this is performed by a flush function associated with transition t_GET_DRINK_duration.

In so doing, in any reachable marking, only the submodel of the current activity contains two non-empty places, one indicating that the activity duration is elapsing (e.g., p_GET_DRINK_durationStart) and the other one meaning that the inter-events time is expiring (e.g., p_GET_DRINK_eventWait). Note that, as a naming convention, any transition named t_EVENT (where EVENT is an event type that may start an activity) or t_ACTIVITY_EVENT (where EVENT is an event type that may occur within ACTIVITY) accounts for an *observable* event, while all the other transitions correspond to *unobservable* events. Finally, note that the general structure of the model is open to modifications in various directions.

For instance, in the submodels of activities, the choice between events might be easily made dependent on the duration of the inter-event time, which would allow a more precise classification without significantly impacting on the analyzability of the model. The viability of such evolutions mainly depends on the significance of the statistics that can be derived form the data set.

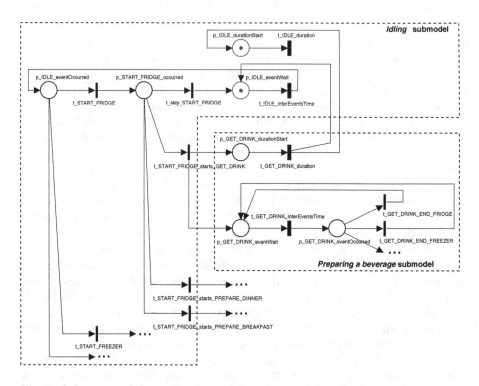

Fig. 3. A fragment of the stochastic model used to evaluate the likelihood measure.

Model Enhancement. The actual topology of the model and its stochastic temporal parameters are derived in automated manner from the statistical indexes extracted in the abstraction of the data set (Sect. 2).

The event types that can start an activity (e.g., in the model of Fig. 3, t_START_FRIDGE_starts_GET_DRINK, t_START_FRIDGE_starts_PREPARE_DINNER, ...) and the discrete probabilities in their random switch are derived from the *starting event type* statistic. The event types that can be observed during each activity (e.g., t_GET_DRINK_END_FRIDGE, t_GET_DRINK_END_FREEZER, ...) or can continue an *Idling* period and the discrete probabilities in their random switch are derived from the *event type* statistics.

The distribution associated with GEN transitions is derived from the *duration* statistic and the *inter-events time* statistic by fitting expected value and Coefficient of Variation (CV) as in [29]: if $0 \leq CV \leq 1/\sqrt{2}$, we assume a shifted exponential distribution with PDF $f(x) = \lambda e^{-\lambda(x-d)}$ over $[d, \infty)$, where σ^2 is the variance, $\lambda = \sigma^{-1}$, and $d = \mu - \sigma$; if $1/\sqrt{2} < CV < 1$, we use a hypo-exponential distribution with PDF $f(x) = \lambda_1\lambda_2/(\lambda_1 - \lambda_2)(e^{-\lambda_2 x} - e^{-\lambda_1 x})$, where $\lambda_i^{-1} = (\mu/2)(1 \pm \sqrt{2CV^2 - 1})$, with $i = 1, 2$; if $CV \approx 1$, we adopt an exponential distribution with $\lambda = 1/\mu$; if $CV > 1$, we consider a hyper-exponential distribution with PDF $f(x) = \sum_{i=1}^{2} p_i\lambda_i e^{-\lambda_i x}$, where $p_i = [1 \pm \sqrt{(CV^2 - 1)/(CV^2 + 1)}]/2$ and $\lambda_i = 2p_i\mu^{-1}$.

3.3   Online Diagnosis and Prediction

We use the transient probability $\pi_\gamma(t)$ that an activity of type $\gamma \in \Gamma$ is being performed at time t as a *measure of likelihood* for γ at t. A prediction $\mathcal{P}(t)$ emitted at time t consists of the set of activity types that may be performed at t, each associated with the likelihood measure, i.e., $\mathcal{P}(t) = \{\langle\gamma, \pi_\gamma(t)\rangle \mid \pi_\gamma(t) \neq 0\}$. Between any two subsequent events $e_n = \langle\sigma_n, t_n\rangle$ and $e_{n+1} = \langle\sigma_{n+1}, t_{n+1}\rangle$, a prediction $\mathcal{P}(t)$ is emitted at equidistant time points in the interval $[t_n, t_{n+1}]$, i.e., $\forall\, t \in \{t_n, t_n + q, t_n + 2q, \ldots, t_{n+1}\}$, with $q \in \mathbb{R}^+$. In the experiments, we assume the activity type with the highest measure of likelihood as the predicted class to be compared against the actual class annotated in the ground truth, i.e., at time t, the predicted activity is $\gamma \mid \pi_\gamma(t) = \max_{a \in \Gamma \mid \langle a, \pi_a(t)\rangle \in \mathcal{P}(t)}\{\pi_a(t)\}$.

As a result of the prescribed model structure and the specific enhancement, the stochastic model subtends a Markov Regenerative Process (MRP) [5,9,10] under enabling restriction, i.e., no more than one GEN transition is enabled in each marking (only the duration of four activities is modeled by a shifted exponential distribution, thus no more than one DET transition is enabled in each marking). According to this, online diagnosis and prediction can be performed by leveraging the regenerative transient analysis of [12]. The solution technique of [12] samples the MRP state after each transition firing, maintaining an additional timer τ_{age} accounting for the absolute elapsed time; each sample, called *transient class*, is made of the marking and the joint PDF of τ_{age} and the times-to-fire of the enabled transitions. Within a given time limit T, enumeration of transient classes is limited to the first regeneration epoch and repeated from every regeneration point (i.e., a state where the future behavior is independent from the past), enabling the evaluation of transient probabilities of reachable markings through the solution of generalized Markov renewal equations.

In the initial transient class of the model, the marking assigns a token to places p_IDLE_durationStart and p_IDLE_eventWait, all transitions are newly enabled, and τ_{age} has a deterministic value equal to zero. After n observed events $e_1 = \langle\sigma_1, t_1\rangle, \ldots, e_n = \langle\sigma_n, t_n\rangle$, let $S_n = \{\langle s_n^i, \omega_n^i\rangle\}$ be the set of possible transient classes s_n^i having probability ω_n^i, where $\sum_{i \mid \langle s_n^i, \omega_n^i\rangle \in S_n} \omega_n^i = 1$. Regenerative transient analysis [12] of the model is performed from each possible transient class $\langle s_n^i, \omega_n^i\rangle \in S_n$ up to any observable event within a given time limit, which is set equal to $48\,h$ to upper bound the time between any two subsequent events. This allows one to evaluate transient probabilities of reachable markings, i.e., $p_m^{n,i}(t) = Prob\{M^{n,i}(t) = m\} \;\forall\, t \leq T, \;\forall\, m \in \mathcal{M}^{n,i}$, where $\mathbb{M}^{n,i} = \{M^{n,i}(t), t \geq 0\}$ is the underlying marking process, and $\mathcal{M}^{n,i}$ is the set of markings that are reachable from s_n^i. Since in any reachable marking only the submodel of the ongoing activity contains non-empty places, transient probabilities of markings are aggregated to derive transient probabilities of ongoing activities $\pi_\gamma(t) = \sum_{\langle s_n^i, \omega_n^i \rangle \in S_n} \omega_n^i \sum_{m \in \mathcal{M}_\gamma^{n,i}} p_m^{n,i}(t) \ \forall \ \gamma \in \Gamma$, where $\mathcal{M}_\gamma^{n,i}$ is the set of markings that are reachable from s_n^i and have non-empty places in the submodel of the activity type γ.

Whenever an event $e_{n+1} = \langle \sigma_{n+1}, t_{n+1} \rangle$ is observed, any tree of transient classes enumerated from class $\langle s_n^i, \omega_n^i \rangle \in S^n$ is explored to determine the possible current classes and their probability. More specifically: (i) the possible current classes are identified as those classes that can be reached after a time $t_{n+1} - t_n$ through a sequence of unobservable events followed by the observed event e_{n+1}; (ii) any possible current class s_{n+1}^j is a regeneration point since, by model construction, each GEN transition is either newly enabled or enabled by a deterministic time (i.e., the timestamp $t_{n+1} - t_n$); and, (iii) the probability ω_{n+1}^j of s_{n+1}^j is obtained as $\lim_{t \to t_{n+1}^-} \zeta_{s_{n+1}^p}(t) \cdot \rho$, where s_{n+1}^p is the last class where the model waits for the arrival of the next event e_{n+1}, $\zeta_{s_{n+1}^p}(t)$ is the probability of being in class s_{n+1}^p at time t, and ρ is the product of transition probabilities of the arcs encountered from s_{n+1}^p to s_{n+1}^j; (iv) in the limit case that s_{n+1}^p is vanishing, ω_{n+1}^j is obtained as the product of transition probabilities of the arcs encountered from the root class to s_{n+1}^j. Hence, the approach is iterated, performing transient analysis from any new current class up to any observable event, still encountering regeneration points after each observed event.

By construction, the approach complexity is linear in the number of observed events. For each observed event $e_n = \langle \sigma_n, t_n \rangle$, the number of transient trees to enumerate is proportional to the number of possible parallel hypotheses $|\mathcal{P}(t_n)|$, i.e., the number of activity types that may be performed at time t_n; moreover, the depth of each transient tree is proportional to the number of events that may occur between two observations (which is bounded in the considered application context), and relatively insensitive to the time elapsed between observed events.

4   Computational Experience

4.1   Experimental Setup

Experiments were performed on the data set [27], using a *dual change-point* representation for sensor events as detailed in Sect. 2. We split data provided by the computed statistics and event logs into training and test sets using a *Leave One Day Out* (LOO) approach, which consists of using each instance of one full day sensor readings for testing and the instances of the remaining days for training. Since, in each test, predictions are emitted starting from the first observed event of the day, we extended online analysis up to the first event of the next day. To avoid inconsistencies in the characterization of *Leaving house*, during which all the event types were observed, we removed from the training sets all events occurring during *Leaving house* that are not of type *start_front door* and *end_front door*. Moreover, whenever the ground truth includes concurrent activities, we considered our prediction correct if the predicted activity type is equal to any of the concurrent activity types.

We performed experiments using two fitting techniques in the evaluation of the duration and the inter-events time statistics: (*i*) only exponential distributions (i.e., *exponential* case); (*ii*) different classes of distributions based on the CV value, as discussed in Sect. 3.2 (i.e., *non-Markovian* case). On a machine with an Intel Xeon 2.67 GHz and 32 GB RAM, the evaluation for a single day took on average 43 s for the exponential case and 18 min for the non-Markovian case.

We evaluated the performance of our approach computing, for each activity class, three measures, derived from the number of true positives (TP), false positives (FP), and false negatives (FN): (*i*) *precision* $Pr = TP/(TP + FP)$, which accounts for the accuracy provided that a specific class has been predicted; (*ii*) *recall* $Re = TP/(TP + FN)$, which represents the ability to select instances of a certain class from a data set; and, (*iii*) *F-measure* $F_1 = 2*Pr*Re/(Pr+Re)$, which is the harmonic mean of precision and recall.

We also compared the outcome of our experiments with the results reported in [27], obtained using a generative model (i.e., an HMM) and a discriminative one (i.e., a CRF) in combination with *offline* inference and the change-point representation. To make this comparison possible, we sampled the result of our prediction using a timeslice of duration $\Delta t = 60$ s and we considered two additional measures: (*i*) *accuracy* $A = 1/N \sum_{i=1}^{N_c} TP_i$, which is the average percentage of correctly classified timeslices, with N being the total number of timeslices, N_c the total number of activity classes, and TP_i the number of TP of class i; and, (*ii*) *average recall* $\bar{Re} = 1/N_c \sum_{i=1}^{N_c} Re_i$, which is the average percentage of correctly classified timeslices per class, with Re_i being the recall of class i.

4.2   Results

Table 4 shows the confusion matrix for the exponential and the non-Markovian cases, which reports in position i, j the number of timeslices of class i predicted as class j; Table 5 shows precision, recall and F_1 score as derived from the confusion matrix. Results show that *Idling*, *Leaving house*, and *Sleeping* are the activities with the highest F_1 score. In terms of F_1 score, the non-Markovian case outperforms the exponential one for all activity classes except for *Preparing a beverage* and *Taking shower*. In terms of precision, the non-Markovian case performs worse only for *Sleeping*. Conversely, in terms of recall, the exponential case outperforms the non-Markovian one for *Preparing a beverage*, *Preparing breakfast*, *Taking shower*, and *Toileting*, and performs worse for *Idling*, *Preparing dinner*, and *Sleeping*. Note that the precision, recall, and F_1 score of *Leaving house* are identical in both cases.

Accuracy and average recall are summarized in Table 6, and compared with results from [27]. As we can see, fitting statistical data according to the CV (non-Markovian case), we achieve the highest accuracy, both w.r.t. our exponential case and w.r.t. HMM and CRF. Nevertheless, the exponential case, HMM, and CRF outperform the non-Markovian case in terms of average recall.

Table 4. Confusion matrix showing the number of timeslices of each class i (first column) predicted as class j (other columns): exponential case/non-Markovian case. Diagonal elements represent TP, while FN (FP) can be read along rows (columns).

	Idling	Leaving house	Preparing a beverage	Preparing breakfast	Preparing dinner	Sleeping	Taking shower	Toileting
Idling	2975/3471	330/330	37/6	82/16	733/400	588/658	131/65	98/28
Leaving house	184/209	22219/22219	1/0	0/0	101/56	25/71	22/4	15/8
Preparing a beverage	8/6	1/1	5/2	0/0	6/11	1/1	0/0	0/0
Preparing breakfast	5/14	0/0	3/2	39/24	15/22	8/8	0/0	0/0
Preparing dinner	83/107	0/0	4/1	39/13	214/221	0/0	0/0	1/0
Sleeping	194/215	0/0	0/0	0/0	0/0	11226/11430	3/1	231/8
Taking shower	94/100	0/0	0/0	0/0	1/1	17/41	105/79	4/0
Toileting	46/60	2/2	0/0	0/0	0/1	36/52	7/8	66/33

Table 5. Precision, recall, and F_1 score achieved for each activity type.

	Exponential			non-Markovian		
	Precision	Recall	F_1	Precision	Recall	F_1
Idling	82.89	59.81	69.49	83.00	69.78	75.82
Leaving house	98.52	98.46	98.49	98.52	98.46	98.49
Preparing a beverage	10.00	23.81	14.09	18.18	9.52	12.50
Preparing breakfast	24.38	55.71	33.91	45.28	34.29	39.02
Preparing dinner	20.00	62.76	30.33	31.04	64.62	41.94
Sleeping	94.33	96.33	95.32	93.22	98.08	95.59
Taking shower	39.18	47.51	42.95	50.32	35.75	41.80
Toileting	15.90	42.04	23.08	42.86	21.15	28.33

Table 6. Accuracy and average recall achieved by the exponential and non-Markovian cases (dual change-point representation and online analysis), compared with those reported in [27] for HMM and CRF (change-point representation and offline analysis).

	Accuracy	Average recall
Exponential case	92.11	60.80
non-Markovian case	93.69	53.96
HMM [27]	80.00	67.20
CRF [27]	89.40	61.70

5   Discussion

Experimentation developed so far achieves results that compare well with the HMM and CRF approaches, with a slight increase in precision and a slight reduction in recall. The proposed approach is open to various possible developments, and the insight on observed cases of success and failure comprises a foundation for refinement and further research on which we are presently working.

In the present implementation, classification of the current activity relies only on past events, which is for us instrumental to open the way to the integration of classification with on-line prediction. However, the assumption of this causal constraint severely hinders our approach in the comparison against the offline classification implemented in [27] through HMM and CRF. Online classification results reported in [27] are unfortunately not comparable due to the different abstraction applied on events, and it should be remarked that in any case they are not completely online as the classification at time t relies on all the events that will be observed within the end of the timeslice that contains t itself, which makes a difference for short duration activities. For the purposes of comparison, our online approach can be relaxed to support offline classification by adding a backtrack from the final states reached by the predictor. This should in particular help the recall of short activities started by events that can be accepted also as the beginning of some longer activity (e.g., Preparing a beverage w.r.t. Preparing dinner). We also expect that this should reduce the number of cases where a time period is misclassified as Idling.

The statistics of durations are now fitted using the basic technique of [29] which preserves only expected value and variation coefficient. Moreover, the deterministic shift introduced in the approximation of hypo-exponential distributions with low CV causes a false negative for all the events occurring before the shift completion, which is in fact observed in various cases. Approximation through acyclic Continuous PHase type (CPH) distributions [13,17,21] would remove the problem and allow an adaptable trade-off between accuracy and complexity. In particular, a promising approach seems to be the method of [6] which permits direct derivation of an acyclic phase type distribution fitting not only expected value and variation coefficient but also skewness.

Following a different approach, the present implementation is completely open to the usage of any generally distributed (GEN) representation of activity durations. This would maintain the underlying stochastic process of the on-line model within the current class, i.e., Markov Regenerative Processes (MRP) that run under enabling restriction [10] and guarantee a regeneration within a bounded number of steps. In this case, on-line prediction could be practically implemented using various tools, including Oris [7,12], TimeNet [31] and Great-SPN [1].

In the present implementation, classification is unaware of the absolute time, which may instead become crucial to separate similar activities, such as for instance Preparing breakfast and Preparing dinner. To overcome the limitation, the model should in principle become non-homogeneous, but a good approximation can be obtained by assuming a discretized partition of the daytime, which can be cast in the on-line model as a sequence of deterministic delays. By exploiting the timestamps, at each event the current estimation of the absolute time is restarted. Under the fair assumption that at least one event is observed in each activity, the underlying stochastic process of the on-line model still falls in the class of MRPs that encounter a regeneration within a bounded number of steps, and can thus be practically analyzed through the Oris Tool [7,12].

References

1. Amparore, E.G., Buchholz, P., Donatelli, S.: A structured solution approach for markov regenerative processes. In: Norman, G., Sanders, W. (eds.) QEST 2014. LNCS, vol. 8657, pp. 9–24. Springer, Heidelberg (2014)
2. Avci, U., Passerini, A.: Improving activity recognition by segmental pattern mining. IEEE Trans. Knowl. Data Eng. **26**(4), 889–902 (2014)
3. Bartocci, E., Bortolussi, L., Sanguinetti, G.: Data-driven statistical learning of temporal logic properties. In: Legay, A., Bozga, M. (eds.) FORMATS 2014. LNCS, vol. 8711, pp. 23–37. Springer, Heidelberg (2014)
4. Berthomieu, B., Diaz, M.: Modeling and verification of time dependent systems using time petri nets. IEEE Trans. Soft. Eng. **17**(3), 259–273 (1991)
5. Bobbio, A., Telek, M.: Markov regenerative SPN with non-overlapping activity cycles. In: International Computer Performance and Dependability Symposium, pp. 124–133 (1995)
6. Bobbio, A., Horváth, A., Telek, M.: Matching three moments with minimal acyclic phase type distributions. Stoch. Models **21**(2–3), 303–326 (2005)
7. Bucci, G., Carnevali, L., Ridi, L., Vicario, E.: Oris: a tool for modeling, verification and evaluation of real-time systems. Int. J. Softw. Tools Technol. Transfer **12**(5), 391–403 (2010)
8. Buchholz, R., Krull, C., Strigl, T., Horton, G.: Using hidden non-markovian models to reconstruct system behavior in partially-observable systems. In: International ICST Conference on Simulation Tools and Techniques, p. 86 (2010)
9. Choi, H., Kulkarni, V.G., Trivedi, K.S.: Markov regenerative stochastic Petri nets. Perf. Eval. **20**(1–3), 337–357 (1994)
10. Ciardo, G., German, R., Lindemann, C.: A characterization of the stochastic process underlying a stochastic Petri net. IEEE Trans. Softw. Eng. **20**(7), 506–515 (1994)
11. Cook, D.J., Augusto, J.C., Jakkula, V.R.: Ambient intelligence: technologies, applications, and opportunities. Pervasive Mob. Comput. **5**(4), 277–298 (2009)
12. Horváth, A., Paolieri, M., Ridi, L., Vicario, E.: Transient analysis of non-Markovian models using stochastic state classes. Perform. Eval. **69**(7–8), 315–335 (2012)
13. Horváth, A., Telek, M.: PhFit: a general phase-type fitting tool. In: Field, T., Harrison, P.G., Bradley, J., Harder, U. (eds.) TOOLS 2002. LNCS, vol. 2324, pp. 82–91. Springer, Heidelberg (2002)
14. Katz, S., Downs, T.D., Cash, H.R., Grotz, R.C.: Progress in development of the index of ADL. The Gerontologist **10**(1 Part 1), 20–30 (1970)
15. Mans, R.S., Schonenberg, M.H., Song, M., van der Aalst, W.M.P., Bakker, P.J.M.: Application of process mining in healthcare - a case study in a dutch hospital. In: Fred, A., Filipe, J., Gamboa, H. (eds.) BIOSTEC 2011. CCIS, vol. 273, pp. 425–438. Springer, Heidelberg (2009)

16. Mitchell, C.D., Jamieson, L.H.: Modeling duration in a hidden Markov model with the exponential family. IEEE Int. Conf. Acoust. Speech Signal Process. **2**, 331–334 (1993)
17. Neuts, M.F.: Matrix Geometric Solutions in Stochastic Models. Johns Hopkins University Press, London (1981)
18. Patterson, D.J., Liao, L., Fox, D., Kautz, H.: Inferring high-level behavior from low-level sensors. In: Dey, A.K., Schmidt, A., McCarthy, J.F. (eds.) UbiComp 2003. LNCS, vol. 2864, pp. 73–89. Springer, Heidelberg (2003)
19. Rashidi, P., Cook, D.J.: Keeping the resident in the loop: adapting the smart home to the user. IEEE Trans. Syst. Man Cybern. Part A: Syst. Hum. **39**(5), 949–959 (2009)
20. Rashidi, P., Cook, D.J., Holder, L.B., Schmitter-Edgecombe, M.: Discovering activities to recognize and track in a smart environment. IEEE Trans. Knowl. Data Eng. **23**(4), 527–539 (2011)
21. Reinecke, P., Krauß, T., Wolter, K.: Phase-type fitting using hyperstar. In: Balsamo, M.S., Knottenbelt, W.J., Marin, A. (eds.) EPEW 2013. LNCS, vol. 8168, pp. 164–175. Springer, Heidelberg (2013)
22. Rogge-Solti, A., Weske, M.: Prediction of remaining service execution time using stochastic petri nets with arbitrary firing delays. In: Basu, S., Pautasso, C., Zhang, L., Fu, X. (eds.) ICSOC 2013. LNCS, vol. 8274, pp. 389–403. Springer, Heidelberg (2013)
23. Trivedi, K.S.: Probability and Statistics with Reliability, Queuing, and Computer Science Applications. John Wiley and Sons, New York (2001)
24. van der Aalst, W., et al.: Process mining manifesto. In: Daniel, F., Barkaoui, K., Dustdar, S. (eds.) BPM Workshops 2011, Part I. LNBIP, vol. 99, pp. 169–194. Springer, Heidelberg (2012)
25. van der Aalst, W.M.P., Reijers, H.A., Weijters, A.J.M.M., van Dongen, B.F., Alves De Medeiros, A.K., Song, M., Verbeek, H.M.W.: Business process mining: an industrial application. Inf. Syst. **32**(5), 713–732 (2007)
26. van Dongen, B.F., de Medeiros, A.K.A., Verbeek, H.M.W.E., Weijters, A.J.M.M.T., van der Aalst, W.M.P.: The ProM framework: a new era in process mining tool support. In: Ciardo, G., Darondeau, P. (eds.) ICATPN 2005. LNCS, vol. 3536, pp. 444–454. Springer, Heidelberg (2005)
27. van Kasteren, T., Noulas, A., Englebienne, G., Kröse, B.: Accurate activity recognition in a home setting. In: Proceedings of the International Conference on Ubiquitous Computing, UbiComp 2008, pp. 1–9. ACM, New York, NY, USA (2008)
28. Vicario, E., Sassoli, L., Carnevali, L.: Using stochastic state classes in quantitative evaluation of dense-time reactive systems. IEEE Trans. Softw. Eng. **35**(5), 703–719 (2009)
29. Whitt, W.: Approximating a point process by a renewal process, I: two basic methods. Oper. Res. **30**(1), 125–147 (1982)
30. Ye, J., Dobson, S., McKeever, S.: Situation identification techniques in pervasive computing: a review. Pervasive Mob. Comput. **8**(1), 36–66 (2012)
31. Zimmermann, A.: Dependability evaluation of complex systems with TimeNET. In: Proceedings of the International Workshop on Dynamic Aspects in Dependability Models for Fault-Tolerant Systems, pp. 33–34 (2010)

Power Trading Coordination in Smart Grids Using Dynamic Learning and Coalitional Game Theory

Farshad Shams[✉] and Mirco Tribastone

IMT Institute for Advanced Studies, Piazza S. Francesco,
19, 55100 Lucca, Italy
{f.shams,m.tribastone}@imtlucca.it

Abstract. In traditional power distribution models, consumers acquire power from the central distribution unit, while "micro-grids" in a smart power grid can also trade power between themselves. In this paper, we investigate the problem of power trading coordination among such micro-grids. Each micro-grid has a surplus or a deficit quantity of power to transfer or to acquire, respectively. A coalitional game theory based algorithm is devised to form a set of coalitions. The coordination among micro-grids determines the amount of power to transfer over each transmission line in order to serve all micro-grids in demand by the supplier micro-grids and the central distribution unit with the purpose of minimizing the amount of dissipated power during generation and transfer. We propose two dynamic learning processes: one to form a coalition structure and one to provide the formed coalitions with the highest power saving. Numerical results show that dissipated power in the proposed cooperative smart grid is only 10 % of that in traditional power distribution networks.

1 Introduction

While "macro-grids" were traditionally viewed as a technology used in remote area power supplies at a high-voltage, a "micro-grid" (MG) is introduced as a collective of geographically proximate, electrically connected loads and generators based on renewable energy technologies at a medium-voltage [1]. In general, an MG may or may not be connected to the wider electricity grid. Figure 1 shows conceptual differences between the traditional grid, which is hierarchical, and a grid including MGs. In traditional electricity transmission, distribution networks act like the branches of the tree, interconnecting loads and the long-distance transmission network. The MG concept offers a path to autonomous, intelligent low-emissions electricity systems, by creating a localized smart grid that allows advanced and distributed control while being compatible with traditional electricity infrastructure. With such promise, MGs are of growing interest to grid operators, as a way of enhancing the performance of electricity systems [2]. However, realizing MGs is not without challenges, among which we will investigate the problem of coordination between MG electricity generators in order to

© Springer International Publishing Switzerland 2015
J. Campos and B.R. Haverkort (Eds.): QEST 2015, LNCS 9259, pp. 54–69, 2015.
DOI: 10.1007/978-3-319-22264-6_4

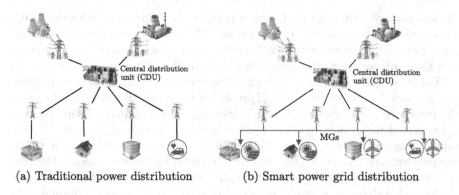

(a) Traditional power distribution   (b) Smart power grid distribution

Fig. 1. Traditional vs. smart power grid electricity distribution.

minimize of the power dissipation over the transmission lines. The minimization of energy dissipation is and will be an important focus in electricity markets considering that power dissipated is accompanied with the cost of energy. References [3–6] investigate the problem of power loss reduction in smart grids. References [3,4] focus on the optimization of various electricity parameters during peak load times.

Game theory [7] is a potential mathematical tool to model smart grid [8–10]. Non-cooperative game theory can model the distributed operations in smart power grids and cooperative/coalitional game theory can model the cooperation among nodes [11,12]. To the best of our knowledge, in the existing literature there are only two game theory based algorithms with aim at minimization of power dissipated. References [5,6] propose coalitional game theory based algorithms which form a set of disjoint coalitions and enable MGs in each coalition to trade power among themselves with aim at minimization the overall power dissipated. The authors in [5] maximize the utility function of the grand coalition (coalition of all MGs), while the authors in [6] maximize Shapley values [13] of MGs. The results of [5] show that the proposed algorithm improves the performance reaching up to 31 % (with 30 MGs) compared to the traditional transmission grids. Unfortunately, the performance of the proposed algorithm in [6] is evaluated with non realistic smart grid network area $10 \times 10\,\mathrm{km}^2$. It is obvious that in small areas the amount of power dissipated is less than in a realistic network area $100 \times 100\,\mathrm{km}^2$. In addition, in References [5,6], the circuit of transmission lines is not realistic (We explain why in Sect. 2).

In this work, we investigate the problem of power trading coordination between MGs in smart power grids. Each MG has either a surplus or deficit quantity of power. Each MG with a surplus amount of power can transfer to the central distribution unit (CDU), and meanwhile it can serve MGs with deficit power. Each MG with a deficit amount of power can receive from the CDU and from MGs with surplus power. With the aim of power loss minimization during power generation and transfer, we will introduce a power trading coordination strategy among MGs. We formulate the problem using coalitional game theory in which MGs form a set of not necessarily disjoint and possibly singleton coalitions.

MGs in each coalition can trade power between themselves and with the CDU. Each non-singleton coalition consists of one MG with surplus power which will serve the needed MG(s) with deficit power. On the other hand, the micro-grid in a singleton coalition only trades directly with the CDU.

For achieving the best power distribution over transmission lines, we propose two dynamic learning algorithms: (1) coalition formation dynamic learning, and (2) power loss minimization dynamic learning. The dynamic learning process (1) is nested in (2) one. In other words, the dynamic learning process (2) iteratively executes learning process (1) until it achieves a fixed point at which the performance can no longer improve. Coalition formation dynamic learning achieves a coalition formation structure and then the complementary power loss minimization dynamic learning leads the MGs to the maximum performance in terms of power saving. We show the stability (the convergence to a fixed-point) of both dynamic processes using the Kakutani fixed point theorem [14]. As our evaluation shows, our approach enables MGs to come to a power trading coordination among themselves that yields a significant power saving compared to the traditional power trading.

The remainder of the paper is structured as follows. Sect. 2 studies the circuit of transmission lines in a smart power grid. Sect. 3 consists of two subsections which contains the smart power network model, and models it as a coalitional game, respectively. The two subsections in Sect. 4 propose an innovative algorithm to form coalition and to maximize the performance in terms of power saving, respectively. Sect. 5 presents some simulation results and finally Sect. 6 concludes the paper.

2   Transmission Line Model

Figure 2a shows the single-phase equivalent[1] circuit of a medium-length (up to 200km) transmission line [1]. On an electricity transmission line from the sending-end s to the receiving-end r, the variables of interest are the (complex-valued) voltages and currents per phase $\overrightarrow{V_s}$, $\overrightarrow{V_r}$, $\overrightarrow{I_s}$, and $\overrightarrow{I_r}$ at the end-line terminals. The variables V_s and I_s denote the amplitudes of $\overrightarrow{V_s}$ and $\overrightarrow{I_s}$, respectively, and likewise for V_r and I_r. As the transmission line model in Fig. 2a has many parameters, it is convenient to replace each line with its π-equivalent shown in Fig. 2b where the impedance $\overrightarrow{Z} = R + i\omega L$ Ω and the admittance $\overrightarrow{Y} = 1/R_1 + i\omega C$ Ω^{-1} with i as the imaginary unit and $\omega = 2\pi f$ as the (sinusoidal) angular frequency in electrical radians at the frequency f.

Suppose the sending-end wants to supply a power amount P_s [W] and transfers it toward the receiving-end. A fraction of P_s will be dissipated during generation and transfer and the receiving-end r can load a power amount $P_r < P_s$. The real power loss is calculated by the following formula:

$$PL_{sr} = P_s - P_r = V_s.I_s.\cos\theta_s - V_r.I_r.\cos\theta_r \quad [\text{W}] \tag{1}$$

---

[1] This circuit is suitable for analysing its symmetrical three-phase operation.

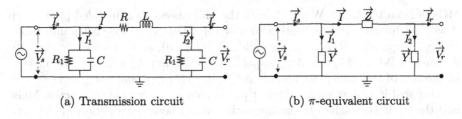

(a) Transmission circuit   (b) π-equivalent circuit

Fig. 2. Single-phase circuit of a medium-length transmission line.

where the parameter θ_s is the "power factor angle", in the range of $(-90°, +90°)$, and that is the phase difference between $\overrightarrow{V_s}$ and $\overrightarrow{I_s}$, and likewise for θ_r. So, the received power at the receiving-end r from the transmission line $s \to r$ can be formulated as $P_r = P_s - PL_{sr}$.

In a transmission line, the amplitude of the voltage at sending-end V_s is known and we suppose $\overrightarrow{V_s}$ to be the reference vector, i.e., $\overrightarrow{V_s} = V_s \angle 0°$. We suppose also, as usual, that the power factor at the sending-end, $\cos\theta_s$, is known with the assumption that the current waveform comes delayed after the voltage waveform, i.e., $\overrightarrow{I_s} = I_s \angle -\theta_s$. For supplying and transferring P_s amount of power, the sending-end regulates the amplitude of the current as $I_s = \frac{P_s}{V_s.\cos\theta_s}$ and synchronizes its phase at $-\theta_s$. Knowing the parameters of the transmission line in the nominal π-circuit in Fig. 2b, the voltage and the current at the receiving-end are calculated by the following equations:

$$\overrightarrow{V_r} = \left(1 + \overrightarrow{Y}.\overrightarrow{Z}\right).V_s - \overrightarrow{Z}.\overrightarrow{I_s}$$
$$\overrightarrow{I_r} = -\overrightarrow{Y}.\left(2 + \overrightarrow{Y}.\overrightarrow{Z}\right).V_s + \left(1 + \overrightarrow{Y}.\overrightarrow{Z}\right).\overrightarrow{I_s} \tag{2}$$

Now, we can calculate the receiving-end's parameters V_r, I_r, and θ_r using (2) and then the amount of power loss is calculated as in (1). It is simple to show that the power loss equation (1) is a concave function of I_s when $L = C = 1/R_1 = 0$ (short-length transmission line model [1] as considered in References [5,6]), otherwise its shape strictly depends upon the values of the parameters.

3 System Model and Problem Formulation

3.1 System Model

We study a smart power grid consisting of a single CDU which is connected to one or more main power plants at a high-voltage. The CDU is connected to M MGs denoted by the set $\mathcal{M}$. At any given time frame [8], each MG $r \in \mathcal{M}$ has a Q_r residual power load which is defined as the difference between the generated power and the overall demand. A positive quantity $Q_r > 0$ determines the surplus power that the MG can transfer to other MGs or to the CDU, whereas a negative quantity $Q_r < 0$ determines the deficit power the MG needs to acquire from other

MGs or from the CDU. When $Q_r = 0$, the MG rmeets its demanded power and it will not interact with any other MG or the CDU. We divide MGs into three groups: "suppliers" ($Q_r > 0$), "demanders" ($Q_r < 0$), and "inactives" ($Q_r = 0$) denoted by $\mathcal{M}^+$, $\mathcal{M}^-$, and $\mathcal{M}^0$, respectively, such that $\mathcal{M} = \mathcal{M}^+ \cup \mathcal{M}^- \cup \mathcal{M}^0$. In the rest of this paper, we assume $\mathcal{M}^0 = \emptyset$, $|\mathcal{M}^+| \geq 1$, and $|\mathcal{M}^-| \geq 1$.

Our goal is to develop an efficient power transfer policy between active MGs and the CDU themselves to minimize the overall power dissipation in the network. We will propose an algorithm which assigns to each supplier $m \in \mathcal{M}^+$ a subset of demanders in $\mathcal{M}^-$ and determines different fractions of Q_m to be transferred to each assigned demander and to the CDU. Doing so, traded power may be transferred on short distance lines. As values of each transmission line components (resistor, inductor, and capacitor) are an inverse function of the distance, the amount of dissipated power will be much lower than that in a traditional distribution. However, a MG can decide to act as a non-cooperative MG which trades only with the CDU. Each MG in demand can be assigned to more suppliers to acquire the whole or a fraction of its own needed power. Two suppliers will not belong to the same coalition, since they do not trade power between themselves.

3.2   Problem Formulation

To study the cooperative behavior of the MGs, we use a coalitional game theory framework [11]. A coalitional game is defined as $\mathcal{G} = (\mathcal{M}, \nu)$ where $\mathcal{M}$ is the player's set (the active MGs), and $\nu : 2^{\mathcal{M}} \longrightarrow \mathbb{R}$ is the characteristic function of each coalition (subset of $\mathcal{M}$) that assigns a real number representing the benefit earned by the coalition. We will propose an algorithm which forms a set of not necessarily disjoint and possibly singleton coalitions denoted by $\mathcal{K} = \{\mathcal{K}_1, \ldots, \mathcal{K}_K\}$. The (unique) MG in a singleton coalition will trade only with the CDU. A non-singleton formed coalition consists of *only* one supplier and one or more demanders, i.e., the unique supplier MG provides the whole or a fraction of the overall power demanded by the assigned demander(s). Each demander in $\mathcal{M}^-$ may belong to more coalitions, i.e., the whole or a fraction of its needed power can be provided by more suppliers. Then, all MGs can trade with the CDU for the rest, if any. Figure 3 shows a sample wherein two cooperative coalitions and one non-cooperative singleton coalition are formed. For instance, MG1 can decide to transfer 10 % of its (surplus) quantity to MG2 and 30 % to MG3 and the rest to the CDU. MG4 can decide to transfer 50 % of its quantity to MG3 and the rest to the CDU. The rest of deficit quantities of MG3 and MG4 will be provided by the CDU. MG5 will be completely served by the CDU.

A coalition structure will be formed only if (i) surplus power quantities of all suppliers are completely loaded, and (ii) deficit power quantities of all demanders are completely served, as the following equations state:

$$
\begin{cases}
\displaystyle\sum_{n \in \mathcal{M}^-} P_{mn} + P_{m0} = Q_m & \forall m \in \mathcal{M}^+ & \text{(3a)} \\[2em]
\displaystyle\sum_{m \in \mathcal{M}^+} (P_{mn} - PL_{mn}) + (P_{0n} - PL_{0n}) = -Q_n & \forall n \in \mathcal{M}^- & \text{(3b)}
\end{cases}
$$

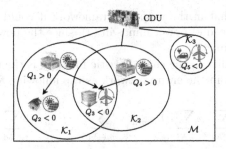

Fig. 3. A sample of formed coalitions for cooperative power distribution.

where the subscript 0 denotes the CDU index. The parameter P_{ij} denotes the amount of loaded power by the sending-end i as $P_{ij} = V_{ij}.I_{ij}.\cos\theta_{ij}$ to transfer over the transmission line $i \to j$. We suppose that the power network is *meshed*, i.e. each sending-end can regulate a different voltage, current, and power factor angle over each connected transmission line. The parameter PL_{ij} is the amount of power dissipated, calculated as in (1) while the sending-end terminal i loads P_{ij} over the transmission line $i \to j$. Condition (3a) guarantees that the power quantity of each supplier $m \in \mathcal{M}^+$ is completely loaded over the connected transmission lines and condition (3b) guarantees that each demander $n \in \mathcal{M}^-$ receives the whole deficit quantity (demanded) power.

We divide the MGs in each formed coalition $\mathcal{K}_k$ into two subsets of suppliers and demanders denoted by $\mathcal{K}_k^+$ and $\mathcal{K}_k^-$, respectively. For a singleton coalition $\mathcal{K}_k$, either $\mathcal{K}_k^+$ or $\mathcal{K}_k^-$ is empty. In each non-singleton coalition $\mathcal{K}_k$, $\mathcal{K}_k^+$ is singleton since the supplier MG is unique. For each coalition $\mathcal{K}_k$ with $\mathcal{K}_k^+ = \{m\}$, we define the characteristic function as the inverse of total dissipated power over the distribution lines incurred by the power generation and transfer, as:

$$\nu(\mathcal{K}_k) = \left(\sum_{n \in \mathcal{K}_k^-} PL_{mn} + PL_{m0} + \sum_{n \in \mathcal{K}_k^-} PL_{0n} \right)^{-1} \tag{4}$$

where in the power minus accounts for the maximization problem, term PL_{mn} refers to the power dissipated incurred by the generation and transfer of P_{mn} from the supplier m to each $n \in \mathcal{K}_k^-$, the term PL_{m0} refers to the power dissipated incurred by the transfer of the power amount P_{m0} from the supplier m to the CDU, and PL_{0n} refers to the power dissipated incurred by the power transfer from the CDU to the demanders in $\mathcal{K}_k^-$. For a singleton coalition which consists of one non-cooperative supplier MG, terms PL_{mn} and PL_{0n} are equal to zero and term PL_{m0} is calculated with setting $P_{m0} = Q_m$. On the other hand, for a singleton coalition which consists of one demander MG, terms PL_{mn} and PL_{m0} are equal to zero and the term PL_{0n} is calculated with setting $P_{0n} = -Q_n + PL_{0n}$, i.e., the demanded power $-Q_n$ will be served only by the CDU. We assign $\nu(\mathcal{K}_k) = -\infty$ when the power distribution conditions (3) do not hold. We assume the unit of currency as the price of unit amount of power. It is worthwhile to note that in the proposed framework it could also be considered coefficient minus instead of power minus in (4), since the power loss is neither a concave nor convex function.

For the grand coalition, the characteristic function is the inverse of the total amount of power loss in the entire network as the following formula:

$$\nu\left(\mathcal{M}\right) = \left(\sum_{m \in \mathcal{M}^+} \sum_{n \in \mathcal{M}^-} PL_{mn} + \sum_{m \in \mathcal{M}^+} PL_{m0} + \sum_{n \in \mathcal{M}^-} PL_{0n} \right)^{-1} \tag{5}$$

We assign $\nu\left(\mathcal{M}\right) = -\infty$ when the power distribution conditions (3) do not hold.

A central question in a coalitional game is how to divide the earnings among the members of the formed coalition. The payoff of each MG (player in the game) can show the power of influence of the MG and it is obvious that a higher payoff is an incentive to cooperate more efficiently. The Shapley value [13] assigns a unique outcome to each MG. Let us denote ϕ_r as the Shapley value of MG $r \in \mathcal{M}$ in the game. For each MG $r \in \mathcal{M}$:

$$\phi_r = \sum_{\substack{\forall \mathcal{K}_k \subseteq \mathcal{M} \\ \text{s.t. } r \in \mathcal{K}_k}} \frac{(|\mathcal{K}_k| - 1)! \cdot (M - |\mathcal{K}_k|)!}{M!} \left(\nu\left(\mathcal{K}_k\right) - \nu\left(\mathcal{K}_k \backslash \{r\}\right) \right) \tag{6}$$

The expression $\nu\left(\mathcal{K}_k\right) - \nu\left(\mathcal{K}_k \backslash \{r\}\right)$ is the marginal payoff of the MG r to the coalition $\mathcal{K}_k$. The Shapley value can be interpreted as the marginal contribution an MG makes, averaged across all permutations of MGs that may occur.

4   Best Response Algorithm

In this section we provide an answer to the question: *How do the MGs form the best coalition structure with aim at minimization of the amount of power dissipated?* Dynamic learning models provide a framework for analyzing the way players set their proper strategies. To address this question, we propose an "adaptive learning algorithm" [15] in which the learners use the same learning algorithm. Typically, these types of algorithms are constructed to iteratively play a game with an opponent, and, by playing this game, to converge a solution. Forming a coalition structure by MGs is equivalent to distribute $Q_r \, \forall r \in \mathcal{M}$ over the transmission lines satisfying conditions (3). Obviously the solution is not unique. In the next subsections, we first propose a dynamic learning process to form a coalition structure where each formed coalition earns a positive bounded payoff and each MG earns a bounded Shapley value. We then introduce another complementary dynamic process which iteratively executes the coalition formation dynamic learning process in order to choose the best coalition structure with minimum power dissipated.

4.1   Coalition Formation Dynamic Learning Process

To realize a coalition formation structure, we use dynamic learning. Our goal is to determine the best amount of power loaded at the sending-end sides of the uni-directional transmission lines denoted by $\mathcal{L} = \{l_{mn}\} \cup \{l_{m0}\} \cup \{l_{0n}\}$, $\forall m \in \mathcal{M}^+$ and $\forall n \in \mathcal{M}^-$ where the symbol l_{ij} denotes the transmission line from the sending-end i to the receiving-end j. We denote $L = |\mathcal{L}| = |\mathcal{M}^+| \cdot |\mathcal{M}^-| + |\mathcal{M}^+| + |\mathcal{M}^-|$ as the number of individuals. We denote the parameters of the sending-end i over the transmission line l_{ij} (to the receiving-end j) by I_{ij}, V_{ij}, P_{ij}, and θ_{ij}. We will propose an algorithm to determine the best value of P_{ij}. The values of V_{ij} and θ_{ij} are known at the sending-end i and it will regulate the electricity current as $I_{ij} = \frac{P_{ij}}{V_{ij} \cos \theta_{ij}}$ to load P_{ij} toward the receiving-end j.

In our dynamic learning process, each transmission line $l_{ij} \in \mathcal{L}$ is an individual learner which, during learning process learns about the best amount of the power P_{ij} to be loaded by the sending-end i to transfer over the transmission line l_{ij}. By doing so: 1) each supplier $m \in \mathcal{M}^+$ will be aware of the amount of power to load over the transmission lines toward all demanders and the CDU, i.e., over l_{m0} and $\{l_{mn}\} \; \forall n \in \mathcal{M}^-$, and 2) the CDU will be aware of the amount of power to load over the transmission lines toward all demanders, i.e., over $\{l_{0n}\}$ $\forall n \in \mathcal{M}^-$. If one supplier m loads a power $P_{mn} > 0$ over the transmission line l_{mn}, the MGs m and n will belong to the same coalition. If one supplier m loads $P_{m0} = Q_m$ over l_{m0}, the supplier m will form a singleton coalition. One demander $n \in \mathcal{M}^-$ will form a singleton coalition if it receives the whole $-Q_n$ from the CDU, i.e., $P_{0n} - PL_{0n} = -Q_n$.

For each P_{ij}, let us denote the maximum power constraint and the best power amount by $\bar{P}_{ij}$ and P_{ij}^*, respectively. We set $\bar{P}_{mn} = \min \{Q_m, -\beta Q_n\}$, $\bar{P}_{m0} = Q_m$, and $\bar{P}_{0n} = -\beta Q_n$ with $\beta > 1$ for $\forall m \in \mathcal{M}^+$ and $\forall n \in \mathcal{M}^-$. About $\bar{P}_{m0}$, it is obvious that loaded power by a supplier cannot be greater than its residual power load. With respect to $\bar{P}_{0n}$, whenever the CDU wants to completely serve one demander n, it must load a power amount $P_{0n} > -Q_n$ to overcome power loss over the transmission line $0 \rightarrow n$ and so to guarantee received power $-Q_n$ at the proper demander side. Now, the setting of $\bar{P}_{mn}$ is obvious. As is apparent, choosing an appropriate value for β strictly depends upon the parameters of the transmission lines and the value of Q_rs. In our simulation in Sect. 5, we set $\beta = 2$, i.e., the CDU should load at most $-2Q_n$ for completely serving the proper demander (the maximum power loss over the line $0 \rightarrow n$ is $-Q_n$).

During the learning process, in each (discrete) time step t every learner $l_{ij} \in \mathcal{L}$ individually and distributively updates its temporary power value P_{ij}^t in a myopic manner, i.e., supposing that all other learners $\mathcal{L} \backslash \{l_{ij}\}$ are inactive and their power values are fixed. In the following, we propose a method to distributively update the power values P_{ij}^t in order to lead it to the best value and we will show that the updating process has a fixed point, i.e., for a given t large enough $P_{ij}^{t+1} = P_{ij}^t \; \forall l_{ij} \in \mathcal{L}$, and then we set $P_{ij}^* = P_{ij}^{t+1}$.

The power value of learners l_{mn} appear in both conditions (3), whereas that of l_{m0} only in (3a) and that of l_{0n} only in (3b). From conditions (3), the following equations are derived:

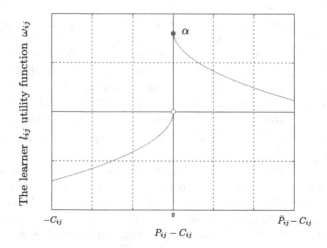

Fig. 4. Learner utility as a function of $P_{ij} - C_{ij}$ with $0 \leq C_{ij} \leq \bar{P}_{ij}$ and $\alpha \gg 0$.

$$
\begin{cases}
P_{ij} = C_{ij}; & \forall\, l_{ij} \in \mathcal{L} \\
C_{m0} = Q_m - \displaystyle\sum_{n \in \mathcal{M}^-} P_{mn} & \forall\, l_{m0} \in \mathcal{L}; \\
C_{0n} = -Q_n - \displaystyle\sum_{m \in \mathcal{M}^+} P_{mn} + PL_{0n} + \sum_{m \in \mathcal{M}^+} PL_{mn} & \forall\, l_{0n} \in \mathcal{L}; \\
C_{mn} = 0.5 \left(Q_m - Q_n - \displaystyle\sum_{n \neq n' \in \mathcal{M}^-} P_{mn'} - \sum_{m \neq m' \in \mathcal{M}^+} P_{m'n} - P_{m0} - P_{0n} + PL_{0n} + \sum_{m' \in \mathcal{M}^+} PL_{m'n} \right) \\
& \hspace{3cm} \forall\, l_{mn} \in \mathcal{L}.
\end{cases}
\tag{7}
$$

According to (7), the rationality of each learner $l_{ij} \in \mathcal{L}$ is focusing on updating the proper value P_{ij} in order to achieve the value of C_{ij} *exactly*, i.e., $P_{mn} = C_{mn}$, $P_{m0} = C_{m0}$, and $P_{0n} = C_{0n}$. At each time step t, the learner l_{ij} will update the proper power value P_{ij}^t in a myopic manner, i.e., supposing that C_{ij}^t is a constant. To this end, we define the following learning utility function for each learner $l_{ij} \in \mathcal{L}$:

$$
\begin{cases}
\omega_{ij} = -\sqrt{|P_{ij} - C_{ij}|} + \alpha \cdot \mathrm{u}\left(P_{ij} - C_{ij}\right); & 0 \leq C_{ij} \leq \bar{P}_{ij} \\
\omega_{ij} = -\alpha & \text{otherwise.}
\end{cases}
\tag{8}
$$

where $\mathrm{u}\left(\cdot\right)$ is the step function, with $\mathrm{u}\left(y\right) = 1$ if $y \geq 0$ and $\mathrm{u}\left(y\right) = 0$ otherwise (see Fig. 4), and α is a constant. When the value C_{ij} is out of the range of $[0, \bar{P}_{ij}]$, the proper learner will gain the minimum possible payoff, $-\alpha$. For a $C_{ij} \in [0, \bar{P}_{ij}]$, if $P_{ij} = C_{ij}$, the learner l_{ij} earns the highest possible payoff $\omega_{ij} = \alpha$. Whereas when $P_{ij} \neq C_{ij}$, the learner l_{ij} gets a payoff lower than α. The factor α is a sufficiently large positive constant that ensures ω_{ij} to be positive when $P_{ij} > C_{ij}$. This is an expedient to let the learners distinguish the value of P_{ij} that is either smaller or larger than C_{ij} only by knowing their own payoffs. In practice, it is obvious that the payoff of each learner is bounded from below by the finite negative number $-\alpha$. Setting an appropriate value to α in order to

Table 1. Myopic decision of the learner l_{ij} at time step t.

```
function  Pᵢⱼᵗ⁺¹ = Powerupdate(lᵢⱼ, t)
    if (ωᵢⱼᵗ = α) or (ωᵢⱼᵗ = −α), then Pᵢⱼᵗ⁺¹ = Pᵢⱼᵗ, exit;
    P̃ᵢⱼ = Pᵢⱼᵗ; //saving the current power
    repeat
        P̂ᵢⱼ = P̃ᵢⱼ; //saving tentative power
        compute ω̃ᵢⱼ; //tentative payoff
        ΔP̃ᵢⱼ = unif [0, ΔPᵢⱼ]; //random power step
        P̃ᵢⱼ = P̃ᵢⱼ − sign(ωᵢⱼᵗ) · ΔP̃ᵢⱼ; //tentative power
    until (ω̃ᵢⱼ > ωᵢⱼᵗ) or (P̃ᵢⱼ > P̄ᵢⱼ) or (P̃ᵢⱼ < 0)
    if (ω̃ᵢⱼ > ωᵢⱼᵗ),   then Pᵢⱼᵗ⁺¹ = P̂ᵢⱼ; //accept
                         else Pᵢⱼᵗ⁺¹ = Pᵢⱼᵗ; //discard
end function.
```

guarantee a positive ω_{ij} when $P_{ij} \geq C_{ij}$ is strictly dependent upon the values of Q_r. Note that the proposed framework could also consider any bounded utility function which increases as its argument moves from $\pm\infty$ to 0. This means that, for any $P_{ij} \neq C_{ij}$, each learner l_{ij} has an incentive to move towards the point zero where $P_{ij} = C_{ij}$ and the learner gains the highest possible payoff. When all learners gain α, then all conditions (3) hold, and obviously $P_{ij} = P_{ij}^*$.

The pseudocode in Table 1 shows how each learner l_{ij} takes its myopic decision during time step t. In this algorithm, sign($\cdot$) is our sign function, with sign(y) = 1 if $y \geq 0$ and sign(y) = −1 otherwise. The parameter $\tilde{\omega}_{ij}$ is the "trial" value of the current payoff of the learner l_{ij} when the tentative power $\tilde{P}_{ij}$ is adopted. As each learning time step, the power step $\Delta\tilde{P}_{ij}$ is the particular outcome (value) of a random variable uniformly distributed between 0 and $\overline{\Delta P}_{ij}$, with $\overline{\Delta P}_{ij} \ll \bar{P}_{ij}$. Optimal values for $\overline{\Delta P}_{ij}$ can be found in order to minimize the computational load of the algorithm, based on experimental results. If $\omega_{ij}^t < 0$, then $P_{ij}^t < C_{ij}$, and the best strategy for the learner l_{ij} is to increase its power so as to increase its payoff. Consequently, the tentative power is a random number in the interval $[P_{ij}^t, \bar{P}_{ij}]$. The learner l_{ij} accepts this value if and only if the utility ω_{ij}^t increases, otherwise it ends up to keep its previous value. If $0 \leq \omega_{ij}^t < \alpha$, the learner l_{ij}'s best strategy is on the contrary to decrease P_{ij}^t, and thus the tentative (random) power level belongs to the interval $[0, P_{ij}^t]$. As is apparent, the convergence speed of the algorithm depends also on the choice of the maximum update step $\overline{\Delta P}_{ij}$.

The process starts at time step $t = 0$ with $P_{ij}^{t=0} = 0$ for all learners in $\mathcal{L}$. Thus, at $t = 0$, we have $C_{m0}^{t=0} = Q_m = \bar{P}_{m0} \; \forall m \in \mathcal{M}^+$ and $C_{0n}^{t=0} = -Q_n < \bar{P}_{0n}$ $\forall n \in \mathcal{M}^-$, i.e., $\omega_{m0}^{t=0}$ and $\omega_{0n}^{t=0}$ are in the range of $(-\alpha, 0)$. Depending on Q_m, $-Q_n$, and β, the values of $C_{mn}^{t=0}$ can be either in or out of the range of $[0, \bar{P}_{mn}]$. At $t = 1$, each learner l_{m0} and l_{0n} may increase the proper power amount which results that some of $C_{mn}^{t=1} < C_{mn}^{t=0}$. After some time steps, some of the learners l_{mn} will have $0 \leq C_{mn}^t \leq \bar{P}_{mn}$ and they can update the proper powers.

Each learner l_{ij} individually decides to adjust its power value P_{ij}^t in a myopic manner while supposing that the value of C_{ij}^t is fixed and all other learners are inactive. The value of C_{ij}^t depends upon the power values of other learners and therefore the decision of adjusting the power value by each learner influences other C values in conflicting and incompatible ways which prevent learners from gaining the expected payoff. At each time step t, adjusting P_{ij}^t changes all C_{m0}^t and C_{0n}^t $\forall m \in \mathcal{M}^+$ and $\forall n \in \mathcal{M}^-$. To reduce the number of occurrences of this event, we modify our algorithm by requesting each learner l_{ij} *not* to update its power value at every time step with a probability $\lambda_{ij} \in [0, 1]$. At each time step t, every learner l_{ij} picks a random number ξ_{ij}^t uniformly distributed in $[0, 1]$. If $\xi_{ij}^t > \lambda_{ij}$, then the learner applies the algorithm and possibly update P_{ij}^{t+1}, otherwise $P_{ij}^{t+1} = P_{ij}^t$. Note that each learner l_{ij} is aware of the value of λ_{ij}. The algorithm is executed in a central computing unit [16], e.g. the CDU, and it can transmit the best amount of power load over each transmission line to MGs.

We show now that our proposed algorithm reaches a fixed point at which $P_{ij}^t = C_{ij}^t$ $\forall l_{ij} \in \mathcal{L}$. We model the evolution of the algorithm as the output of a Markov chain with state space $\Psi = \{\psi = (\boldsymbol{\omega}) | \boldsymbol{\omega} \in [-\alpha, \alpha]^L\}$. At time step t, our system is in the state $\psi^t = (\boldsymbol{\omega}^t)$, where $\boldsymbol{\omega}^t$ is the set $\{\omega_{ij}^t\}$ $\forall l_{ij} \in \mathcal{L}$. For all time steps t, since ω_{ij} is continuous and bounded in $[-\alpha, \alpha]$, $\boldsymbol{\omega}^t$ is bounded in the L-dimensional space $[-\alpha, \alpha]^L$. So, $\boldsymbol{\omega}^t$ is compact and Lebesgue measurable. The evolution of the Markov chain is then dictated by the strategy of the learning process. The strategy of each learner l_{ij} is to find the best power amount P_{ij}^t that leads to an increase its own payoff ω_{ij}^t. In practice, each learner l_{ij} autonomously decides whether to change its power value P_{ij}^t, making its payoff better off, or to keep the power at the same power level (when its payoff is equal to α or $-\alpha$). The transition from the state $\psi = (\boldsymbol{\omega})$ to a new state $\check{\psi} = (\check{\boldsymbol{\omega}})$ occurs if and only if the new state $\check{\psi}$ "dominates" the state ψ, i.e., compared to $(\boldsymbol{\omega})$, no learner gets worse off in $(\check{\boldsymbol{\omega}})$. The CDU computes C_{ij}^{t+1} $\forall l_{ij}$ using the values of P_{ij}^{t+1} and then computes all utility payoffs $\boldsymbol{\omega}^{t+1}$. If the transition from the state ψ^t to the state ψ^{t+1} occurs, then the CDU communicates to all learners sending them the proper C_{ij}^{t+1}, otherwise announce them to keep the previous power amount, i.e., $P_{ij}^{t+1} = P_{ij}^t$. The Markov process asymptotically tends towards a stable power distribution state at which no learner has any incentive to change its power value. In other words, all learners get their maximum payoffs, $\boldsymbol{\omega} = \{\alpha\}^L$, and consequently $P_{ij}^t = C_{ij}^t$ $\forall l_{ij} \in \mathcal{L}$, and then, obviously, no learner has any incentive to update its power value. Our algorithm guarantees tending the Markov chain towards a fixed point state with probability one when $t \to \infty$. Obviously, the stable state is not unique and according to the way the learners generate random numbers, the algorithm leads them to one of the possible solutions and then the power values are no longer updated.

Theorem 1. *The coalition formation learning process converges a stable state.*

Due to space limitations, we skip the proof here. Note that, during the learning process for achieving stable state, the payoffs of all coalitions are equal to $-\infty$ since the conditions (3) are not still satisfied. Once all learners achieve the coalition formation stable state, the conditions hold a possible coalition structure $\mathcal{K}$ is formed, but the power dissipated is not necessarily the minimum one. In the next subsection, we propose another "complementary" Markov model which, in each of its time steps τ, executes coalition formation dynamic process and calls the achieved stable state $\psi = (\boldsymbol{\omega})$. The evolution during time τ leads to a network with the minimum amount of power dissipated.

4.2   Power Loss Minimization Dynamic Learning Process

At the stable state of the coalition formation process, conditions (3) hold, a possible coalition structure $\mathcal{K}$ is formed, and each coalition earns a limited payoff. The payoff of each coalition is the inverse of total power dissipated and so higher payoffs for coalitions means lower amount of power dissipated. Obviously, the coalitions' payoffs strictly depend upon the distribution of power over transmission lines, i.e., the achieved coalition formation stable state $\psi = (\boldsymbol{\omega})$. In general, the payoffs of coalitions and MGs is neither a convex nor a concave function of the loaded power over transmission lines. Hence, it is not possible to find an analytical solution for finding the absolute best payoffs of coalitions. With aim at minimization of the overall power dissipated, we propose a complementary dynamic learning process. In a similar fashion as the coalition formation dynamic learning, we model the evolution of these learning processes as a Markov model $\Pi = \left\{ \pi = (\boldsymbol{\omega}, \rho) \mid \boldsymbol{\omega} = \{\alpha\}^L, \rho \subset \mathbb{R}^M \right\}$ wherein $\psi = (\boldsymbol{\omega})$ is the achieved coalition formation stable state. For the second input ρ, we will consider one of the characteristics of the network which is limited and Lebesgue measurable. In the simulation results, we will consider the following two values for ρ and we will compare the performance of them:

(i) We will set $\rho = \nu(\mathcal{M}) \in \mathbb{R}$. In fact, in this case we disregard the coalition structure and MGs payoffs, but we aim at maximization of $\nu(\mathcal{M})$ which is, according to (5), limited and Lebesgue measurable.

(ii) We will set $\rho = \boldsymbol{\phi} \in \mathbb{R}^M$ where $\boldsymbol{\phi} = \{\phi_r\}$ denotes the MGs Shapley values. In this case, the structure of the coalitions and their payoffs are important.

In the simulation results we will show that maximizing Shapley values outperforms maximizing $\nu(\mathcal{M})$ in terms of power saving. At each time step τ, the central computing unit, first, executes the coalition formation dynamic learning. Once the coalition formation learning achieves a stable state, the complementary Markov model evolution is in the state $\pi^\tau = (\boldsymbol{\omega}^\tau, \rho^\tau)$, i.e., either $\pi^\tau = (\boldsymbol{\omega}^\tau, \nu(\mathcal{M})^\tau)$ if we choose (i), or $\pi^\tau = (\boldsymbol{\omega}^\tau, \boldsymbol{\phi}^\tau)$ if (ii), where $\boldsymbol{\omega}^\tau$ is the stable state of the coalition formation dynamic learning. At the next time step $\tau + 1$, the CDU re-executes the coalition formation dynamic learning picking different random numbers and achieves a new stable state $\boldsymbol{\omega}^{\tau+1}$ with a different power distribution and coalition structure. The transition from π^τ to $\pi^{\tau+1}$

occurs if and only if $\rho^{\tau+1}$ dominates ρ^τ, otherwise the Markov model Π stays at the same state π^τ. If we set $\rho = \nu(\mathcal{M})$ the transition occurs when the grand coalition earns a higher payoff, whereas with $\rho = \phi$ it occurs when no MG will get worse off. At time step $\tau + 1$, if the coalition formation learning process happens to achieve the same ω^τ, again the complementary Markov model stays at the same state π^τ. Since the values of ρ are limited, with a same discussions in Theorem 1, it is easy to show that the evolution of Π achieves a fixed point for ρ, i.e., for a given time step τ large enough, $\rho^{\tau+1} = \rho^\tau$ for a $\omega^{\tau+1} \neq \omega^\tau$. At the fixed point of the complementary dynamic learning process, in case (i) the grand coalition earns the highest possible payoff, and in case (ii) all MGs in $\mathcal{M}$ earn the highest possible Shapley value ϕ_r. This means that the amount of power dissipated is the lowest possible. Obviously, this algorithm can not guarantee the absolute minimum power loss, since the shape of the power loss function strictly depends upon the parameters of the transmission line. For this, the power loss minimization stable point is not necessarily unique and the proposed learning process leads micro-grids to one of the stable points.

5   Numerical Results

In this section, we show the performance of the proposed algorithm. Throughout the simulations, we make use of the following transmission lines parameters: $\omega C = 4.518 \ \mu\Omega^{-1}/\text{km}$, $\omega L = 367 \ \text{m}\Omega/\text{km}$, $R = 37 \ \text{m}\Omega/\text{km}$, and $R_1 = 1 \ \Omega$ [1, p.68]. Power transfer among MGs themselves is done at a medium voltage 100 kV, while between the central transmission unit and the MGs that is done at 345 kV [1, p.68]. The MGs are uniform randomly distributed within an area with radius 100 km and the CDU located at the center. For each MG $r \in \mathcal{M}$, the amount of residual power Q_r is supposed to be a normal random distribution with zero mean and a standard deviation uniformly distributed in the range [100, 500] MW [2]. In the following set of evaluations, based on numerical optimizations, we consider the following parameters in the function Powerupdate: the power update step $\overline{\Delta P}_{mn} = \bar{P}_{mn}/5$, $\overline{\Delta P}_{m0} = \bar{P}_{m0}/10$, and $\overline{\Delta P}_{0n} = \bar{P}_{0n}/10$; the parameter that reduces the probability of conflicting decisions among learners $\lambda_{mn} = 0.7$, $\lambda_{m0} = 0.85$, and $\lambda_{0n} = 0.85$ for $\forall \ m \in \mathcal{M}^+$ and $\forall \ n \in \mathcal{M}^-$. All results are obtained by averaging over $2,000$ random realizations of a network with different positions and different values of Q_rs.

Figure 5 reports a snapshot of the achieved coalition formation dynamic learning process stable state in a network consisting of $M = 6$ MGs randomly located in an area with residual quantities Q_rs equal to $\{+150, +350, +200, -150, -200, -450\}$ MW, respectively. After one execution of coalition formation dynamic learning process, the power distribution at the coalition formation stable state is as follows: $P_{16} = 0.43Q_1$, $P_{25} = 0.38Q_2$, $P_{26} = 0.42Q_2$, $P_{34} = 0.53Q_3$, $P_{36} = 0.47Q_3$ and consequently the coalition structure is $\mathcal{K}_1 = \{1, 6\}$, $\mathcal{K}_2 = \{2, 5, 6\}$, $\mathcal{K}_3 = \{3, 4, 6\}$, and $\mathcal{M}$. The achieved coalitions payoffs are $\{107, 14.9, 29, 9.5\} \ \text{nW}^{-1}$, respectively, and the Shapley values of the MGs are $\{4.426, -0.430, 0.038, 0.038, -0.430, 5.888\} \ \text{nW}^{-1}$, respectively. The total

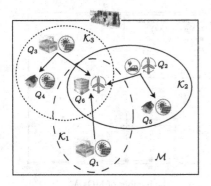

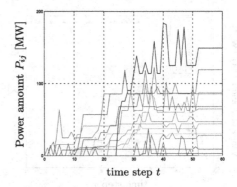

time step t

Fig. 5. Snapshot of a formed coalition structure.

Fig. 6. Learners behaviour during coalition formation process.

amount of power dissipated, $1/\nu(\mathcal{M})$, is 59 % of that of the traditional non-cooperative model which is calculated as a network wherein each MG trades only with CDU as in Fig. 1a. Note that here we report the results at the stable point of one execution of coalition formation dynamic learning process only.

Figure 6 exhibits the behaviour of the learners during coalition formation dynamic process. At $t = 0$, all $P_{ij} = 0$ and only the learners of P_{m0} and P_{0n} $\forall m \in \mathcal{M}^+$ and $\forall m \in \mathcal{M}^+$ may increase the proper power, since they have a negative payoff not equal to $-\alpha$. As can be seen, after some time steps, some other learners earn a utility not equal to $-\alpha$, and so may update their power value. Despite conflicts between simultaneous and myopic decisions, Fig. 6 shows the convergence of P_{ij} the stable point after 52 time steps.

Figure 7 reports the average number of time steps τ for achieving the stable point of the power loss minimization process in the network scenario with the above mentioned (fixed) residual quantities. The dashed line reports the grand coalition's payoff and other curves report the Shapley values of the MGs during the dynamic process as a function of τ. As can be seen, when we choose $\rho = \nu(\mathcal{M})$ as the input of the Markov state $\pi = (\omega, \rho)$, the algorithm achieves stable state after 7 time steps, while this happens in $\tau = 12$ with $\rho = \phi$. It is worthwhile to emphasize that, at each time step τ, first coalition formation dynamic learning process is executed (Fig. 6), and then the inputs of the state π^τ are updated.

Figure 8 depicts the average of power loss fraction that is defined as the proportion of the overall amount of dissipated power to the overall loaded power. The dashed line reports the traditional non-cooperative scheme with $PL = \sum_{m \in \mathcal{M}^+} PL_{m0} + \sum_{n \in \mathcal{M}^-} PL_{0n}$ by setting $P_{m0} = Q_m$ and $P_{0n} = -Q_n$, and the solid reports the proposed cooperative scheme with $PL = 1/\nu(\mathcal{M})$. In other words, the total power dissipated in the traditional non-cooperative scheme is calculated as a network with M singleton coalitions. The average of the loss fraction in the non-cooperative scheme is 11 % and that is not very sensitive to the number of MGs. However, in our cooperative scheme the average of the ratio decreases with the increasing number of MGs and its average is 0.83 % with

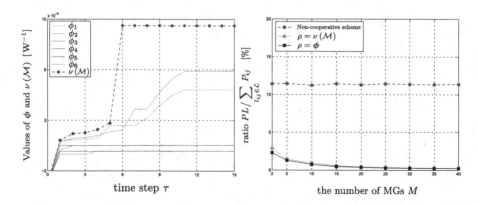

Fig. 7. Shapley values and grand coali- **Fig. 8.** Average of loss fraction (power dis-
tion's payoff during power loss minimiza- sipated as a proportion of power loaded) as
tion dynamic learning. a function of M.

$\rho = \nu(\mathcal{M})$, and 0.69 % with $\rho = \phi$. Considering the range of the quantities (in
$\pm[100, 500]$ MW), the power dissipated in non-cooperative scheme is significant
and that reduces significantly in the cooperative scheme. As can be seen, with
applying the proposed cooperative algorithm, the power dissipated will become
10 % of the traditional distribution networks. As a result, setting $\rho = \phi$ in which
the coalition structure and all coalitions' payoffs are involved outperforms setting $\rho = \nu(\mathcal{M})$ in which only the grand coalition payoff is considered, at the cost
of more computational complexity. Figure 8 shows also that any value greater
than 1.2 is a sufficient value for β in the power constraints $\bar{P}_{mn}$ and $\bar{P}_{0n}$, as we
set $\beta = 2$.

6   Conclusion

In this work, we investigate the problem of power trading coordination in a smart
power grid consisting of a set of micro-grids (MGs) each of which has a quantity
of either surplus or deficit amount of power to sell or to acquire, respectively.
The proposed coalitional game theory based approach allows micro-grids (MGs)
to form a set of coalitions and trade power between themselves in order to
achieve the minimum power dissipated during power generation and transfer.
In this context, we propose a dynamic learning process to form a set of non
necessarily disjoint and possibly singleton coalitions. In the proposed dynamic
learning process each transmission line, as a learner, adjusts the best power load
such that: (i) all suppliers to be completely loaded, and (ii) all demanders to
be completely served. A complementary dynamic learning process is proposed
to lead the coalition structure to the best performance in terms of power saving.
Numerical results show power dissipated in the proposed cooperative smart grid
is 10 % of that in the traditional non-cooperative networks.

Acknowledgement. This work is supported by the EU project QUANTICOL, 600708.

References

1. Machowski, J., Bialek, J.W., Bumby, J.R.: Power System Dynamics: Stability and Control, 2nd edn. Wiley, New York (2008)
2. Coster, E.J.: Distribution Grid Operation Including Distributed Generation. Eindhoven University of Technology, The Netherlands (2010)
3. Ochoa, L., Harrison, G.: Minimizing energy losses: Optimal accommodation and smart operation of renewable distributed generation. IEEE Trans. Power Sys. **26**(1), 198–205 (2011)
4. Tenti, P., Costabeber, A., Mattavelli, P., Trombetti, D.: Distribution loss minimization by token ring control of power electronic interfaces in residential microgrids. IEEE Trans. Ind. Electron **59**(10), 3817–3826 (2012)
5. Saad, W., Han, Z., Poor, H.: Coalitional game theory for cooperative micro-grid distribution networks. In: IEEE International Conference on Communications Workshops (ICC), pp. 1–5, Kyoto, June 2011
6. Wei, C., Fadlullah, Z., Kato, N., Takeuchi, V.: GT-CFS: a game theoretic coalition formulation strategy for reducing power loss in micro grids. IEEE Trans. Parallel Distrib. Sys. **25**(9), 2307–2317 (2014)
7. Osborne, M.J., Rubinstein, A.: A Course in Game Theory. MIT Press, Cambridge (1994)
8. Mohsenian-Rad, A.-H., Wong, V., Jatskevich, J., Schober, R., Leon-Garcia, A.: Autonomous demand-side management based on game-theoretic energy consumption scheduling for the future smart grid. IEEE Trans. Smart Grid **1**(3), 320–331 (2010)
9. Vytelingum, P., Ramchurn, S., Voice, T., Rogers, A., Jennings, N.: Agent-based modeling of smart-grid market operations. In: IEEE Power and Energy Society General Meeting, pp. 1–8, Detroit, July 2011
10. Wang, Y., Saad, W., Han, Z., Poor, H., Basar, T.: A game-theoretic approach to energy trading in the smart grid. IEEE Trans. Smart Grid **5**(3), 1439–1450 (2014)
11. Shams, F., Luise, M.: Basics of coalitional games with applications to communications and networking. EURASIP J. Wirel. Commun. Networks **1**, 2013 (2013)
12. Saad, W., Han, Z., Poor, H., Basar, T.: Game-theoretic methods for the smart grid: an overview of microgrid systems, demand-side management, and smart grid communications. IEEE Signal Process. Mag. **29**(5), 86–105 (2012)
13. Shapley, L.S.: A value for n-person games. contribution to the theory of games. Ann. Math. Stud. **2**, 28 (1953)
14. Kakutani, S.: A generalization of Brouwer's fixed point theorem. Duke Math. J. **8**(3), 457–459 (1941)
15. Yahyasoltani, N.: Dynamic learning and resource management under uncertainties for smart grid and cognitive radio networks. Ph.D. dissertation, Department of Computer Engineering, University of Minnesota, USA (2014)
16. Galli, S., Scaglione, A., Wang, Z.: For the grid and through the grid: the role of power line communications in the smart grid. Proc. IEEE **99**(6), 92–951 (2011)

PCA-Based Method for Detecting Integrity Attacks on Advanced Metering Infrastructure

Varun Badrinath Krishna, Gabriel A. Weaver$^{(\boxtimes)}$,
and William H. Sanders

Information Trust Institute, Department of Electrical
and Computer Engineering, University of Illinois at Urbana-Champaign,
1308 West Main Street, Urbana, IL 61801, USA
{varunbk,gweaver,whs}@illinois.edu
http://iti.illinois.edu

Abstract. Electric utilities are in the process of installing millions of smart meters around the world, to help improve their power delivery service. Although many of these meters come equipped with encrypted communications, they may potentially be vulnerable to cyber intrusion attempts. These attempts may be aimed at stealing electricity, or destabilizing the electricity market system. Therefore, there is a need for an additional layer of verification to detect these intrusion attempts. In this paper, we propose an anomaly detection method that uniquely combines Principal Component Analysis (PCA) and Density-Based Spatial Clustering of Applications with Noise (DBSCAN) to verify the integrity of the smart meter measurements. Anomalies are deviations from the normal electricity consumption behavior. This behavior is modeled using a large, open database of smart meter readings obtained from a real deployment. We provide quantitative arguments that describe design choices for this method and use false-data injections to quantitatively compare this method with another method described in related work.

Keywords: Smart · Meter · Grid · Anomaly · Detection · Principal · Component · Analysis · Data · Cyber-physical · AMI · PCA · SVD · DBSCAN · Electricity · Theft · Energy · Computer · Communication · Network · Security

1 Introduction

The *Advanced Metering Infrastructure (AMI)* provides a means for communication between electric utilities and consumers. Smart meters are increasingly replacing traditional analog meters to enable the automated reading of electricity consumption and the detection of voltage variations that may lead to outages. For example, by 2018, the Illinois-based Commonwealth Edison Company will have installed 4 million smart meters in all homes and businesses in Northern Illinois [5].

© Springer International Publishing Switzerland 2015
J. Campos and B.R. Haverkort (Eds.): QEST 2015, LNCS 9259, pp. 70–85, 2015.
DOI: 10.1007/978-3-319-22264-6_5

AMI is perceived to provide other benefits, beyond describing the state of the electric distribution grid. For example, smart meters have been rolled out by electric utilities such as BC Hydro to detect electricity theft [2]. In 2010, however, the Cyber Intelligence Section of the FBI reported that smart meters were hijacked in Puerto Rico, causing electricity theft amounting to annual losses for the utility estimated at $400 million [6].

In [20], we show that an attacker may be able to destabilize a real-time electricity market system by compromising the electricity price relayed to the Automated Demand Response (ADR) interfaces. Equivalently, it may be possible to destabilize the system by compromising smart meter consumption readings, causing suppliers to modify the electricity price accordingly. Electricity theft and destabilization of electricity markets are just two of several attacker goals that illustrate the need for effective intrusion detection systems. Other attacker models are discussed in [3].

It must be noted that smart meters, such as those manufactured by GE, are equipped with encrypted communication capabilities and tamper-detection features. However, reliance on those mechanisms is not a sufficient defense against cyber intrusions that exploit software vulnerabilities. In their *Cyber Risk Report 2015*, HP Security Research states that the enterprises most successful in securing their environments employ complementary protection technologies [8]. Such technologies work best in the context of the assumption that breaches will occur. By using all tools available and not relying on a single product or service, defenders place themselves in a better position to prevent, detect, and recover from attacks.

The anomaly detection methods presented in this paper assume that an attacker has compromised the integrity of smart meter consumption readings, and aim to mitigate the impact of such an intrusion. How the attacker can get into a position where he is capable of modifying communication signals is not a focus of this paper and is discussed in [9,13], and [14]. Our aim is to verify the data reported to the utility by modeling the normal consumption patterns of consumers and looking for deviations from this model.

Our proposed method leverages Principal Component Analysis (PCA) [15]; anomaly detection methods that leverage PCA have been proposed in [4,11, 18,19]. However, these papers focus on classifying anomalies, such as network volume anomalies, that manifest themselves as spikes in the data. Electricity consumption behavior, however, tends to be naturally spiky. Therefore, these methods fail to detect actual anomalies in consumption, such as extended changes in consumption patterns.

We propose a method that leverages the Density-Based Spatial Clustering of Applications with Noise (DBSCAN) algorithm and show that this algorithm [1,7], when combined with PCA, effectively detects anomalies in electricity consumption data. There are three advantages to using this method. First, by extracting the principal components that retain the maximum amount of variance in the data, we extract underlying consumption trends that repeat on a daily or weekly basis. Principal components that account for lower variance essentially represent noise in the consumption behavior, and this noise is filtered out. Second, the first two principal components allow us to visualize a massive dataset in a 2-dimensional space. Anomaly detection can then be performed in a way that can be visually verified. Third, the anomaly detection is performed in a space that spans the consumptions of all consumers. Therefore, it becomes significantly harder for an attacker to reverse-engineer and circumvent this detector, as he would need full information of all the consumers' smart meters in the network.

Online anomaly detection is an important feature that would enable better adoption of our method. For this feature, we leverage the technique in [17]. The results of our method could feed into recent cyber physical vulnerability assessment techniques such as [21], or be incorporated into a stand-alone tool.

Our model of consumption patterns is based on a large, open dataset that is described in Sect. 2. We propose and delineate our own anomaly detection method in Sect. 3, and evaluate this method against other well-known methods in Sect. 4. We conclude in Sect. 5.

2   Description of the Dataset

The dataset we use was collected by Ireland's Commission for Energy Regulation (CER) as part of a trial that aimed at studying smart meter communication technologies. It is the largest, publicly available dataset that we know of, and access details are provided in the Acknowledgments section of this paper. The fact that the dataset is public makes it possible for researchers to replicate and extend this paper's results.

The dataset is an anonymized collection of readings from 6,408 consumers, collected at a half-hour time resolution, for a period of up to 74 weeks. Of the 6,408 consumers, we restrict our analysis to the largest subset that contains the same 74 weeks, by calendar date. This restriction results in a set of 2,982 consumers, of which 2,374 were residential, 253 were small and medium enterprises (SMEs), and 355 were unclassified by CER.

3   Data-Driven Detection Strategies

In this section, we analyze the CER smart meter dataset and model electricity consumption patterns to aid in the detection of integrity attacks. We discuss two distinct detection strategies. The first is based on the average detector proposed in [12]. Given the limitations of that technique, we devised an alternative method, which is based on Principal Component Analysis (PCA). We discuss it in detail and quantify its effectiveness in Sect. 4.

The electricity consumption patterns in the CER dataset guide our anomaly detection methods. The authors of [12] admit that they arbitrarily evaluate detection strategies and use consumption models (such as the Auto-Regressive Moving-Average model) that do not capture actual electricity consumption patterns. In contrast, our detection methods stem from our analysis of consumption patterns in a dataset obtained from a real, large-scale deployment.

3.1 Assumptions and Notations Used in Detection

The detection strategies presented in this section look for anomalies in the smart meter readings that are reported to the utility. We assume that the meters are correctly measuring current, but the readings being communicated may have been compromised.

In the context of this paper, a detection strategy is a centralized online algorithm that would typically run at the utility control center and is defined as follows. The input is a set of new smart meter consumption readings that are reported to the utility. We refer to it as the *input set*, and it may contain one or more readings for each consumer. We refer to the output as the *classification* of the set, which is binary: normal or suspected attack. Note that the classification is based on an input set, and not on a single reported reading. So it is possible for the detection algorithm to classify individual readings as normal, but classify a combined set of such readings as anomalous. This may happen because of a deviation of the combined readings from the expected combined pattern.

We divided the 74 weeks of consumption data obtained from the CER dataset into two sets: a *training set* of the first 60 weeks and a *test set* of the remaining 14 weeks. Note that anomalies in the training set are not labeled, so we do not have ground truth on which readings are anomalous. As such, our algorithm is unsupervised, and our training set serves to build a model of the consumption patterns while accounting for the possibility of anomalies in it.

It is reasonable to assume that the training set obtained from CER is free from integrity attacks. However, there are anomalous consumption behaviors in the dataset. These anomalies might reflect periods when consumers were, for example, traveling, leading to abnormally low consumption, or hosting parties, leading to abnormally high consumption. Such events lead to false positives if the detection strategy classifies them as suspected attacks. The test set is used in Sect. 4 to evaluate false positives and false negatives reported by the detection algorithms using models built from the training set.

We use the following matrix notations in this section. $A_{(i,j)}$ refers to the element in matrix A at the intersection of row i and column j. $A_{(i,:)}$ refers to the row vector at row i, and $A_{(:,j)}$ refers to the column vector at column j.

3.2 Use of Averages to Detect Anomalies

The Average Detector was shown to be effective relative to the other methods proposed in [12]. This detector is formulated as follows. Let $D_c(t)$ represent the total consumption of consumer c during time period t. Given that our dataset contains smart meter measurements at a half-hour granularity, t refers to a particular half-hour. The consumption reported to the utility is denoted by $D'_c(t)$.

$D_c(t)$ is non-deterministic, and the true value at a certain time t is unknown to us, since our only knowledge of the value is through the reported reading $D'_c(t)$. As $D'_c(t)$ may be manipulated by integrity attacks on smart meter communications, we need to devise methods to validate $D'_c(t)$ for each t.

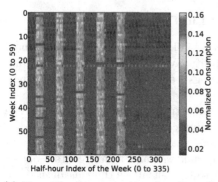
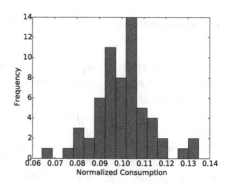

(a) Normalized consumption for all weeks for SME 1028

(b) Distribution of normalized consumption at half-hour index 71

Fig. 1. Normalized consumption of SME Consumer 1028. The five green/blue vertical bands in (a) represent time periods of higher electricity consumption that correspond to weekdays (Color figure online).

For each consumer c, we define an $M \times N$ matrix H_c in which each row represents one week, and there are M weeks. Each column represents the time of the week, and there are N times. In addition, we carefully align the weeks such that the first column of each matrix H_c refers to the first time period of a Monday, and the last column refers to the last time period of a Sunday. In our training set, H_c is a 60×336 matrix, as there are 60 weeks in the training data and 336 half-hours in a week.

Note that while consumption patterns typically repeat every week, the absolute consumption depends on the week. If the week is during winter in a country whose climate is like Ireland's, chances are that an occupant of the house will turn on the heating system. In contrast, the heating system will most likely be turned off during a week in the summer. To ensure that weeks can be compared on even grounds, we normalize each row in the matrix H_c, corresponding to the week, by dividing the row by its $l2$-norm. Figure 1 illustrates the normalized H_{1028} matrix (consumer identity $c = 1028$). The fact that this consumer has nearly zero consumption on weekends indicates that it is likely not a residential consumer. Indeed, the CER dataset has labeled 1028 as an SME.

We define the half-hour index (HHI) of the time t as a mapping $HHI : t \to \{0, 1, ..., 335\}$. For example, if we want to know whether $D'_c(t)$ is anomalous when t is December 2, 2014 at 12 p.m., we determine that this date is a Tuesday and that the time of the week corresponds to an HHI of 71. Figure 1(b) represents $P(D_{1028}|HHI(t) = 71)$. We do not make any assumptions on the underlying distribution, as we do not have the data necessary to construct a valid distribution. In previous work, we showed that a normal distribution is observed when D_c is conditioned on multiple parameters, such as $HHI(t)$, solar irradiation, external temperature, and building occupancy [10].

The detection algorithm for an input set is performed on a per-user basis as follows. For each consumer c_k, we calculate the average of the *input set* $IS_k = \{D'_{c_k}(t_1), D'_{c_k}(t_2)...\}$, where the times t_j may index a single time point ($j = 1$), a day ($j \in [1, 48]$) or a week ($j \in [1, 336]$), etc. This produces a single average value $avg(IS_k)$ for the new data. We then compare this average to the averages taken over the same time points in all previous weeks for consumer c_k. For example, if IS_k contains the set of all consumption points on a Tuesday, we compare the average $avg(IS_k)$, with averages of every Tuesday in the history of the dataset. If $avg(IS_k)$ is less than the lowest (or greater than the highest) average seen in past Tuesdays, we say that the input set is anomalous. If IS_k is a singleton set containing, say, the consumption at 12 p.m. on a Tuesday, we compare it against a set of consumptions at 12 p.m. on previous Tuesdays. In this case, $avg(IS_k)$ is the same as the single value in IS_k, so the notation remains valid.

3.3   Detecting Anomalies with Principal Component Analysis

The drawback of the average detector is that an attacker can circumvent it by ensuring that the average of the input set for a consumer $avg(IS_k)$ does not change significantly. Specifically, the elements of the input set can vary in a manner that is not consistent with the typical consumption patterns, but this change of pattern will not be quickly detected if the average is kept within reasonable bounds. Therefore, there is a need for a method that analyzes the variation in the consumption pattern as a collection of meter readings, as opposed to individual meter readings. For this purpose, we propose using Principal Component Analysis (PCA), and to detect deviations from the pattern we propose the use of a clustering technique.

PCA reveals the underlying trends in the smart meter data, across thousands of consumers, by reducing the dimensionality of the data, while retaining most of the data's variance. As such, it provides us with a way we can collapse a vector of electricity consumption readings in a high-dimensional space into one in a lower-dimensional space. This greatly aids anomaly detection methods, which can be intuitively executed in the lower-dimensional space, without loss of significant information. PCA not only immediately reveals clusters in data, but also is sensitive to changes in consumption patterns that may indicate integrity attacks.

The PCA Mechanism. We demonstrate the mechanism of PCA by constructing two different matrices (A &B) from our entire training set. A has $M_A = 20,160$ rows, one for each half-hour of the 60-week period of study, and $N_A = 2,982$ columns, one for each consumer. In this example, we can think of the consumption of each consumer across all 60 weeks as a column vector in a 20,160-dimensional space. There are 2,982 such column vectors. Using PCA, we will collapse these column vectors from $M_A = 20,160$ dimensions into two dimensions. Due to high correlation, two data points are sufficient to capture the patterns of each consumer, relative to the patterns of other consumers. Let P_A be the matrix of dimension $2 \times M_A$ that transforms A of dimension $M_A \times N_A$ to Y_A of dimension $2 \times N_A$. Then,

$$P_A A = Y_A \tag{1}$$

For calculation and notation convenience, we pre-process A by subtracting each row by the mean for that row and dividing the entire matrix by $N_A = 2,982$. We are interested in the covariance between the M_A rows (or readings per consumer) in A. For the corresponding AA^T covariance matrix, $P_A = U_{(0:1,:)}^T$, where U is obtained from the Singular Value Decomposition (SVD) of $A = U\Sigma V^T$. Here, the columns of U are the eigenvectors of the covariance matrix AA^T, and Σ^2 (the eigenvalues) represent the amount of variance retained by the principal components. This is illustrated in Fig. 2. Together, the two components in P_A retain 63.63 % of the variance in A, and the marginal variance retained by each further component is negligible.

There are two advantages to retaining only the first two components. First, maximum variance is retained by these components, so discarding further components effectively discards the noise in the consumption patterns. Second, it allows us to visualize a vector of 20, 160 dimensions in a two dimensional space. This then facilitates anomaly detection in this 2D space, as we will discuss later.

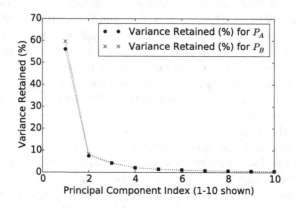

Fig. 2. Variance (%) retained by principal components of matrices A & B.

Construction of PCA Biplots. We transform A into the P_A space by taking the product $P_A A = Y_A$, where Y_A is the $2 \times N_A$ PCA score matrix. The two rows of Y_A are called the *Principal Component 1 Score* and the *Principal Component 2 Score*. The scatter plot of the two scores is the PCA biplot shown in Fig. 3(a). Points that lie close together in this 2D space describe consumers, or columns in A, whose consumption patterns are similar. Given that the comparison is over 60 weeks at a half-hour granularity, the large extent to which the consumers cluster together in the biplot was unexpected, and indicates that most of the consumers in the dataset have highly similar consumption patterns.

In Fig. 3(a), we used the labels in the CER dataset to distinguish the points in the biplot by consumer type. These labels revealed an interesting behavior where most residential consumers were found to cluster together in the 2D space. This indicated that their consumptions were similar to each other. However, SMEs varied greatly, which might reflect the unique electricity consumption requirements of their businesses.

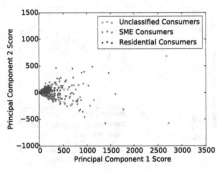

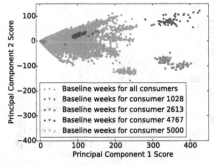

(a) Observed clustering of consumers by type in the P_A space

(b) Observed clustering of consumers and their consumption weeks in the P_B space

Fig. 3. Principal Component Analysis biplots describing the structure and similarities within the dataset.

In order to capture the relationship among consumers' individual patterns across different weeks, we reshaped A to get another matrix B; it contains $M_B = 48 * 7 = 336$ rows (one for each half-hour of the week) and $N_B = 2,982 * 60 = 178,920$ columns (one for each week of each consumer in the 60-week period). Again, we reduced the dimension of each week from $M_B = 336$ dimensions to 2 dimensions, retaining 68 % of the variance in B as shown in Fig. 2. The corresponding principal component matrix, P_B, is 2×336, and the PCA score matrix, $Y_B = P_B B$, is $2 \times N_B$.

Note that although A and B contain the same number of elements, their Principal Component Scores are of different dimensions and describe completely different characteristics of the data. The scores in Y_B tell us how similar the 60 weeks of consumption are in the training set across all consumers and they are plotted in Fig. 3(b). We observe a dense clustering of points corresponding to each consumer in the Y_B matrix, which captures how similar the consumption weeks are for each consumer, in comparison to weeks of other consumers. This can easily be seen in Fig. 3(b), where we have colored the points corresponding to four very different consumers and their consumption weeks.

A closer look at the weeks for consumer 1028 in Fig. 3(b) revealed a single blue triangle at around $(70, -15)$ in the plot that is significantly distant from the others. It corresponds to Week Index 23 in Fig. 1(a), which is clearly anomalous and probably a vacation week. There are other anomalous points that are distant from the dense cluster. As the anomalies are inherent in the dataset, we assume that they are natural anomalies, and not the consequence of attacks. Attacks, which modify the consumption signals in a manner that changes their pattern, cause a shift in the location of the original point (corresponding to a week) to a completely new one on the biplot, as shown in Fig. 4(b).

Clustering Points in the Principal Component Space. A natural density-based clustering of points in the principal component space is observed in Fig. 3. Therefore, we employ the Density-Based Spatial Clustering of Applications with

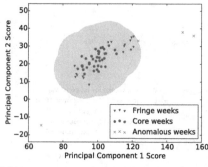 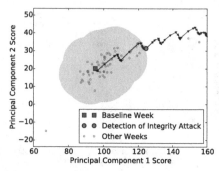

(a) Decision boundary and types of points in the DBSCAN clustering results

(b) Movement of baseline point with increase in attack duration

Fig. 4. Principal Component Analysis biplots for Consumer 1028 capturing (a) the decision boundary for anomalous points and (b) the movement of a baseline week of consumption in the principal component space with increase in duration of an integrity attack (Random Scale Attack, discussed in Sect. 4.3).

Noise (DBSCAN) algorithm [7] to determine which points correspond to regular weeks and which points correspond to anomalous weeks. An inherent benefit of DBSCAN lies in the fact that it works well for irregular geometries of dense clusters, and that it does not assume any underlying probability density of the points. The non-convex boundaries and treatment of dense clusters make DBSCAN better suited to our application, as opposed to other hierarchical, centroid-based, and distribution-based clustering methods.

The DBSCAN algorithm has two configurable parameters: *eps* and *MinPts*. These are used to obtain dense neighborhoods. In a two-dimensional Euclidean space, such as our principal component space, the circular region of radius *eps* centered at a point is referred to as the *eps neighborhood* of the point. A *core point* is a point that contains *MinPts* points within its *eps* neighborhood. All points that lie within the *eps* neighborhood of a core point are considered members of a dense cluster.

In our specific case, we are clustering 60 points that correspond to the weeks of consumption, for each consumer in our training set. These points were extracted from Y_B. We define *MinPts* as the number of points that achieves a simple majority (which in this case is 31). As a result, a single continuous cluster corresponding to normal weeks is produced, because any two *eps* neighborhoods containing *MinPts* points must intersect at at least one point.

Points that lie within the *eps* neighborhood of a core point, but are not core points themselves, are referred to as *fringe points*, as they usually lie at the fringes of the dense neighborhood. The algorithm considers all points that are neither core points nor fringe points to be "noise". This noise is how we define anomalous points in the dataset.

Figure 4(a) illustrates the result of the DBSCAN algorithm for consumer 1028. The green points are core points, and the blue triangles are fringe points. Since we chose *MinPts* to represent a simple majority, the circular *eps* neighborhoods overlap to form a single dense region, indicated by the yellow region in Fig. 4(a). All points within this region are considered to be normal. Anomalies, which may be caused by attacks, are points that do not lie in this region. The red crosses correspond to natural anomalies, which the algorithm would flag as false positives, as they lie outside the yellow "safe" region.

Clearly, the detection takes place in the 2D space spanned by P_B's basis vectors, which span the weekly consumptions of all consumers. In order to reverse-engineer the PCA detector for the purpose of circumventing it, the attacker would need to recreate this 2D space by gaining access to the meters of all consumers. In contrast, he would only need to compromise the smart meter of a single victim in order to reverse-engineer the Average Detector for that victim. *Therefore, the PCA-based detection method is more secure, because circumventing it requires full knowledge of all consumers' smart meter readings.*

Although the DBSCAN authors provide recommendations in [7] on how to set *eps*, these methods are not scalable. Specifically, they suggest calculating a list of core distances for each point and observing a knee-point at which a threshold should be set for *eps*. Given that there are 2,982 sets of points in our dataset (one for each consumer), eyeballing knee-points for each set is not feasible, so we needed to find an alternative. OPTICS, described in [1], can be used to determine cluster memberships for a single set of points containing multiple clusters. However, this method is not suitable for our study, in which we are defining a single cluster per set in 2,982 sets of points.

We set *eps* based on S_n, a measure proposed by statisticians in [16]. S_n looks at a typical distance between points, which makes it a good estimator of *eps*. In contrast, the Median Absolute Deviation (MAD) and the Mahalanobis distance measure the distance between points and a centroid, which is not how *eps* is defined. And, unlike the standard deviation, S_n is robust to outliers. Further, S_n is applicable to asymmetric geometries of points, like those in Fig. 3(b).

4   Evaluation of Detection Methods

We used data-driven simulation methods to evaluate the performance of our PCA-based detection method for multiple consumers in our dataset. We present a quantitative comparison of the performance of this method with that of the Average Detector method.

4.1   Runtime Memory Cost of Implementing the Detectors

The memory cost of the PCA-based method depends on the size of the input set that needs to be verified. For each consumer, if the input set is a week of readings at a half-hour time resolution, the principal component matrix would have a 2×336 dimension. The dimensionality of the principal component matrix remains the same as that of the input set. It does not increase as more input sets are evaluated, but can be updated without further memory costs, as shown in [17]. Also, it is not a function of how many consumers are in the system and thus occupies $O(1)$ memory.

In comparison, the Average Detector occupies $O(N)$ memory, scaling with the number of consumers in the system. For each consumer, this detector needs to access the minimum and maximum of the averages of each input set it ever processed. Therefore, this detector maintains 2 scalar values (a maximum and a minimum) per consumer.

In our evaluation, we do not consider the scenario where input sets are restricted to readings obtained from a single day. However, we briefly describe the procedure for evaluating such a scenario, because it might be useful in practice. In this scenario, we would separately perform PCA for each day of the week to obtain the two principal components. Therefore, our model for each day of the week would be a 2×48 matrix, as there are 48 half-hours in a day. Depending on the input set's day of the week, we would then rotate the input set into its corresponding principal component space. Following this, we would use the DBSCAN algorithm to detect whether the input set was an outlier. Note that we would need 7 principal component matrices in this case, one for each day of the week; the total memory requirement remains 2×336.

4.2 Evaluation Methodology

In order to evaluate both methods, we injected attacks that modify the smart meter readings. Our objective was to test the robustness of the two methods to such modifications, so we gradually varied the attack duration and recorded both false positives as well as false negatives. In this case, a false positive occurs when the input set is classified as a suspected attack, when it was actually not altered. A false negative occurs when the input set was compromised but the detection method classifies it as normal behavior.

Let $Attacked : IS_k \rightarrow \{0,1\}$ be a function that takes the value 1 when the Input Set for a consumer IS_k is compromised and 0 when IS_k is not compromised. We measured the *false positive rate* (FPR) and *false negative rate* (FNR) defined in terms of probabilities as follows:

$$FPR = P(Classification = Suspected\_Attack \quad | \ Attacked(IS_k) = 0)$$
$$FNR = P(Classification = Normal\_Behavior \quad | \ Attacked(IS_k) = 1) \tag{2}$$

In order to make a fair comparison between the two methods, we standardized the size of the input set to half-hour values over a week. Therefore, the input set contained 336 readings. We could have equivalently limited the standardized size of the input set to a day, but the stealthy attacks that we injected would take multiple days to be detected. A stealthy attack in this sense is one where the readings are not significantly altered, and examples are described in Sect. 4.3.

Our injection approach was broken down into two tests. In both tests, we constructed an initial input set containing the elements of any previous week in the training data that corresponded to a core point according to DBSCAN. This set is guaranteed to be free from anomalies and was used to complete the input set with 336 readings. Beginning with the first half-hour element of this input set, we modified the consumptions in chronological order. The modifications were made differently for the two tests, as described next.

The first test was a test for false negatives and the readings were modified using the attack injection methods explained in Sect. 4.3. The ideal result would have been the classification of all input sets as suspected attacks. By modifying the input sets in a chronological sequence, we varied the *attack duration* from one half-hour period to the entire week. For each attack duration, a new input set was created. The *time to detection* (TTD) for a detector is captured by the input set that corresponded to the shortest detectable attack duration. Better detectors have smaller TTDs.

The second test was a test for false positives. In this case, the input set was constructed using readings from the *test set*, which contained 14 weeks from the CER dataset (see Sect. 3.1). For each half-hour index of the input set, we randomly picked a reading from the test set with the same half-hour index. This random choice was made from a discrete uniform distribution with range $[1, 14]$. The ideal result would have been the classification of all input sets as normal, since $P(Attacked(IS_k)) = 0$ for all of them.

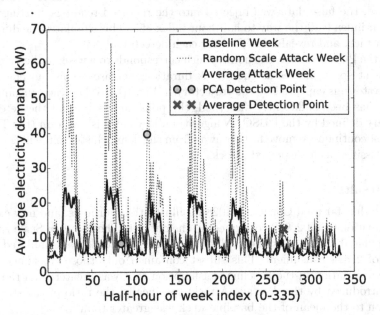

Fig. 5. Physical Manifestation of Random Scale ($\alpha = 0.5, \beta = 3$) and Average ($\gamma = 0.5$) Attacks on Consumer 1028. The attacks are launched at time 0. The PCA Detector has a TTD of 114 half-hours for the Random Scale Attack, and 84 half-hours for the Average Attack. The Average Detector has a TTD of 270 half-hours for the Random Scale Attack, and cannot detect the Average Attack.

4.3   Attack Injection Methods

In order to evaluate false negatives, we discuss two specific types of attack injection methods that modify the baseline week of readings (input set):

Random Scale Attack: At each half-hour time point, the consumption signal is multiplied by a uniform random variable $R \sim Unif(\alpha, \beta)$ where $0 < a < b$. The consumption is under-reported when R is a fraction below 1, and over-reported when it is a fraction above 1. If on average the values are over-reported, it can cause instability as shown in [20]. If they are under-reported, however, they can lead to electricity theft, as shown in [12].

Average Attack: At each half-hour time point, the consumption signal is replaced by the average value of the baseline week, multiplied by a uniform random variable $G \sim Unif(1 - \gamma, 1 + \gamma)$ where $\gamma \in [0, 1]$. The reported consumption effectively oscillates around the average. An attacker may use this method in a time-of-use electricity pricing scheme by under-reporting consumption readings when the price is high and over-reporting them when the price is low, while maintaining the average for a given day. This assumes that the electric utility uses redundant meters to verify aggregate consumptions at the end of the day.

Figure 5 illustrates the physical manifestation of these attacks. As previously described, the false data was injected into the reported readings starting from half-hour index 0 all the way to half-hour index 355. This simulates the duration of the attack, and reveals the TTD for each detector.

For the PCA detector, the point, which corresponds to a week in the principal component space, moves as the attack duration increases. This movement, in discrete steps, is captured by the trajectory in Fig. 4(b). When the consumption pattern has been sufficiently disturbed, the point moves beyond the detection boundary defined by the DBSCAN algorithm. Beyond this duration (the TTD), the point continues to move farther away from the dense cluster and will continue to be classified as a suspected attack.

4.4   Results

We have thus far used Consumer 1028 for illustration purposes, and now extend our evaluation to all $2,982$ consumers in the dataset. The Random Scale Attack was simulated on all consumers while the Average Attack was simulated on a subset of $2,814$ consumers. This subset contained only those consumers who exhibited variation in the baseline consumption that was greater than the variation introduced by the Average Attack; specifically, the ratio of the standard deviation to the mean of the baseline week was greater than γ.

For each consumer, we created $2 * 336 * 1,000$ input sets, for a combination of 2 tests (false positive and false negative), 336 discrete attack durations (for 336 half-hours in a week), and $1,000$ trials to capture the range of the uniform random variables. Increasing the number of trials from 100 to $1,000$ resulted in a 5.4% increase in the range of means observed for a uniform random sample of 336 values. Further increasing the number of trials from $1,000$ to $100,000$ increased the range by just 6.2%. We therefore decided to use $1,000$ trials to reduce the cost of the simulation without losing a large fraction of the range of the uniform random variables.

Given the large size of the simulation (it took around $3,840$ CPU hours to complete), we limited the attack parameter space to the values given in Fig. 5.

We then calculated the FNR and TTD for the two different types of attacks. Table 1 captures the metrics across all consumers.

To save space in Table 1, we introduce some new notation. *RS Attack* is the Random Scale Attack and *A Attack* is the Average Attack. *dPCA* denotes the event that an attack was successfully detected by the PCA detector within the 336 half-hour time frame in all 1,000 trials. This event applies to each consumer. $P(dPCA)$ is a probability that denotes the fraction of consumers for whom this event held true. *dAVG* is the corresponding event for the Average (AVG) detector. Result 1 (in Table 1) tells us that for 84.9% of consumers, the PCA detector was successful in detecting the Random Scale Attack. The PCA detector performs better against both attacks, as indicated in the *Win* column. Although the PCA detector did not perform as well as we had hoped for the Average Attack, it was a significant improvement on the AVG Detector.

Note that the detectors work in many of the 1,000 trials conducted for each consumer, but we wanted to test robustness under the stochastic attacker behavior. Therefore, our results conservatively captured only the consumers for whom the detectors worked in all 1,000 trials.

Results 2, 3, and 4 describe the TTD for the PCA and AVG detectors. The best case (min), average case (mean), and worst case (max) TTDs for the PCA detector are lower than the corresponding values for the AVG detector for most consumers, which again makes the PCA detector better. Note that the probability of having a higher TTD can be inferred from the probability of having equal and lower TTDs, which are given in the table. Result 5 compares the probability of having lower FNRs. Note that the remaining probability is accounted for by the case where the FNRs are equal.

On false positive tests, the AVG detector outperformed the PCA detector overall. In fact, the AVG detector had a perfect result: $P(AVG_{FPR} = 0) = 1.0$ and $P(AVG_{FPR} = 1) = 0.0$. For the PCA detector, however, $P(PCA_{FPR} = 0) = 0.637$ and $P(0 < PCA_{FPR} < 1) = 0.363$. This means that for 63.7% of

Table 1. Evaluation results for false negative tests

Metric	RS Attack		A Attack	
	Value	Win	Value	Win
1. $P(dAVG)$	0.635	PCA	0.040	PCA
$P(dPCA)$	0.849		0.081	
2. $P(mean(PCA_{TTD}) < mean(AVG_{TTD})|dPCA\ \&dAVG)$	0.652	PCA	0.520	PCA
$P(mean(PCA_{TTD}) = mean(AVG_{TTD})|dPCA\ \&dAVG)$	0.001		0.000	
3. $P(min(PCA_{TTD}) < min(AVG_{TTD})|dPCA\ \&dAVG)$	0.689	PCA	0.480	TIE
$P(min(PCA_{TTD}) = min(AVG_{TTD})|dPCA\ \&dAVG)$	0.025		0.040	
4. $P(max(PCA_{TTD}) < max(AVG_{TTD})|dPCA\ \&dAVG)$	0.513	PCA	0.600	PCA
$P(max(PCA_{TTD}) = max(AVG_{TTD})|dPCA\ \&dAVG)$	0.121		0.040	
5. $P(PCA_{FNR} < AVG_{FNR})$	0.630	PCA	0.079	PCA
$P(PCA_{FNR} > AVG_{FNR})$	0.232		0.033	

consumers, false positives were not detected in all $1,000$ trials. For the remaining consumers, the consumption patterns changed dramatically in the test set, leading to at least one false positive reported by the PCA detector in the $1,000$ trials. However, the consumption did not increase or decrease beyond the AVG detector thresholds, leading to the success of the AVG detector.

In summary, we have shown that the PCA detector probabilistically outperforms the AVG detector on false negative tests. We suspect that the PCA detector can be improved to reduce the false positive rate for the 36.3% of consumers. This might be achieved by correlating their consumption pattern deviations with deviations observed in the patterns of other consumers. In a vacation weak, for example, the PCA detector would suspect an attack due to deviation in consumption patterns. However, if other consumers show deviations for the same week, it provides evidence against classification as a suspected attack.

5 Conclusion

In this paper, we proposed a PCA-based anomaly detection method that utilities can use to detect integrity attacks on smart meter communications in an Advanced Metering Infrastructure. We provided quantitative arguments describing design choices for this method and presented a quantitative evaluation of the method with respect to the Average Detector proposed in related work.

In future work, we intend to use the framework developed in this paper to build a tool that can allow us to perform a more comprehensive evaluation of detection strategies under different attack parameters. Also, we plan to investigate the false positives of the PCA method by correlating simultaneous anomalies across multiple consumers.

Acknowledgements. This material is based upon work supported by the Department of Energy under Award Number DE-OE0000097. The smart meter data used in this paper is accessed via the Irish Social Science Data Archive - www.ucd.ie/issda. The providers of this data, the Commission for Energy Regulation, bear no responsibility for the further analysis or interpretation of it. We thank Shweta Ramdas, Jeremy Jones and Tim Yardley for their support.

References

1. Ankerst, M., Breunig, M.M., Kriegel, H.P., Sander, J.: OPTICS: ordering points to identify the clustering structure. ACM SIGMOD Rec. **28**(2), 49–60 (1999)
2. Hydro, B.C.: Smart metering program (2014). https://www.bchydro.com/energy-in-bc/projects/smart_metering_infrastructure_program.html
3. Berthier, R., Sanders, W.H., Khurana, H.: Intrusion detection for advanced metering infrastructures: requirements and architectural directions. In: Proceedings of IEEE SmartGridComm 2010, pp. 350–355. IEEE (2010)
4. Brauckhoff, D., Salamatian, K., May, M.: Applying PCA for traffic anomaly detection: problems and solutions. In: Proceedings of IEEE INFOCOMM 2009 (2009)

5. ComEd: Smart meter (2015). https://www.comed.com/technology/smart-meter-smart-grid/Pages/smart-meter.aspx
6. Cyber Intelligence Section: Smart grid electric meters altered to steal electricity, May 2010. http://krebsonsecurity.com/2012/04/fbi-smart-meter-hacks-likely-to-spread/
7. Ester, M., Kriegel, H.P., Sander, J., Xu, X.: A density-based algorithm for discovering clusters in large spatial databases with noise. In: Proceedings of KDD 1996. vol. 96, pp. 226–231 (1996)
8. HP Security Research: Cyber Risk Report 2015 (2015)
9. Jiang, R., Lu, R., Wang, L., Luo, J., Changxiang, S., Xuemin, S.: Energy-theft detection issues for advanced metering infrastructure in smart grid. Tsinghua Sci. Technol. **19**(2), 105–120 (2014)
10. Jung, D., Badrinath Krishna, V., Temple, W.G., Yau, D.K.: Data-driven evaluation of building demand response capacity. In: Proceedings of IEEE SmartGridComm 2014, pp. 547–553. IEEE (2014)
11. Lakhina, A., Crovella, M., Diot, C.: Diagnosing network-wide traffic anomalies. In: Proceedings of ACM SIGCOMM 2004. ACM, New York (2004)
12. Mashima, D., Cárdenas, A.A.: Evaluating electricity theft detectors in smart grid networks. In: Balzarotti, D., Stolfo, S.J., Cova, M. (eds.) RAID 2012. LNCS, vol. 7462, pp. 210–229. Springer, Heidelberg (2012)
13. McLaughlin, S., Holbert, B., Zonouz, S., Berthier, R.: AMIDS: A multi-sensor energy theft detection framework for advanced metering infrastructures. In: Proceedings of SmartGridComm 2012. pp. 354–359, November 2012
14. McLaughlin, S., Podkuiko, D., Miadzvezhanka, S., Delozier, A., McDaniel, P.: Multi-vendor penetration testing in the advanced metering infrastructure. In: Proceedings of ACSAC 2010, pp. 107–116. ACM, New York (2010)
15. Pearson, K.: LIII. on lines and planes of closest fit to systems of points in space. Philos. Mag. **2**(11), 559–572 (1901). Series 6
16. Rousseeuw, P.J., Croux, C.: Alternatives to the median absolute deviation. J. Am. Stat. Assoc. **88**(424), 1273–1283 (1993)
17. Sarwar, B., Karypis, G., Konstan, J., Riedl, J.: Incremental singular value decomposition algorithms for highly scalable recommender systems. In: Fifth International Conference on Computer and Information Science. Citeseer (2002)
18. Shyu, M.L., Chen, S.C., Sarinnapakorn, K., Chang, L.: A Novel Anomaly Detection Scheme Based on Principal Component Classifier, DTIC (ADA465712) (2003)
19. Shyu, M.L., Chen, S.C., Sarinnapakorn, K., Chang, L.: Principal component-based anomaly detection scheme (2006)
20. Tan, R., Badrinath Krishna, V., Yau, D.K., Kalbarczyk, Z.: Impact of integrity attacks on real-time pricing in smart grids. In: Proceedings of ACM CCS 2013, pp. 439–450. ACM, New York (2013)
21. Vellaithurai, C., Srivastava, A., Zonouz, S., Berthier, R.: CPIndex: cyber-physical vulnerability assessment for power-grid infrastructures. IEEE Trans. Smart Grid **6**(2), 566–575 (2015). doi:10.1109/TSG.2014.2372315

Tools

U-Check: Model Checking and Parameter Synthesis Under Uncertainty

Luca Bortolussi[1,2,3], Dimitrios Milios[4(✉)], and Guido Sanguinetti[4,5]

[1] Modelling and Simulation Group, University of Saarland, Saarbrücken, Germany
[2] Department of Mathematics and Geosciences, University of Trieste, Trieste, Italy
[3] CNR/ISTI, Pisa, Italy
[4] School of Informatics, University of Edinburgh, Edinburgh, Scotland
dmilios@inf.ed.ac.uk
[5] SynthSys, Centre for Synthetic and Systems Biology,
University of Edinburgh, Edinburgh, Scotland

Abstract. Novel applications of formal modelling such as systems biology have highlighted the need to extend formal analysis techniques to domains with pervasive parametric uncertainty. Consequently, machine learning methods for parameter synthesis and uncertainty quantification are playing an increasingly significant role in quantitative formal modelling. In this paper, we introduce a toolbox for parameter synthesis and model checking in uncertain systems based on Gaussian Process emulation and optimisation. The toolbox implements in a user friendly way the techniques described in a series of recent papers at QEST and other primary venues, and it interfaces easily with widely used modelling languages such as PRISM and Bio-PEPA. We describe in detail the architecture and use of the software, demonstrating its application on a case study.

1 Introduction and Motivation

Tools and methodologies from formal analysis are increasingly playing a central role in science and engineering. Recent years have seen a veritable explosion in the number of novel application domains for quantitative analysis of dynamical systems, from systems biology, to smart cities, to epidemiology. A common feature of these novel application domains is the presence of uncertainty: while expert opinion may inform modellers about the presence of specific interactions (the *structure* of the model), it seldom is sufficient to quantify precisely the kinetic parameters underpinning the system dynamics. Extending formal analysis methods to handle models with parametric uncertainty is therefore rapidly becoming a major area of development in formal modelling.

L. Bortolussi—Work partially supported by EU-FET project QUANTICOL (nr. 600708) and by FRA-UniTS.
D. Milios—Work supported by European Research Council under grant MLCS 306999.

© Springer International Publishing Switzerland 2015
J. Campos and B.R. Haverkort (Eds.): QEST 2015, LNCS 9259, pp. 89–104, 2015.
DOI: 10.1007/978-3-319-22264-6_6

Unsurprisingly, the trend towards modelling uncertain systems has led to a convergence between ideas from machine learning and formal modelling. While most efforts are focussing on the problem of identifying parameter values that may match particular specifications in terms of observations or global properties (*parameter synthesis*, [1,4,8,10,12]), more recent efforts aim at characterising and exploiting the dependence of system properties on the parametrisation, and embedding the concept of uncertainty in formal modelling languages [7,9,15]. Despite the considerable interest such approaches are generating, user-friendly implementations of machine learning methodologies for formal analysis are currently lacking.

In this paper, we present U-check, a toolkit for formal analysis of models with parametric uncertainties based on *Gaussian Processes* (GPs), a flexible class of prior distributions over functions underpinning many Bayesian regression and optimisation algorithms [21]. GPs are at the core of several novel developments in formal analysis [2,3,7–9,18]; here we focus on three particular tasks: estimating the parametric dependence of the truth probability of a linear temporal logic formula; synthesising parameters from logical constraints on trajectories, and identifying parameters that maximise the robustness (quantitative satisfaction score [2,13]) of a formula. Our tools are based on Java and interface with popular formal modelling programming languages such as PRISM [17] and Bio-PEPA [11]; we also offer support for hybrid models specified in the SimHyA modelling language (for stochastic hybrid systems) [5]. U-check is available to download at https://github.com/dmilios/U-check.

The Problems Tackled by U-Check. U-check is a tool to perform model checking, parameter estimation, and parameter synthesis for uncertain stochastic models. We consider a parametric family of stochastic processes $\mathcal{M}_\theta$, indexed by parameters θ in a bounded subset $\mathcal{D} \subseteq \mathbb{R}^m$, usually a hyperrectangle. The parametric dependence is introduced to reflect the impossibility of removing uncertainty from model specification. We will refer to the pair $(\mathcal{M}_\theta, \mathcal{D})$ as un *uncertain stochastic model*. Given a model $\mathcal{M}_\theta$, we are often interested in understanding some features of its global behaviour, which can be specified as a set of formal properties, for instance using a linear temporal logic formalism as MiTL or STL [19,20]. U-check solves one of the following problems:

- **Smoothed Model Checking.** Given an uncertain stochastic model $(\mathcal{M}_\theta, \mathcal{D})$, and a linear property φ, smoothed model checking provides a statistical estimate of the satisfaction probability of the formula as a function of the model parameters (*satisfaction function*), $p_\varphi(\theta)$; the tool also returns pointwise confidence bounds. The estimate is obtained in a Bayesian framework combining simulation to generate trajectories, a monitoring routine to very φ on the so obtained trajectories, and Gaussian Process-based statistical inference; details are provided in [7].
- **Parameter Estimation from Qualitative Observations.** This algorithm takes as input an uncertain stochastic model $(\mathcal{M}_\theta, \mathcal{D})$, n linear time properties $\varphi_1, \ldots, \varphi_n$, and N observations of the joint satisfaction value of such

properties. Then, using an active learning optimisation algorithm based on Gaussian Processes [22], it computes the maximum likelihood (or the maximum a-posteriori) estimate $\hat{\theta}$ of parameters θ that best explain the observed dataset [8].
- **Robust Parameter Synthesis.** Given an uncertain stochastic model $(\mathcal{M}_\theta, \mathcal{D})$ and a property φ, the algorithm identifies the parameters set θ^* that maximises the expected robustness (satisfaction score) of φ, again exploiting a Gaussian Process-based optimisation algorithm [2].

2 Background Material

U-check takes as input an uncertain stochastic model $(\mathcal{M}_\theta, \mathcal{D})$ specified as a stochastic model $\mathcal{M}$ and of a range of values for (a subset of) its parameters. The primary focus of U-check is on Continuous-Time Markov Chain models [14], that are simulated by the standard stochastic simulation algorithm [16].

2.1 Population CTMC

A Continuous time Markov Chain (CTMC) $\mathcal{M}$ is a Markovian (i.e. memoryless) stochastic process defined on a finite or countable state space S and evolving in continuous time [14]. We will specifically consider population models of interacting agents [6], which can be easily represented by

- a vector of population variables $X = (X_1, \ldots, X_n)$, counting the number of entities of each kind, and taking values in $S \subseteq \mathbb{N}^n$;
- a finite set of reaction rules, describing how the system state can change. Each rule η is a tuple $\eta = (r_\eta, s_\eta, f_\eta)$. r_η (respectively s_η) is a vector encoding how many agents are consumed (respectively produced) in the reaction, so that $v_\eta = s_\eta - r_\eta$ gives the net change of agents due to the reaction. $f_\eta = f_\eta(X, \theta)$ is the rate function, associating to each reaction the rate of an exponential distribution, depending on the global state of the model and on a d dimensional vector of *model parameters*, θ. Reaction rules are easily visualised in the chemical reaction style, as

$$r_1 X_1 + \ldots r_n X_n \xrightarrow{f(X,\theta)} s_1 X_1 + \ldots s_n X_n$$

2.2 Property Specification

In this work, properties systems are expressed as properties of their trajectories via *Metric Interval Temporal Logic* (MiTL) [20]. Formally, the syntax of a MiTL formula φ is given by the following grammar:

$$\varphi ::= \mathbf{tt} \mid \mu \mid \neg\varphi \mid \varphi_1 \wedge \varphi_2 \mid \varphi_1 \mathbf{U}_{[T_1,T_2]} \varphi_2 \tag{1}$$

where $\mathbf{tt}$ is the true formula, conjunction and negation are the standard boolean connectives, and there is only one temporal modality, the time-bounded until

$\mathbf{U}_{[T_1, T_2]}$. Further connectives can be easily derived from the terms of the grammar above. For example, temporal modalities like time-bounded eventually and always can be defined as: $\mathbf{F}_{[T_1, T_2]}\varphi \equiv \mathtt{tt}\,\mathbf{U}_{[T_1, T_2]}\varphi$ and $\mathbf{G}_{[T_1, T_2]}\varphi \equiv \neg\mathbf{F}_{[T_1, T_2]}\neg\varphi$.

Given a system with n population variables, a trajectory will be a real-valued function $\mathbf{x}(t)$, $\mathbf{x} : [0, T] \to \mathbb{R}^n$. An atomic proposition μ transforms a function $\mathbf{x}(t)$, to a boolean signal $\mathbf{s}_\mu(t) = \mu(\mathbf{x}(t))$, where $\mathbf{s} : [0, T] \to \{\mathtt{tt}, \mathtt{ff}\}$. The truth of a formula φ with respect to a trajectory $\mathbf{x}$ at time t is given by the standard satisfiability relation $\boldsymbol{x}, t \models \varphi$. For instance, the rule for the temporal modality states that $\boldsymbol{x}, t \models \varphi_1\mathbf{U}_{[T_1, T_2]}\varphi_2$ if and only if $\exists t_1 \in [t + T_1, t + T_2]$ such that $\boldsymbol{x}, t_1 \models \varphi_2$ and $\forall t_0 \in [t, t_1]$, $\boldsymbol{x}, t_0 \models \varphi_1$, while $\boldsymbol{x}, t \models \mu$ if and only if $\mathbf{s}_\mu(t) = \mathtt{tt}$.

A CTMC $\mathcal{M}_\theta$ is characterised by a distribution of random trajectories. In this context, a MiTL formula φ can be associated with the probability $Pr(\boldsymbol{x}, 0 \models \varphi | \mathcal{M}_\theta)$, which is the probability that the formula is satisfied at time zero by a trajectory $\boldsymbol{x}$ sampled from $\mathcal{M}_\theta$[1].

We also offer evaluation of MiTL formulae under quantitative semantics, which returns a measure of robustness for a given trajectory [2]. The quantitative specification function ρ returns a value $\rho(\varphi, \mathbf{x}, t) \in \mathbb{R} \cup \{-\infty, +\infty\}$ that quantifies the robustness degree of φ by the trajectory $\mathbf{x}$ at time t subject to perturbations. In the stochastic setting, the robustness will be a real-valued random variable R_φ which captures the distribution of robustness degrees over the trajectory space.

2.3   Statistical Methodologies

We provide here a very brief intuition about the statistical methodologies employed by U-check. A full discussion is provided in the cited papers [7,8] and is beyond the scope of this tool paper.

The fundamental idea behind U-check is that the (intractable) functional dependence of a formula's truth probability on the model parameters can be abstracted through the use of statistical methods. We leverage the fact that the satisfaction probability of a temporal logic formula over an uncertain CTMC is a smooth function of the parameters, which was proved in [7]. Smoothness has important practical repercussions: knowing the value of a smooth function at a point $\mathbf{x}$ is informative about the value of the function in a neighbourhood of the point $\mathbf{x}$ through the Lipschitz property implied by smoothness. Intuitively, U-check exploits this transfer of information between neighbouring points to devise effective algorithms to explore (and optimise) the parametric dependence of truth probabilities.

More formally, the starting point for U-check is to obtain estimates of a truth probability at a set of initial points $\mathbf{x}_1, \ldots, \mathbf{x}_N$ via a standard simulation-based statistical model checking algorithm. These values are treated as *noisy observations* of the unknown function (the truth probability). We then proceed in a Bayesian framework, place a Gaussian Process (GP) prior over the function values at *all* parameter values and combine this with the observations to obtain

---

[1] We assume implicitly that T is sufficiently large so that the truth of φ at time 0 can always be established from $\boldsymbol{x}$.

a *posterior* estimate. GPs are infinite dimensional generalisations of Gaussian distributions [21]; crucially, they can easily encode smoothness and inherit many favourable computational properties from the finite dimensional case. GPs therefore enable us to construct a statistical surrogate of the satisfaction function; this is the central idea behind all of the techniques implemented in U-check. The specific algorithms used for optimisation and smoothed model checking are different, and are described in detail in the main references of the paper.

3   Software Architecture

One of the main requirements of U-check is to be a multi-platform tool, which can also be easily incorporated as a library in other software projects. For this reason, the entire system has been implemented in Java. U-check depends on separate projects that have been developed independently, namely the PRISM project, the Bio-PEPA project, and the SimHyA project, which offer functionality to load models written in the respective languages. Other external libraries that our tool depends on are jBLAS, which is a linear algebra library for Java, and Apache Commons Math, which we use for local optimisation routines.

The software components of U-check, along with their dependencies are summarised in Fig. 1. The main functionality of the tool is implemented by two main components: Learning From Formulae and Smoothed Model Checking. The former depends on the GP optimisation component which implements the non-convex optimisation algorithms required. Both components depend on the Gaussian Process framework and the Model Checking component, which offers routines to perform statistical model checking for CTMCs. Finally, the U-check CLI component offers the functionality of all of the components involved in a common command line interface.

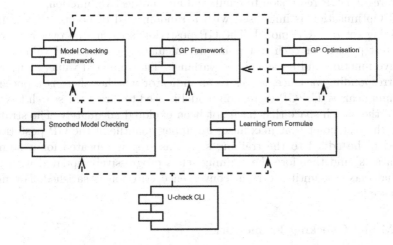

Fig. 1. Component diagram for U-check

3.1   Gaussian Processes Framework

This module is responsible for the GP regression tasks, on which the main methodologies that we cover rely. **Regular GP Regression** assumes that the data are noisy observations of the latent function, which is then analytically approximated by a series of linear algebra calculations. We offer two ways of handling the observation noise either as a constant defined by the user, or using an automatically calibrated heteroskedastic noise model. **Probit Regression** is used in the case of smoothed model checking, where the objective is to emulate satisfaction probability as a function of the parameters. The output of the emulated function has to be strictly in the interval $[0, 1]$, hence regular regression is no longer appropriate and has to be combined with a probit transformation.

Both regression approaches require the specification of a covariance function; we use the *Radial Basis Function* (RBF) kernel due to its theoretical properties [7]. For the RBF kernel, there is an isometric version labelled as `rbfiso`, and a version that supports a different lengthscale parameter for each dimension of the input space. The latter is labelled as `rbfard` and it can be combined with hyperparameter optimisation to achieve *automatic relevance determination*. To automatically determine the hyperparameters of the kernel, we offer two alternatives: a heuristic that relies on the range of the training observations, or a local hyperparameter optimisation using the marginal likelihood of the observed data. For this local optimisation task (using the heuristic as default initialisation), we use the optimisation toolkit of Apache Commons.

3.2   GP Optimisation Framework

This module constitutes an implementation of the GP optimisation algorithm, which is described in [8], as a generic framework for non-convex optimisation of noisy objective functions. The module depends on the GP framework, as it utilises regular GP regression to emulate a given objective function.

GP Optimisation is initialised with a random grid of points, that is used as training set in a GP model. The GP posterior is calculated over a random set of test points; for each test point we calculate the estimated value of the objective function and its associated variance. The emulated value, along with the corresponding variance, serves as an indicator whether there is a potential maximum nearby. The GP regression model is used to direct the search towards areas of the search-space that have not been explored adequately. The strategy is that the test point that maximises an upper quantile of the GP posterior is selected to be added to the training set. This step is repeated for a number of iterations, and therefore the training set is progressively updated with new potential maxima, until a certain convergence criterion is satisfied. For more details see [8].

3.3   Model Checking Framework

This component is responsible for parsing and evaluating MiTL properties. The successful parsing of a MiTL formula will result in an abstract syntax tree which can be evaluated over a specified trajectory. A trajectory can be either a random sample from a CTMC model, or a solution to a system of ordinary differential equations (ODEs). The formulae can be evaluated either in terms of the standard boolean semantics, which is used for smoothed model checking and parameter inference from qualitative data, or the quantitative semantics, which is used for robust parameter synthesis.

The module also involves stochastic simulation routines for CTMCs, which are used by default to evaluate the satisfaction probabilities of MiTL formulae. The simulation capabilities can be overridden, if the model checking module is used as a library. Regarding the modelling component, we have defined an interface that accepts different implementations; in the current version we offer implementations based on PRISM, Bio-PEPA and SimHyA.

3.4   Smoothed Model Checking

This component combines the Gaussian process framework with the model checking capabilities, in order to construct analytic approximations to the satisfaction probability as a function of the model parameters. The structure is outlined in the UML class diagram of Fig. 2. The specification of the model and the properties to be verified is responsibility of `MitlModelChecker` class. The `ModelInterface` specifies the method signatures for loading models, setting the model parameters, and generating trajectories. The model checking framework is agnostic of the modelling and simulation details; it is therefore easily expandable to different kinds of implementations. The form of the trajectories to be generated is part of the interface (the `Trajectory` class), so that these are compatible with the `MiTL` class, which implements an abstract syntax tree of MiTL expressions.

The `SmoothedModelChecker` class is responsible for the main functionality of the module, which makes use of `MitlModelChecker` and `GPEP` to construct an analytic approximation of the satisfaction function. The respective class, i.e. `AnalyticApproximation`, is essentially a trained probit regression GP model. The `performSmoothedModelChecking` method uses this analytic result to estimate the satisfaction probability for a number of points in the parameter space. The ranges of the parameters to be explored are specified by the `Parameter` class, while the `SmmcOptions` class controls the configuration options of a smoothed model checking experiment, further discussed in Sect. 4.

3.5   Learning from Formulae

This module depends on the model checking and the GP optimisation framework. The module is responsible for the application of the GP optimisation algorithm to parameter synthesis, which is achieved by appropriate objective functions.

The structure of the module is outlined in Fig. 3. The `GPOptimisation` class implements the GP optimisation algorithm; it relies on `RegressionGP`, while its options are controlled by `GpoOptions`. As already discussed in Sect. 3.4, `MitlModelChecker` is responsible for the specification of models, properties

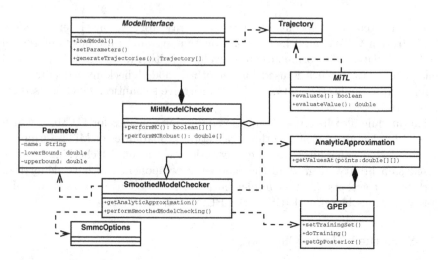

Fig. 2. UML class diagram for the smoothed model checking component

and the simulation algorithms. The `LearnFromFormulae` class performs the actual parameter synthesis. It has to be initialised with an object of type `MitlModelChecker`, a set of parameters and their corresponding prior distributions. The latter are represented by the abstract `Prior` class, whose implementations offer different options, including uniform, exponential, Gaussian and gamma distributions. Other options are specified by `LFFOptions`, which involve options regarding the simulation algorithms used, and the entire set of options in `GpoOptions`. The `GpoResult` class contains the result of the optimisation process; that involves the optimal solution found, along with a covariance matrix that captures the uncertainty of the approximated optimum. In a Bayesian setting, this is interpreted as a Gaussian approximation of the posterior distribution of the parameters.

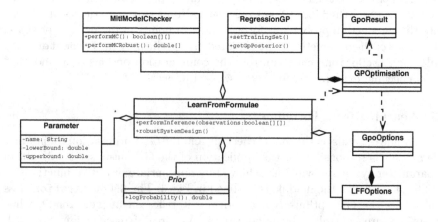

Fig. 3. UML class diagram for the learning from formulae component

3.6   U-Check CLI

This module is the implementation of the command-line tool for model checking and parameter synthesis under uncertainty. It offers a common API to provide the functionality of both Learning From Formulae and Smoothed Model Checking components. It is responsible for linking the other components of U-check with the PRISM, SimHyA and Bio-PEPA libraries, providing implementations to the specified interfaces. It also provides functionality of reading the required experiment options from a configuration file, whose structure is outlined in the next section.

4   Configuration Options

We describe the practical usage of U-check on an example in the next section. Here we summarise the configuration options of the program, avoiding excessive details for readability; a comprehensive description of the options is given in the user manual associated to the code release. The command line interface of our tool dictates that the program is provided with a configuration file, which lists a number of options in the form of assignments as follows:

`OPTION = VALUE`

where `VALUE` can be either a number, a truth value, or a string, depending on the nature of the option. Only a few configuration options are necessary to produce results, while the rest are optional and their corresponding default value will be used if no explicit assignment is made.

4.1   Experiment Configuration

The following options control the main setting of each experiment, which involves the definition of the model, the properties and the mode of operation.

- `modelFile`: A file that contains the specification of a population CTMC described either in PRISM, Bio-PEPA or the SimHyA language.
- `propertiesFile`: A file that contains the specification of one or more MiTL properties.
- `observationsFile`: A file that contains a $n \times m$ matrix, whose rows correspond to n independent observations of the system in question. A row is a single observation and contains m values (0 or 1), one for each one of the MiTL formulae specified in the properties file.
- `mode`: It can be either `inference`, `robust` or `smoothedmc`.

Parameter uncertainty is expressed as a range of possible parameter values, which is provided by the user in addition to model specification. Each parameter has to be associated with an interval, which is specified with an assignment of the form:

`parameter NAME = [A, B]`

Optionally, each parameter can be associated with a prior distribution using an assignment of the form:

```
prior NAME = uniform(A, B) | exponential(MU)
           | gamma(ALPHA, BETA) | gaussian(MU, S2)
```

Only independent univariate priors are supported. If no prior is explicitly declared, then a uniform prior will be assumed. The prior information is only utilised in the parameter inference scenario.

In order for an experiment to progress, `modelFile`, `propertiesFile`, `mode` and at least one parameter are required to be specified in the configuration file. In the case of parameter inference from qualitative data, `observationsFile` is also required to be specified.

4.2 Simulation Options

The options in this category control the parameters of the simulation process. The most important of these are `endTime`, which sets the time up to which the system will be simulated, and `runs` which controls the number of the independent simulation runs per parameter value.

It is also possible to use simulation engines other than stochastic simulation via the `simulator` option. This can take one of three values: `ssa` for Gillespie's stochastic simulation algorithm, `odes` for mean-field approximation, and `hybrid`. The `odes` and the `hybrid` simulation are based on the implementation of SimHyA [5], and they support SimHyA models only. Moreover, the mean-field approximation by ODEs can only be used for robust parameter synthesis.

4.3 GP Options

The options in this category determine the properties of the GP regression models used in the tool. The size of the training set for the GP is defined by the `initialObservations` option. For the parameter synthesis operations, it affects the initialisation of the GP optimisation algorithm. For smoothed model checking, it controls the initial evaluations of the satisfaction function via statistical model checking. In general, increasing this value is expected to increase the approximation quality for the GP regression model. However, an excessively large value for this parameter implies that the entire process degenerates to naïve parameter space exploration via statistical model checking. It is recommended to begin with a relatively small value (for example 100, which is the default value), and progressively increase until the results are estimated with adequate confidence.

The `numberOfTestPoints` option controls the size of the test set for the GP process. In case of parameter synthesis, that is the set of points where the GP posterior is estimated at each step, in order to find a new potential global maximum. Increasing the size of this set will increase the chances of discovering a new potential maximum. For smoothed model checking, the test set contains the points at which we explore the satisfaction function. Alternatively, the test set can be specified via the `testPointsFile` option: a set of parameter values is directly specified in a csv file that contains a $n \times d$ matrix, each row of which is a d-dimensional point in the parameter space.

The `kernel` option defines the kind of the covariance function used: either `rbfiso` for isometric RBF kernel, or `rbfard` for RBF kernel that supports *automatic relevance determination*. Other options control the hyperparameters of the kernel; by default these are internally optimised via a local optimisation process.

GP Optimisation Options. For the modes of operation that involve parameter optimisation, there are a number of options available that control the convergence properties of the GP optimisation algorithm. The optimisation process is considered to have converged, if a certain number of added points with no significant improvement is reached, defined by the `maxAddedPointsNoImprovement` option. An improvement is considered significant if: $f_{n+1} > f_n * \alpha$, where α is the improvement factor, also set by the user. Convergence is alternatively assumed if a number of failed attempts to find a new local optimum is reached, which is set via `maxFailedAttempts`.

5 Case Study

We shall demonstrate the use of U-check on a rumour-spreading model, whose PRISM specification is outlined in Fig. 4. There are three modules that correspond to the population variables of a PCTMC; these are spreaders, ignorants and blockers. A spreader and an ignorant may interact via the `spreading` action, which means that the ignorant is converted to a spreader. If two spreaders interact via the `stop_spreading1` action, then one of them will become blocker and therefore stop spreading the rumour. Finally, the `stop_spreading2` action dictates that a blocker may convert a spreader to blocker.

All rates of this model follow the law of mass action, with kinetic constants `k_s` and `k_r`. We shall measure the probability that the rumour has not reached the entire population, assuming that there has been some initial outbreak. We consider the following MiTL property, which states that there will be still ignorants between time 3 and 5, while the population of spreaders has climbed above 50% of the population before time 1. Note that the nested globally term indicates that the spreader population has to remain above 50 for at least 0.02 time units.

$$\varphi_1 = \mathbf{G}_{[3,5]} ignorants > 0 \ \wedge \ \mathbf{F}_{[0,1]}(\mathbf{G}_{[0,0.02]} spreaders > 50) \tag{2}$$

An example of property specification as used in U-check is shown in Fig. 5. A property file contains a list of constant declarations, followed by one or more modal expressions.

5.1 Smoothed Model Checking

We demonstrate the application of smoothed model checking, considering uncertain parameters `k_s` and `k_r`. The performance of the approach in terms of

```
ctmc
const double k_s=0.05;
const double k_r=0.02;

module spreaders
  spreaders : [0..100] init 10;
  [spreading] true -> spreaders : (spreaders'=spreaders+1);
  [stop_spreading1] true -> spreaders * (spreaders - 1)
                                  : (spreaders'=spreaders-1);
  [stop_spreading2] true -> spreaders
                                  : (spreaders'=spreaders-1);
endmodule

module ignorants
  ignorants : [0..100] init 100-10;
  [spreading] true -> ignorants : (ignorants'=ignorants-1);
endmodule

module blockers
  blockers : [0..100] init 0;
  [stop_spreading1] true -> 1 : (blockers'=blockers+1);
  [stop_spreading2] true -> blockers : (blockers'=blockers+1);
endmodule

module base_rates
  [spreading]       true -> k_s : true;
  [stop_spreading1] true -> k_r : true;
  [stop_spreading2] true -> k_r : true;
endmodule
```

Fig. 4. PRISM model specification for a rumour-spreading system

approximation quality and efficiency compared to naïve parameter exploration
has been analysed in [7]. Given a model file specification rumour.sm, and a
property file rumour.mtl that contains the formula in (2), the contents of the
configuration file to perform smoothed model checking will be the following:

```
modelFile = rumour.sm
propertiesFile = rumour.mtl
mode = smoothedmc
parameter k_s = [0.0001, 2]
parameter k_r = [0.0001, 0.5]
endTime = 5
runs = 10
initialObservtions = 100
numberOfTestPoints = 625
```

According to the initialObservtions option, U-check will evaluate the satisfaction probability on a grid of 100 regularly distributed parameters values
between 0.0001 and 2 for k_s and between 0.0001 and 0.5 for k_r correspond-
ingly. The numberOfTestPoints option means that the GP posterior will be
evaluated on a grid of 625 points.

```
// constant declarations
const int threshold = 3.75

// MiTL properties
G[3,5] ignorants > 0 & F[0,1] (G<=0.02 spreaders>=threshold)
```

Fig. 5. An example of a properties file

The results are written in a text file named MODEL.csv, where MODEL is the name of the input model. The output file contains the grid of input points along with the associated predictions and confidence intervals. The program also produces a script file named load_MODEL.m, that allow easy manipulation of the results under either matlab or octave. The tool also supports limited plotting capabilities via the gnuplot program. If only one parameter is explored, the satisfaction function is plotted along with the confidence bounds obtained by the GP. In the two dimensional case, the satisfaction function is depicted as a 2-D map. Out-of-the-box visualisation of higher-dimensional data is not currently supported. An example of plots produced automatically by U-check can be seen in Fig. 6.

5.2   Robust Parameter Synthesis

We next use U-check to maximise the robustness of the property in (2). In this case, the initialObservations and the numberOfTestPoints options control the GP training and test sets in the optimisation context.

```
modelFile = rumour.sm
propertiesFile = rumour.mtl
mode = robust
parameter k_s = [0.0001, 2]
parameter k_r = [0.0001, 0.5]
endTime = 5
runs = 100
initialObservations = 100
numberOfTestPoints = 50
```

We quote the part of the program output that contains information regarding the solution obtained. For k_s, the most robust value is 0.341, while for k_r we have optimal value equal to 0.107. Note however that the standard deviations calculated are significantly large compared to the estimates. This is an indication that the optimum obtained is unstable. In this case, this is due to the fact that the robustness of the property in question is not very sensitive to the parameters.

```
# Gaussian Process Optimisation --- Results
Solution:       [0.3410609996706949, 0.10736369126944924]
Standard Dev: [3.6248813872572243, 0.7830513406533174]
Covariance matrix:
[13.13976507168386, 2.838061284295024;
 2.838061284295024, 0.6131694020989578]
```

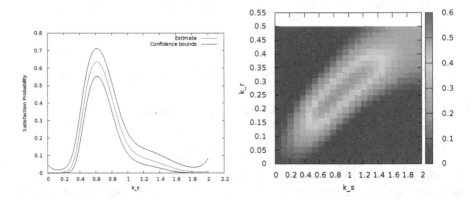

Fig. 6. Emulated satisfaction probability of φ_1 as function of the parameters. Left: We vary k_s only, while k_r is fixed to 0.2. Right: We vary both k_s and k_r

5.3   Inference from Qualitative Data

Finally, we present an example of performing parameter inference from qualitative data. We consider two additional MiTL properties:

$$\varphi_2 = \mathbf{F}_{[0,1]}\,blockers \geq spreaders$$
$$\varphi_3 = \mathbf{G}_{[1.4,2]}\,spreaders \leq 25$$

The φ_2 property states that at some point before time 1 the population blockers surpasses the spreader population, while φ_3 states that the spreaders are always fewer that 25 between time 1.4 and 2. We have considered the rumour-spreading model with parameters k_s= 0.05 and k_r= 0.02, and we have produced a synthetic file of observations named `rumour.dat`, by performing model checking on random trajectories of the original model. U-check then requires the following input:

```
modelFile = rumour.sm
propertiesFile = rumour.mtl
observationsFile = rumour.dat
mode = inference
parameter k_s = [0.0001, 2]
parameter k_r = [0.0001, 0.5]
prior k_s = exponential(0.5)
prior k_r = exponential(0.1)
endTime = 5
runs = 100
```

As for robust parameter synthesis, the results of the optimisation process involve the optimal value obtained for each parameter and the corresponding standard deviation. Note that we have a particularly good fit for both parameters; the estimate for k_s is 0.050, and for k_r is 0.0243, which are very close to the original values, with a low standard deviation.

```
|| # Gaussian Process Optimisation --- Results
|| Solution:      [0.05098993039106938, 0.02430176520400998]
|| Standard Dev:  [0.007751880832319672, 0.0019051431370011513]
|| Covariance matrix:
|| [6.009165643848513E-5, 0.0;
||  0.0, 3.629570372462587E-6]
```

6 Conclusions

Uncertainty is increasingly recognised as an unavoidable companion in many applications of formal methods. This has motivated an increasing cross-fertilisation of ideas between machine learning and quantitative formal modelling. In this paper we describe U-check, a novel tool which implements a number of Gaussian Process based methods for formal analysis of uncertain stochastic processes. Our aim is to offer a set of tools that can be used by modellers without an in-depth knowledge of statistical machine learning. To our knowledge, U-check is the first such tool available; to further facilitate adoption of the tool, U-check can take as input models formulated in widely used modelling languages such as PRISM and Bio-PEPA.

The principal conceptual innovation of the methods implemented in U-check is the smoothness of the satisfaction function for a MiTL formula as a function of the model parameters. This enables us to transfer information across neighbouring observations, yielding potentially very significant computational savings: while we cannot comprehensively review results here, Smoothed Model Checking was shown in [7] to yield computational savings of an order of magnitude on non-trivial systems biology models. Similarly, the smoothness of the satisfaction function enables us to deploy a provably convergent algorithm for (robust) parameter synthesis, with both theoretical guarantees and computational advantages, as shown in non-trivial case studies in [2,8].

While in our work we primarily focus on examples from biology, uncertain stochastic processes are the norm in many other areas of application of formal methods, from smart cities to cyber-physical systems. Future work will explore increasing support for hybrid systems, as well as supporting other formal analysis methodologies such as reachability computations [9] and property synthesis [3].

References

1. Andreychenko, A., Mikeev, L., Spieler, D., Wolf, V.: Approximate maximum likelihood estimation for stochastic chemical kinetics. EURASIP J. Bioinform. Syst. Biol. **2012**(1), 1–14 (2012)
2. Bartocci, E., Bortolussi, L., Nenzi, L., Sanguinetti, G.: On the robustness of temporal properties for stochastic models. Proc. of HSB **125**, 3–19 (2013)
3. Bartocci, E., Bortolussi, L., Sanguinetti, G.: Data-driven statistical learning of temporal logic properties. In: Legay, A., Bozga, M. (eds.) FORMATS 2014. LNCS, vol. 8711, pp. 23–37. Springer, Heidelberg (2014)

4. Bartocci, E., Grosu, R., Katsaros, P., Ramakrishnan, C.R., Smolka, S.A.: Model repair for probabilistic systems. In: Abdulla, P.A., Leino, K.R.M. (eds.) TACAS 2011. LNCS, vol. 6605, pp. 326–340. Springer, Heidelberg (2011)

5. Bortolussi, L., Galpin, V., Hillston, J.: Hybrid performance modelling of opportunistic networks. EPTCS **85**, 106–121 (2012)

6. Bortolussi, L., Hillston, J., Latella, D., Massink, M.: Continuous approximation of collective systems behaviour: a tutorial. Perform. Eval. **70**, 317–349 (2013)

7. Bortolussi, L., Milios, D., Sanguinetti, G.: Smoothed model checking for uncertain continuous time Markov chains. CoRR, abs/1402.1450 (2014)

8. Bortolussi, L., Sanguinetti, G.: Learning and designing stochastic processes from logical constraints. In: Joshi, K., Siegle, M., Stoelinga, M., D'Argenio, P.R. (eds.) QEST 2013. LNCS, vol. 8054, pp. 89–105. Springer, Heidelberg (2013)

9. Bortolussi, L., Sanguinetti, G.: A statistical approach for computing reachability of non-linear and stochastic dynamical systems. In: Norman, G., Sanders, W. (eds.) QEST 2014. LNCS, vol. 8657, pp. 41–56. Springer, Heidelberg (2014)

10. Češka, M., Dannenberg, F., Kwiatkowska, M., Paoletti, N.: Precise parameter synthesis for stochastic biochemical systems. In: Mendes, P., Dada, J.O., Smallbone, K. (eds.) CMSB 2014. LNCS, vol. 8859, pp. 86–98. Springer, Heidelberg (2014)

11. Ciocchetta, F., Hillston, J.: Bio-PEPA: a framework for the modelling and analysis of biological systems. Theoret. Comput. Sci. **410**(33–34), 3065–3084 (2009)

12. Donaldson, R., Gilbert, D.: A model checking approach to the parameter estimation of biochemical pathways. In: Heiner, M., Uhrmacher, A.M. (eds.) CMSB 2008. LNCS (LNBI), vol. 5307, pp. 269–287. Springer, Heidelberg (2008)

13. Donzé, A., Maler, O.: Robust satisfaction of temporal logic over real-valued signals. In: Chatterjee, K., Henzinger, T.A. (eds.) FORMATS 2010. LNCS, vol. 6246, pp. 92–106. Springer, Heidelberg (2010)

14. Durrett, R.: Essentials of Stochastic Processes. Springer, New York (2012)

15. Georgoulas, A., Hillston, J., Milios, D., Sanguinetti, G.: Probabilistic programming process algebra. In: Norman, G., Sanders, W. (eds.) QEST 2014. LNCS, vol. 8657, pp. 249–264. Springer, Heidelberg (2014)

16. Gillespie, D.T.: Exact stochastic simulation of coupled chemical reactions. J. Phys. Chem. **81**(25), 2340–2361 (1977)

17. Kwiatkowska, M., Norman, G., Parker, D.: PRISM 4.0: verification of probabilistic real-time systems. In: Gopalakrishnan, G., Qadeer, S. (eds.) CAV 2011. LNCS, vol. 6806, pp. 585–591. Springer, Heidelberg (2011)

18. Legay, A., Sedwards, S.: Statistical abstraction boosts design and test efficiency of evolving critical systems. In: Margaria, T., Steffen, B. (eds.) ISoLA 2014, Part I. LNCS, vol. 8802, pp. 4–25. Springer, Heidelberg (2014)

19. Maler, O., Nickovic, D.: Monitoring temporal properties of continuous signals. In: Lakhnech, Y., Yovine, S. (eds.) FORMATS 2004 and FTRTFT 2004. LNCS, vol. 3253, pp. 152–166. Springer, Heidelberg (2004)

20. Ouaknine, J., Worrell, J.B.: Some recent results in metric temporal logic. In: Cassez, F., Jard, C. (eds.) FORMATS 2008. LNCS, vol. 5215, pp. 1–13. Springer, Heidelberg (2008)

21. Rasmussen, C., Williams, C.: Gaussian Processes for Machine Learning. MIT Press, Cambridge (2006)

22. Srinivas, N., Krause, A., Kakade, S., Seeger, M.: Information-theoretic regret bounds for Gaussian process optimisation in the bandit setting. IEEE Trans. Inf. Th. **58**(5), 3250–3265 (2012)

mapfit: An R-Based Tool for PH/MAP Parameter Estimation

Hiroyuki Okamura[✉] and Tadashi Dohi

Department of Information Engineering, Graduate School of Engineering,
Hiroshima University, 1–4–1 Kagamiyama,
Higashi-Hiroshima 739–8527, Japan
{okamu,dohi}@rel.hiroshima-u.ac.jp

Abstract. In this paper, we present a tool for estimating parameters of phase-type distribution (PH) and Markovian arrival process (MAP) on the statistical analysis package R. PH and MAP are useful for the analysis of non-Markovian models approximately. By approximating general distributions and point processes with PH and MAP, the non-Markovian models is reduced to continuous-time Markov chains (CTMCs) that can be solved by analytical approaches. The significant features of our tool are (i) PH/MAP fitting from grouped data and (ii) PH fitting from probability density functions.

Keywords: Phase-type distribution · Markovian arrival process · Parameter estimation · Grouped data · R

1 Introduction

This paper presents a PH/MAP fitting tool. There are two main features of our tool; PH/MAP fitting from grouped data and PH fitting from probability density functions (p.d.f.s). The grouped data is also called count data where the number of events for each time interval is recorded. In practice, we frequently encounter grouped data for measurements of real system. It is essentially important to estimate PH/MAP parameters from grouped data. In recent years, Okamura et al. [5,7] developed PH/MAP fitting algorithms from grouped data by applying the EM (expectation-maximization) algorithm. Our tool implements these algorithms, though all of the existing PH/MAP fitting tools cannot deal with grouped data directly.

The second feature is to implement PH fitting algorithm from p.d.f.s directly. This is also significant for PH/MAP approximation in the analysis of non-Markovian models. However, most of existing tools do not support PH fitting algorithm from p.d.f.s. In our tool, we implement PH fitting algorithm from p.d.f.s by utilizing a numerical quadrature based on a double exponential formula.

© Springer International Publishing Switzerland 2015
J. Campos and B.R. Haverkort (Eds.): QEST 2015, LNCS 9259, pp. 105–112, 2015.
DOI: 10.1007/978-3-319-22264-6_7

2   Related Work

There exist several PH/MAP fitting tools. *EMpht*[1] is a C program for PH fitting with EM algorithm by Asmussen et al. [1], which is one of the oldest tools for PH fitting. *EMpht* can also handle both PH fitting algorithms from p.d.f. and samples. In addition, it implements the PH fitting from censored data. However, the user interface is not well designed, and the computation speed is not fast. *PhFit*[2] consists of a Java-based interference and computation engines written by C language. This tool deals with continuous and discrete PHs. Furthermore, PH fitting from a density function is implemented. *momentmatching*[3] is a set of MATLAB files to execute three moment matching [8]. *BuTools*[4] are program packages for Mathematica and MATLAB/Octave. The tool mainly provides PH/MAP fitting with the moment matching. *G-FIT*[5] is a command line tool of the EM algorithms for the hyper-Erlang distribution [9,13]. *jPhaseFit*[6] is a library for Java to handle PHs. In the library, PH fitting with moment matching and EM algorithm are implemented [10]. In particular, the EM algorithm for hyper-Erlang distribution [13] is implemented in the tool. *HyperStar*[7] is a Java-based GUI tool to estimate hyper-Erlang distributions and to draw their graphs.

3   PH Fitting

3.1   Overview

PH distribution is defined as the time to absorption in an absorbing CTMC (continuous-time Markov chain). The p.d.f. and cumulative distribution function (c.d.f.) are mathematically given as the expressions using matrix exponential. Let α and Q denote a probability (row) vector determining an initial state and an infinitesimal generator dominating transitions in transient states, respectively. Since the c.d.f. is given as the probability that the current state of underlying CTMC has already been absorbed, the c.d.f. of PH distribution is given by

$$F(x) = 1 - \alpha \exp(Qx)\mathbf{1}, \tag{1}$$

where $\mathbf{1}$ is a column vector whose entries are 1. Also the p.d.f. can be obtained by taking the first derivative of the c.d.f.

$$f(x) = \alpha \exp(Qx)\xi, \tag{2}$$

where $\xi = -Q\mathbf{1}$. The purpose of PH fitting is to estimate PH parameters α and Q so that the estimated PH distribution fits to observed data. There are two

---

[1] http://home.math.au.dk/asmus/pspapers.html.

[2] http://webspn.hit.bme.hu/~telek/tools.htm.

[3] http://www.cs.cmu.edu/~osogami/code/momentmatching/index.html.

[4] http://webspn.hit.bme.hu/~telek/tools/butools/butools.html.

[5] http://ls4-www.cs.tu-dortmund.de/home/thummler/mainE.html.

[6] http://copa.uniandes.edu.co/?p=141.

[7] http://www.mi.fu-berlin.de/inf/groups/ag-tech/projects/HyperStar/index.html.

different approaches; moment matching (MM) method and maximum likelihood estimation (MLE). The MM method is to find PH parameters whose moments match to the moments obtained from empirical data or distribution functions. On the other hand, MLE is to find PH parameters maximizing the probability that the data is drawn from the model as samples.

3.2 Data for PH Fitting

The parameter estimation algorithm generally depends on data forms. Our tool deals with several kinds of data for PH fitting; IID (independent and identically-distributed) samples, weighted IID samples, grouped data with missing and truncated values. The grouped data is same as the frequency table, i.e., a table of the number of samples falling in each time interval. In this paper, IID samples are called the point data, and the time points dividing the intervals are called break points. Moreover, the tool can handle missing data. The missing values are expressed by NA. Also the truncated data can be represented as the number of samples in the interval to infinity. Table 1 (a) shows an example of the grouped data on the break points $0, 10, 20, \ldots, 100$ where the data have missing values in $[10, 20]$ and $[40, 50]$. Furthermore, the last 5 samples are truncated at 100. Similarly, Table 1 (b) shows an example of another grouped data. In Table 1 (b), although samples are truncated at 100, we do not know the exact number of truncated samples.

Table 1. Examples of grouped data for PH fitting (Breaks: $0, 10, 20, \ldots, 90, 100$).

Grouped data (a)		Grouped data (b)	
Time interval	Counts	Time interval	Counts
$[0, 10]$	1	$[0, 10]$	1
$[10, 20]$	NA	$[10, 20]$	NA
$[20, 30]$	4	$[20, 30]$	4
$[30, 40]$	10	$[30, 40]$	10
$[40, 50]$	NA	$[40, 50]$	NA
$[50, 60]$	30	$[50, 60]$	30
$[60, 70]$	10	$[60, 70]$	10
$[70, 80]$	12	$[70, 80]$	12
$[80, 90]$	4	$[80, 90]$	4
$[90, 100]$	0	$[90, 100]$	0
$[100, \infty]$	5	$[100, \infty]$	NA

3.3 Models and Methods

PH distributions are classified to sub-classes by the structures of α and Q, and the parameter estimation algorithms depend on the class of PH distribution. The tool deals with the following classes of PH distribution:

- general PH: The PH distribution in which there are no constraints on α and Q. In the tool, this is referred to as 'ph' class.
- canonical form 1 (CF1): One of the minimal representations of acyclic PH distribution. The matrix Q becomes a bidiagonal matrix whose entries are sorted in increasing order [3]. In the tool, this is referred to as 'cf1' class.
- hyper-Erlang distribution: One of the representations of acyclic PH distribution. The distribution consists of a mixture of Erlang distributions. In the tool, this is referred to as 'herlang' class.

The parameters of ph class have slots α, Q and ξ, which are defined as members of S4 class in R. To represent the matrix Q, we use Matrix package which is an external package of R. The slots of cf1 class are inherited from the ph class. In addition to inherited slots, cf1 has a slot rate to store the absolute values of diagonal elements of Q. The herlang class has the slots for the mixed ratio, shape parameters of Erlang components, rate parameters of Erlang components. Both of cf1 and herlang classes can be transformed to ph class by using 'as' method of R.

The R functions for PH parameter estimation are;

- phfit.point: MLEs for general PH, CF1 and hyper-Erlang distribution from point and weighted point data. The estimation algorithms for general PH and CF1 are the EM algorithms proposed in [6]. The algorithm for hyper-Erlang distribution is the EM algorithm with shape parameter selection described in [9,13].
- phfit.group: MLEs for general PH, CF1 and hyper-Erlang distribution from grouped data. The estimation algorithms for general PH and CF1 are the EM algorithms proposed in [7]. The algorithm for hyper-Erlang distribution is the EM algorithm with shape parameter selection, which is originally developed as an extension of [9,13].
- phfit.density: MLEs for general PH, CF1 and hyper-Erlang distribution from a density function defined on the positive domain $[0, \infty)$. The function phfit.density calls phfit.point after making weighted point data. The weighted point data is generated by numerical quadrature. In the tool, the numerical quadrature is performed with a double exponential (DE) formula [12].
- phfit.3mom: MM methods for acyclic PH distribution with the first three moments [2,8], which are ported from *momentmatching*[8] and *BuTools*[9].

The functions phfit.point, phfit.group and phfit.density select an appropriate algorithm depending on the class of a given PH distribution. These functions return a list including the estimated model (ph, cf1 or herlang), the maximum log-likelihood (llf), Akaike information criterion (aic) and other statistics of the estimation algorithm. Also, the function phfit.3mom returns a ph class whose first three moments match to the given three moments.

---

[8] http://www.cs.cmu.edu/~osogami/code/momentmatching/index.html.
[9] http://webspn.hit.bme.hu/~telek/tools/butools/butools.html.

4   MAP Fitting

4.1   Overview

MAP is a stochastic point process whose arrival rates are dominated by a CTMC. The CTMC expresses the internal state of MAP called a phase process. MAP is generally defined by an initial probability vector α and two matrices for infinitesimal generators D_0, D_1 without and with arrivals, respectively. Note that $D_0 + D_1$ becomes the infinitesimal generator of phase process. Similar to PH fitting, the purpose of MAP fitting is to find MAP parameters α, D_0 and D_1 so that the estimated MAP fits to observed data. In MAP fitting, there are also two approaches; MM method and MLE. The MM method for MAP is to determine MAP parameters with marginal/joint moments and k-lag correlation. MLE is to find MAP parameters maximizing the log-likelihood function. We implement MLE approaches in the tool.

4.2   Data for MAP Fitting

The tool deals with two kinds of data; point and grouped data in MAP fitting. The point data for MAP fitting is a time series for arrivals. The grouped data for MAP fitting consists of break points and counts which are made from a time series for arrivals. Table 2 shows examples of point data and grouped data in MAP fitting. In the table, the grouped data are made from the point data by counting the number of arrivals in every 5 seconds. Note that missing values cannot be treated in MAP fitting of this version of tool.

4.3   Models and Methods

In the tool, there are three classes (models) for MAP, which require different parameter estimation algorithms.

- general MAP: MAP with no constraints on parameters. This class is referred to as 'map' in the tool. Also, some of sub-classes of general MAP such as a Markov modulated Poisson process (MMPP) are defined by a specific structure of map. For instance, MMPP can be generated by an 'mmpp' command.
- HMM (hidden Markov model) with Erlang outputs (ER-HMM): One of MAP representations where Erlang outputs correspond to time intervals between two successive arrivals [4]. In the tool, this class is referred to as 'erhmm'.
- MMPP with grouped data: MMPP with approximate parameter estimation. This is referred to as 'gmmpp' in the tool, and is essentially same as mmpp except for parameter estimation algorithm. In the parameter estimation of gmmpp, it is assumed that at most one phase change is allowed in one observed time interval [5].

The map class consists of the slots α, D_0 and D_1. The gmmpp class also has the slots alpha, D0 and D1. The erhmm class has an initial probability vector for HMM states (alpha), a probability transition matrix for HMM states (P), the

Table 2. Examples of point and grouped data for MAP fitting.

Point data		Grouped data (Breaks: $0, 5, 10, 15, 20, 25, \ldots$)	
Arrival No.	Time (sec)	Time interval	Counts
1	1.340	[0, 5]	3
2	1.508	[5, 10]	1
3	4.176	[10, 15]	1
4	8.140	[15, 20]	2
5	11.036	[20, 25]	4
6	15.072	⋮	⋮
7	17.892		
8	20.604		
9	22.032		
10	24.300		
⋮	⋮		

shape parameters for Erlang distribution (**shape**) and the rate parameters for Erlang distribution (**rate**). The S4 class **erhmm** can be transformed to **map** by using 'as' method.

The R functions for MAP parameter estimation are;

- **mapfit.point**: MLEs for general MAP and ER-HMM from point data. The estimation algorithm for general MAP is the EM algorithm introduced by [11]. The algorithm for ER-HMM is the fast EM algorithm proposed in [4].
- **mapfit.group**: MLEs for general MAP and **gmmpp** from grouped data. The estimation algorithms for general MAP and **gmmpp** are presented in [5]. Note that **erhmm** cannot handle grouped data.

The functions **mapfit.point** and **mapfit.group** select an appropriate algorithm depending on the class of a given MAP. These functions return a list including the estimated model (**map**, **erhmm** or **gmmpp**), the maximum log-likelihood (**llf**), Akaike information criterion (**aic**) and other statistics of the estimation algorithm.

5   Example

In this section, we introduce examples of the usage of PH fitting based on IID samples from Weibull distribution. When **wsample** is a vector for IID samples of Weibull distribution, we make grouped data by using the function **hist** as follows.

```
> h.res <- hist(wsample, breaks="fd", plot=FALSE)
> h.res$counts
 [1]  4  9 13 14 20 16  9  8  3  3  1
> h.res$breaks
 [1] 0.0 0.2 0.4 0.6 0.8 1.0 1.2 1.4 1.6 1.8 2.0 2.2
```

Then we can get estimated CF1 parameters from grouped data with the following command:

```
> phfit.group(ph=cf1(5), counts=h.res$counts, breaks=h.res$breaks)
```

Next we present the PH fitting from a density function. Since the density function of Weibull distribution is given by a function dweibull in R, the PH fitting from a density function can be done with the following command:

```
> phfit.density(ph=cf1(5), f=dweibull, shape=2, scale=1)
```

The last two arguments are parameters of dweibull function. User-defined functions can also be used as density functions.

Finally we demonstrate MAP fitting with grouped data. The data used in this example is the traffic data; BCpAug89 [10] which consists of time intervals for packet arrivals. Although the original BCpAug89 involves the inter-arrival times of 1 million packets, we use only the first 1000 arrival times in this example. Similar to PH fitting, the grouped data can be made from the inter-arrival time data BCpAug89 by using hist function, i.e.,

```
> BCpAug89.group<-hist(cumsum(BCpAug89), breaks=seq(0,2.7,0.01),
                       plot=FALSE)
```

In the above, break points are set as a time point sequence from 0 to 2.7 by 0.01, which is generated by a seq function. Using the grouped data, we have the estimated parameters for general MAP, MMPP and MMPP with approximate estimation (gmmpp).

```
> mapfit.group(map=map(5), counts=BCpAug89.group$counts,
               breaks=BCpAug89.group$breaks)
> mapfit.group(map=mmpp(5), counts=BCpAug89.group$counts,
               breaks=BCpAug89.group$breaks)
> mapfit.group(map=gmmpp(5), counts=BCpAug89.group$counts,
               breaks=BCpAug89.group$breaks)
```

6   Conclusion and Future Work

In this paper, we have presented the R-based tool for PH/MAP fitting. The significant two features of our tool are (i) PH/MAP fitting from grouped data and (ii) PH fitting from p.d.f.s. In fact, many measurements are provided as grouped data, and thus it is important to develop PH/MAP fitting algorithms from grouped data. In addition, since the presented package is installed in R, we easily use features of R. For instance, we can draw a graph with estimated PH/MAP by using common functions of R. This is one of the advantages of developing a tool as a package of R. *mapfit* has already been published in CRAN. Anyone can download the tool through CRAN with a command; install.packages() in R.

---

[10] http://ita.ee.lbl.gov/html/contrib/BC.html.

References

1. Asmussen, S., Nerman, O., Olsson, M.: Fitting phase-type distributions via the EM algorithm. Scand. J. Stat. **23**(4), 419–441 (1996)
2. Bobbio, A., Horváth, A., Telek, M.: Matching three moments with minimal acyclic phase type distributions. Stoch. Models **21**(2/3), 303–326 (2005)
3. Cumani, A.: On the canonical representation of homogeneous Markov processes modelling failure-time distributions. Microelectron. Reliab. **22**, 583–602 (1982)
4. Okamura, H., Dohi, T.: Faster maximum likelihood estimation algorithms for Markovian arrival processes. In: Proceedings of 6th International Conference on Quantitative Evaluation of Systems (QEST2009), pp. 73–82 (2009)
5. Okamura, H., Dohi, T., Trivedi, K.S.: Markovian arrival process parameter estimation with group data, in submission
6. Okamura, H., Dohi, T., Trivedi, K.S.: A refined EM algorithm for PH distributions. Perform. Eval. **68**(10), 938–954 (2011)
7. Okamura, H., Dohi, T., Trivedi, K.S.: Improvement of EM algorithm for phase-type distributions with grouped and truncated data. Appl. Stochast. Models Bus. Ind. **29**(2), 141–156 (2013)
8. Osogami, T., Harchol-Balter, M.: Closed form solutions for mapping general distributions to minimal PH distributions. Perform. Eval. **63**(6), 524–552 (2006)
9. Panchenko, A., Thümmler, A.: Efficient phase-type fitting with aggregated traffic traces. Perform. Eval. **64**, 629–645 (2007)
10. Perez, J.F., Riano, G.: jPhase: an object-oriented tool for modeling phase-type distributions. In: In SMCtools 2006: Proceeding from the 2006 Workshop on Tools for Solving Structured Markov Chains, p. 5. ACM (2006)
11. Rydén, T.: An EM algorithm for estimation in Markov-modulated poisson processes. Comput. Stat. Data Anal. **21**(4), 431–447 (1996)
12. Takahasi, H., Mori, M.: Double exponential formulas for numerical integration. Publ. RIMS Kyoto Univ. **9**, 721–741 (1974)
13. Thümmler, A., Buchholz, P., Telek, M.: A novel approach for phase-type fitting with the EM algorithm. IEEE Trans. Dependable Secure Comput. **3**(3), 245–258 (2006)

A Compression App for Continuous Probability Distributions

Michael Bungert[1], Holger Hermanns[1], and Reza Pulungan[2][✉]

[1] Saarland University – Computer Science, Saarbrücken, Germany
mbungert@depend.cs.uni-saarland.de, hermanns@cs.uni-saarland.de
[2] Jurusan Ilmu Komputer Dan Elektronika, Universitas Gadjah Mada,
Yogyakarta, Indonesia
pulungan@ugm.ac.id

Abstract. This paper presents an Android app supporting the construction and compact representation of continuous probability distributions. Its intuitive drag-and-drop approach considerably eases an often delicate modelling step in model-based performance and dependability evaluation, stochastic model checking, as well in the quantitative study of system-level concurrency phenomena. To compress the size of the representations, the app connects to a web service that implements an efficient compression algorithm, which constitutes the core technological innovation behind the approach. The app enables interested users to perform rapid experiments with this technology. From a more general perspective, this approach might pioneer how web service and app technology can provide a convenient vehicle for promoting and evaluating computer aided verification innovations.

1 Quantitative Verification at Your Fingertips

Touch-enabled devices have reached the mainstream. Touchscreens are nowadays common assets to smartphones, PDAs, and laptop computers, and are finding their way into the medical field as well as into heavy industry. Many types of information appliances are driving the demand and acceptance of common touch-enabledness allowing an intuitive, rapid, and accurate interaction with the user. Recognition of 3D gestures might become the next emerging technology making the keyboard more and more obsolete [13].

Does this trend impact quantitative modelling and verification research?

It appears difficult to imagine a future where research innovations will be conceived and documented without the use of a classical keyboard-enabled computer. The present paper does not intend to speculate about this. Instead it discusses, by means of a concrete example, if and how current and future research innovations can have a greater outreach by dedicated efforts for making them gesture or touch enabled.

This work is partly supported by the DFG as part of SFB/TR 14 AVACS, by the EU FP7 grants 295261 (MEALS) and 318490 (SENSATION), and by the CAS/SAFEA International Partnership Program for Creative Research Teams.

© Springer International Publishing Switzerland 2015
J. Campos and B.R. Haverkort (Eds.): QEST 2015, LNCS 9259, pp. 113–121, 2015.
DOI: 10.1007/978-3-319-22264-6_8

Another aspect of tool-oriented research is the deployment of software arte-facts. Historically, software started being deployed by installation from external storage media, then it became downloadable from the internet. In line with the greater adoption of the Software-as-a-Service paradigm [4] we nowadays see com-mercial distribution platforms such as Google Play, iOS App Store and Windows Store providing opportunities for third parties to deploy their artefacts, albeit with a special focus on mobile technology.

Does this trend impact quantitative modelling and verification research?

The deployment of commercial software via a commercial store or as a service appears in principle to be free of specific obstacles. It is just another distribu-tion strategy that needs to be looked at from the economic perspective. We are in this paper focussing on research prototypes, not on quality products. The former are usually distributed free-of-charge for empirical evaluation purposes. Often they turn out to be difficult to install. This is an impediment to the take-up of the underlying sometimes brilliant idea by the community, and beyond. Making installation and experimentation possible with a fingertip seems a great alternative.

Installation difficulties are no serious concern for more mature academic software artefacts with long-term commitment behind them, such as SPIN, UPPAAL, PRISM, NuSMV, CADP, or Z3, to name a few. Still, some of the main developers of the latter tools are often facing scientific publications that make comparison with their tools without using the latest available version. Dis-tributing updates via the functionality provided by the store seems a natural way forward in this respect.

The ease-of-installation and ease-of-use attributes associated to many apps might be directly linked to the fact that almost all existing apps are actually very simple algorithmically. Quantitative verification innovations are the oppo-site. This puts some boundaries on what can be rolled out via a store. Other boundaries are imposed by the fact that apps on mobile devices are severely limited in battery and computing power. However, if the algorithmic innovation can be hosted separately from the device, these problems can be overcome by wrapping the functionality in a web service, and hosting that in the cloud. In fact several of the mature tool mentioned above have a built-in separation of computing service and GUI, implemented via sockets, and therefore appear not overly difficult to migrate to a web service.

This paper elaborates on these ideas for a concrete research innovation. It is concerned with the modelling of stochastic timed behaviour as needed for faithful stochastic model checking [2], statistical model checking [14] and dis-crete event simulation [8]. In all three contexts, the model representation size is an impediment to the solution efficiency[1]. The innovation we consider is a preprocessing step that combines a highly convenient and intuitive modelling of time dependencies with an aggressive compression algorithm concerning the internal representation size.

---

[1] For statistical model checking and discrete event simulation the problem lies in the solution time needed, while for stochastic model checking it is both time and space.

Table 1. Operations on probability distributions

	con		*min*		*max*		
Composition Operators	Sequence	;	Choice	⊕	Interleaving Parallel ‖		[17]
Attack Trees	SEQ	⟁	OR	⟁	AND	⟁	[1]
Dynamic Fault Trees	COLD SPARE	⊏⊐	OR	⟁	AND	⟁	[10,5]

2 Stochastic Time-Dependent Behavior Modelling

The study of the time-dependent behavior of a system can reveal distinguishing aspects of many systems. Often, the timing of events is described using continuous probability distributions, and internally often represented by Markov chains or related discrete-state structures [3]. When modelling the time-dependent behavior, one can identify a few operations on probability distributions that are common and widely used in different, only loosely related areas. These recurring operations are: *con*volution, *min*imum, and *max*imum. Table 1 gives an illustrative overview of their use in the context of dependability (fault trees), quantitative security (attack trees), and applied concurrency theory (composition operators). Given two events, the convolution corresponds to the sum of the timing occurrences of the two events, the minimum corresponds to the earliest, while the maximum corresponds to the latest of the timing occurrences.

There are several ways to represent continuous probability distributions, and one of them, which links directly to Markov chains, uses acyclic phase-type (APH) distributions. APH distributions are topologically dense [12]; hence, any continuous probability distribution can be approximated arbitrarily closely by an APH distribution. Several effective tools exist that can tightly fit small APH distributions to arbitrary distributions or empirical data [11,18]. Moreover, APH distributions are closed under the three operations mentioned above. The class comprises important classes of distributions including exponential, hyperexponential, hypoexponential, and Erlang distributions.

3 Acyclic Phase-Type Compression

An APH distribution can be seen as the probability distribution of the time to absorption in a finite acyclic *continuous-time Markov chain* (CTMC) [15]. CTMCs are usually represented by their generator matrix and initial distribution. The *size* of a CTMC is the dimension of its generator matrix. Convolution, maximum, and minimum correspond to certain tensor operations on matrices , which tend to grow with the product of the sizes involved (except for convolution, which is linear), because they are basically cross-product constructions.

However, an APH distribution has many distinct CTMC representations of the same size, but also of drastically different size. In practice, there is a need to work with the smallest possible representation, especially when applying minimum and maximum operators in succession, since this induces an exponential blow-up of the representations.

To combat this problem, we have devised an effective technique to compress the size of the result of the operations, yielding minimality in almost all cases. This is based on a polynomial-time algorithm [16,17] compressing representation sizes. The compression achievable goes beyond concepts like lumpability [6], since the algorithm exploits properties of the Laplace-Stieltjes transform. The algorithm has time complexity $\mathcal{O}(n^3)$, where n is the size of the original representation. The compression works especially well if the representation has many duplicate matrix entries, and remarkably, this is precisely what the minimum and maximum operators induce. In practice, this can turn an exponential growth (in the number of operands) of the representation size into a linear growth, and the resulting APH representation is almost surely minimal [16,17]. The resulting representation is in *Cox* form, a linear structure where each state (apart from the absorbing state) has a transition to the next state, possibly a transition to the absorbing state, but no other transitions.

To exploit these potentially very useful algorithmic achievements, a mechanism to generate and manipulate APH representations, together with the compression algorithm, has earlier been implemented within a web service, called APHzip. The web service supports a single operation called aphmin, and is deployed using the SOAP 1.1 protocol over HTTP. APHzip accepts as input an expression written in a prefix notation that follows the grammar

$$P ::= \mathbf{exp}(\lambda) \mid \mathbf{erl}(k, \lambda) \mid \mathbf{cox}(\mu, \lambda, P) \mid \mathbf{con}(P, P) \mid \mathbf{min}(P, P) \mid \mathbf{max}(P, P),$$

where $\lambda \in \mathbb{R}_+$ and $\mu \in \mathbb{R}_{\geq 0}$ are *rates*, and $k \in \mathbb{Z}_+$. Here, **exp** and **erl** represent exponential and Erlang distributions, the basic blocks of more complex APHs constructed by using operators convolution (**con**), minimum (**min**), and maximum (**max**). **cox** is an operator used to produce Cox representations. This operator semantically works as follows: given an APH P, $\mathbf{cox}(\mu, \lambda, P)$ is a new APH obtained by adding a new state having a transition with rate μ to the absorbing state and a transition with rate λ to P. Repeated application of this operator enables us to produce any Cox representation, and hence represent any APH distribution (and hence approximate any distribution with arbitrary precision) as APHzip expressions.

Provided with a valid expression, APHzip parses it, calls the compression algorithm (in a divide-and-conquer fashion) and then returns the resulting compressed APH representation [17].

4   The App and Its Functionality

In order to further enhance modelling and analysis convenience, but also to experiment with new forms of intuitive graphical modelling on mobile devices,

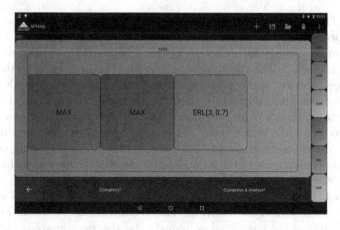

Fig. 1. Screenshots of the APHzip Android app using a Google Nexus 10 tablet

we have implemented a prototype Android app, available for Android 4.0 or later. It uses an intuitive drag-and-drop mechanism, rooted in the observation that nestings of n-ary operators can be represented as typed (or better, colored) sets. Therefore, the app displays every operator type by a distinct box with a distinct color, and similarly for the basic blocks (*i.e.*, **exp** and **erl**), here called *delays*. The graphical design of the app is shown in Fig. 1. A toolbar on the right offers the different operators and delays, and is the basis of the drag-and-drop mechanism. The toolbar, from bottom to top, echoes the grammar presented earlier, from left to right. Expressions are built by simply dragging an operator or a delay and dropping it into an already existing box representing another operator. To navigate through the different levels, the user simply needs to click on an operator to display its operands. The "←" button allows the user to navigate one level upwards. The breadcrumbs on top of the screen also allows direct navigation through the different levels. The buttons on the top toolbar allow the user to create, save, load, and discard an expression, and to change app settings, respectively from left to right.

The app ensures that expressions are consistent with the n-ary version of the APHzip grammar: Subexpressions that are not yet consistent are displayed with a grey background. Consistent expressions enable the "Compress!" button at the bottom of the screen. After clicking this button the app as a frontend is interfacing with the web-service backend.

The actual compression algorithm is executed by APHzip on the server. This separation is the key for easy installation and use, and prevents that more computing resources are required than a mobile device could deliver. So the app at this point requires an active Internet connection to send the expression and to receive the result. Since the Android SDK does not offer a possibility to use SOAP connections, a third party library is used for this purpose, called `ksoap2-android`. Notably, in contrast to the web service, the app provides the possibility to use the convolution, minimum, and maximum operators as n-ary operators. This is not supported by the web-service backend. Thus, before sending the expression to the server, the app runs a heuristics locally on the device to transform from n-ary to binary operators. The heuristics is discussed in further detail in Sect. 5. The resulting expression is sent as a string to the APHzip server. Once received, the resulting, compressed representation is displayed to the user, together with information about the computation time and the size savings. Optionally, the user can provide an email address (as an option of the app settings), to which the result is to be sent once computed. (The app itself does not access any private user data.) This option comes in handy for larger models: For instance, compressions from 81 million to about 500 states are achievable by the web service, but take about 5 minutes server time [17].

Another button that is enabled whenever a consistent expressions is at hand is the "Compress &Analyse!" button. This button enables an additional functionality on top of the compression, namely the transient analysis of the APH distribution represented by the expression after it has been compressed. Transient analysis basically computes the cumulative distribution function of the APH distribution (*i.e.*, the probability until absorption in the APH distribution) on a set of time-points, provided by parameter *bound* (the upper bound time-point computed) and parameter *points* (the number of equidistance points in $[0, bound]$ computed).

5   Compression Acceleration Heuristics

Convolution, minimum, and maximum are all commutative and associative operators on probability distributions, and thus they naturally generalize to n-ary operators. However, in the APHzip web service they are implemented as binary operators only. The reason behind this is practical: operators with more operands can result in prohibitively large representation sizes without the chance to put in intermediate compression steps that keep the representation small. In fact, to keep the representation small, each application of a binary operator in APHzip is directly followed by compression. Composition and compression thus alternates in a divide-and-conquer fashion.

From a modeller perspective, it is however more convenient to work with n-ary operators. It is surely easy to turn this into an arbitrary nesting of binary operators under the hood. But unfortunately, the divide-and-conquer approach is sensitive to the nesting it faces. Even though the final size is the same, it may well happen that a certain nesting leads to an intermediate state explosion that renders the entire construction impossible, while another ordering gives a linear growth with medium-size intermediate results.

This resembles the setting of *smart reduction* [9], where compositional minimization of transition systems is discussed. And indeed, as in this case, finding the optimal nesting is an NP-hard problem. Inspired by the smart reduction approach, we have developed a heuristic method to break expressions with n-ary operators down to expressions consisting only of binary operators, such that the resulting computation times and memory consumptions are minimal.

For each n-ary operator op $\in \{\mathbf{con}, \mathbf{min}, \mathbf{max}\}$, the heuristics performs two steps: (1) It reorders the operands according to different, yet easy-to-compute, criteria described below, and (2) the operator op is broken down into its left-associative version:

$$\text{op}(x_1, x_2, \ldots, x_n) \Rightarrow \text{op}(\text{op}(\ldots(\text{op}(x_1, x_2), x_3), \ldots), x_n).$$

For the minimum and maximum operators, the heuristics uses two criteria. First, the operands are put into descending order according to their operand sizes. After that, same size operands are swapped, so that operands with most duplicate matrix entries are put together first. For the convolution operator, the operands are not reordered because the computation times and space requirements are (provably) not affected. The time complexity of the heuristic algorithm is $\mathcal{O}(n^2)$, where n is the number of operands.

This heuristics turns out to be decisive to achieve small computation times. It exploits that the compression algorithm is applied to every single subexpression. The largest subexpressions are minimized first which keeps the state space small, and inside the compression algorithm, many duplicate matrix entries tend to induce considerable compression. So the heuristics ensures that the state space is compressed on an early level, which leads to improved computation times and memory savings. For instance, for a simple expression with four operands like $\mathbf{max}(\mathbf{erl}(2, 0.2), \mathbf{erl}(7, 0.2), \mathbf{erl}(10, 0.2), \mathbf{erl}(65, 0.2))$, the time saving achieved by the heuristics is almost one order of magnitude. In fact, albeit the approach being a heuristics, in all examples we were able to explore exhaustively, the heuristics produces the time-optimal ordering.

6 Discussion

This paper has reported on advances in tool support for acyclic phase-type compression: We have presented details of an Android app, designed to provide a very easy-to-use interface to the compression technology delivered by the APHzip web service. Furthermore, we have sketched effective heuristics for supporting n-ary operators instead of just binary operators for modelling.

As far as we are aware, this is the first time that an app is used to enable easy experimentation with computer aided verification or quantitative evaluation advances. This fits particularly well to the simple setting of APHzip and integrates very well with its web service. We however believe that some of the principal features of this approach are of interest in their own right: (1) The learning curve for using a technology can become very flat if it is interfaced by easy-to-use gesture-based tools. (2) Rapid and anonymous experimentation can be supported by providing a trivial-to-install app via a store or website, which is (3) separated from the core technology running behind a web service.

At the same time, there are, of course, certain disadvantages that come with the fact that the web service might get overloaded by floods of users (or reviewers) submitting ever larger models. Furthermore, we think that the web-service idea is attractive for rapid experimentation, but not necessarily the method of choice for serious usage, where for instance one may not want to disclose models under study in any way, so as to protect intellectual property. In this context, an *amanat* [7], a dedicated server with trusted protocols, can provide a solution.

The APHzip app can be obtained directly from Google Play (look for "aphzip") or from http://depend.cs.uni-saarland.de/tools/aphzip, where also background literature, the textual (web based) APHzip interface, the WSDL description of the web service, a YOUTUBE video demonstrating its use, and several examples to experiment with can be found.

References

1. Arnold, F., Hermanns, H., Pulungan, R., Stoelinga, M.: Time-dependent analysis of attacks. In: Abadi, M., Kremer, S. (eds.) POST 2014 (ETAPS 2014). LNCS, vol. 8414, pp. 285–305. Springer, Heidelberg (2014)
2. Baier, C., Haverkort, B.R., Hermanns, H., Katoen, J.-P.: Model-checking algorithms for continuous-time Markov chains. IEEE Trans. Software Eng. **29**(6), 524–541 (2003)
3. Baier, C., Haverkort, B.R., Hermanns, H., Katoen, J.-P.: Performance evaluation and model checking join forces. Commun. ACM **53**(9), 76–85 (2010). Sept
4. Benlian, A., Hess, T., Buxmann, P.: Drivers of SaaS-adoption an empirical study of different application types. Business & Information Systems Engineering **1**(5), 357–369 (2009)
5. Boudali, H., Crouzen, P., Stoelinga, M.: A rigorous, compositional, and extensible framework for dynamic fault tree analysis. IEEE Trans. Depend. Sec. Comput. **7**(2), 128–143 (2010)
6. Buchholz, P.: Exact and ordinary lumpability in finite Markov chains. Journal of Applied Probability **31**, 59–75 (1994)
7. Chaki, S., Schallhart, C., Veith, H.: Verification across intellectual property boundaries. ACM Trans. Softw. Eng. Methodol. **22**(2), 15 (2013)
8. Courtney, T., Gaonkar, S., Keefe, K., Rozier, E., Sanders, W.H.: Möbius 2.3: an extensible tool for dependability, security, and performance evaluation of large and complex system models. In: DSN, pp. 353–358. IEEE (2009)
9. Crouzen, P., Lang, F.: Smart reduction. In: Giannakopoulou, D., Orejas, F. (eds.) FASE 2011. LNCS, vol. 6603, pp. 111–126. Springer, Heidelberg (2011)
10. Dugan, J.B., Bavuso, S.J., Boyd, M.A.: Dynamic fault-tree models for fault-tolerant computer systems. IEEE Trans. Reliab. **41**(3), 363–377 (1992)
11. Horváth, A., Telek, M.: PhFit: a general phase-type fitting tool. In: Field, T., Harrison, P.G., Bradley, J., Harder, U. (eds.) TOOLS 2002. LNCS, vol. 2324, pp. 82–91. Springer, Heidelberg (2002)
12. Johnson, M.A., Taaffe, M.R.: The denseness of phase distributions. School of Industrial Engineering Research Memoranda, pp. 88–20. Purdue University (1988)
13. Kaplan, F.: Are gesture-based interfaces the future of human computer interaction? In: ICMI 2009, pp. 239–240. ACM (2009)
14. Legay, A., Delahaye, B., Bensalem, S.: Statistical model checking: an overview. In: Barringer, H., Falcone, Y., Finkbeiner, B., Havelund, K., Lee, I., Pace, G., Roşu, G., Sokolsky, O., Tillmann, N. (eds.) RV 2010. LNCS, vol. 6418, pp. 122–135. Springer, Heidelberg (2010)
15. Neuts, M.F.: Matrix-Geometric Solutions in Stochastic Models: an Algorithmic Approach. Dover, Baltimore (1981)

16. Pulungan, R., Hermanns, H.: Acyclic minimality by construction–almost. In: QEST, pp. 63–72. IEEE Computer Society (2009)
17. Pulungan, R., Hermanns, H.: A construction and minimization service for continuous probability distributions. STTT **17**(1), 77–90 (2015)
18. Thümmler, A., Buchholz, P., Telek, M.: A novel approach for phase-type fitting with the EM algorithm. IEEE Trans. Depend. Sec. Comput. **3**(3), 245–258 (2006)

Petri Nets, Process Algebra
and Fault Trees

Computing Structural Properties
of Symmetric Nets

Lorenzo Capra[1], Massimiliano De Pierro[2], and Giuliana Franceschinis[3](✉)

[1] Dipartimento di Informatica, Univ. di Milano, Milano, Italy
[2] Dipartimento di Informatica, Univ. di Torino, Torino, Italy
[3] DiSIT, Univ. del Piemonte Orientale, Alessandria, Italy
giuliana.franceschinis@di.unipmn.it

Abstract. Structural properties of Petri Nets (PN) have an important role in the process of model validation and analysis. When considering Stochastic PNs, comprising stochastic timed and immediate transitions, structural analysis becomes a fundamental step in net-level definition of probabilistic parameters. High Level PN (HLPN) structural analysis still poses many problems and is often based on the unfolding of the HLPN model: this approach prevents the exploitation of model behavioural symmetries. A more effective alternative approach consists in providing a language, along with an associated calculus, making it possible to derive expressions defining structural relations among node instances of a HLPN model in a symbolic and parametric form: this has been proposed in the literature for Symmetric Nets (SN). The goal of the present paper is to summarize the language defined to express SNs' structural relations and to formalize the derivation of a basic set of such relations; in particular the algorithms to compute the Structural Mutual Exclusion relation and the symmetric and transitive closure of Structural Conflict are an original contribution of this paper. Examples of applications are also included. The algorithms required to support the calculus for symbolic structural relations computation have been recently completed and implemented in a tool called SNexpression.

1 Introduction

Structural properties of Petri Nets (PN) have an important role in the process of model validation and analysis, since they can answer interesting questions on the model potential behaviour, that can also be exploited to improve the efficiency of state-space based methods. When considering Stochastic PNs, including stochastic timed and immediate transitions, structural analysis becomes a fundamental step in net-level definition of probabilistic parameters.

When dealing with high level Petri nets (HLPN) it is desirable to exploit the opportunities offered by this class of formalisms, among which the ability to represent systems in a more compact and parametric way and to make regularities in the model structure explicit. Some HLPN formalism have been devised to make some form of symmetry easier to be exploited at the analysis level: an example of formalism in this class is Symmetric Nets (SN) [4].

J. Campos and B.R. Haverkort (Eds.): QEST 2015, LNCS 9259, pp. 125–140, 2015.
DOI: 10.1007/978-3-319-22264-6_9

Structural properties of classical PNs express relations between nodes in the model, while in HLPNs one wants to express relations between *node instances*, preferably using symbolic and parametric expressions, as proposed in [3], where a language and a calculus have been introduced to this purpose. To the best of our knowledge, only a few papers in the literature propose concrete results in this line: in [5] the problem of computing Extended Conflict Sets (ECS) with the aim of detecting confusion in coloured generalized stochastic PNs is considered (in analogy with what is done for Generalized Stochastic PNs [1]): the formulae expressing the required structural relations are provided, but the effectiveness of the approach requires to constrain the form of arc functions, so it only applies to the subclass of Unary Regular Nets. In [7] a structural calculus is proposed (along with some applications) for Safe Coloured Petri Nets: it is based on the representation of color mappings as equivalent constraint systems; the calculus operators however work only on functions mapping onto sets. In [6] a symbolic approach is used to implement behaviour-preserving model reductions on a subclass of SNs, Quasi Well-Formed Nets; besides limitations on the model syntax, further restrictions are posed on the form of arc functions when trying to symbolically resolve the composition operator. Finally in [8] symbolic expressions are used to compute stubborn-sets of Coloured Petri Nets (with the aim of building a reduced state space, yet preserving the desired properties); an adapted version of structural conflict and causal connection relations are used, together with a *reverseMap* operator similar to the transpose: the considered arc functions are structured, and the basic color maps can either belong to a restricted set or be user defined, however only when the former type is used it is possible to obtain precise stubborn sets, otherwise very coarse sets are computed making the method less effective.

The goal of the present paper is to summarize the language defined in [3] to express the structural properties of SN models (Sect. 2), and to complete the specification of a number of interesting structural properties, providing examples of applications and showing their usefulness (Sect. 3). In particular two symbolic structural relations not considered in [3] are developed in details, namely Structural Mutual Exclusion, and the Symmetric Structural Conflict, whose reflexive and transitive closure leads to the derivation of the Extended Conflict Sets of (stochastic) SN models in symbolic form: these new developments are based on the extension to multisets of all operators of the calculus (except composition), and the completion of the rewriting rules required to deal with all cases of composition (of functions mapping onto sets). Another original contribution of this paper is the parametric nature of the calculus, which can be applied without fixing the color classes size. All the computations presented in the paper have been carried out using the SNexpression tool, which implements the rewriting rules used for deriving the structural properties expressions, and which also directly supports the computation of structural properties on a SN model (automatically producing the expressions from the model specification, and applying the operators to obtain the result). In Sect. 4 the main results presented in the paper are summarized, and the comparison with the related work is further elaborated; ongoing and future developments are also outlined.

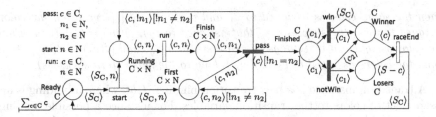

Fig. 1. The SN model of a relay race

2   Basic Definitions and Notation

In this section the SN formalism [4] is quickly recalled, then the definitions and the notation needed in the next sections are introduced.

2.1   The Symmetric Nets Formalism

The SN formalism is introduced through an example. The focus is on the color inscriptions appearing in the model, which are the basis of the calculus introduced later. Let us consider the model in Fig. 1 describing the dynamics of a relay race. The net structure is a bipartite graph whose nodes are places (circles) and transitions[1] (rectangles).

Places are state variables associated with a *color domain*, defining the variable type, expressed as a Cartesian product of *basic color classes*, disjoint and finite non empty sets (denoted with capital letters A, B, ..., Z) which may be partitioned into two or more *static subclasses*, and may be ordered. Static subclasses are denoted by the class name with a numeric subscript (e.g. A_1). In this model there are two basic color classes: $C = \{IT, FR, D, ES\}$, encoding the competing teams identifiers, and $N = \{0, 1, 2, 3\}$ encoding the athletes identifiers; N is an ordered class, hence a successor function, denoted !, is defined on it and induces a circular order among its elements ($!(i) = (i + 1)\%4$). A pair (2-tuple) $\langle c, n \rangle, c \in C, n \in N$ represents athlete n of team c. Each place can contain a multiset of tuples belonging to its color domain: this is called its *marking*.

Transitions are events that cause state changes. Also transitions have a color domain since they describe parametric events; the parameters are *variables* denoted with small letters with a subscript. The letter used for a variable implicitly defines its type, which is the color class denoted by the corresponding capital letter; subscripts are thus used to distinguish parameters of the same type associated with the same transition. Transition **pass** in the net of Fig. 1 has tree parameters $c : C, n_1, n_2 : N$. Transitions can have guards, expressed in terms of predicates on transition variables.

Definition 1 (Guards syntax). *Guards in SN models are boolean expressions whose terms are* basic predicates: *the set of basic predicates is: $[v_1 = v_2]$ (true*

---

[1] In the Stochastic SN formalism white transitions are timed, black transitions are immediate and have priority over the timed ones.

when the same color is assigned to v_1 and v_2), $[v_1 =! v_2]$ (true when the color assigned to v_1 is the successor of that assigned to v_2), $[d(v_1) = C_q]$ (true when the color assigned to v_1 belongs to static subclass C_q, and $[d(v_1) = d(v_2)]$ (true when the colors assigned to v_1 and v_2 belong to the same static subclass).

A transition instance is a pair transition-binding (t, b), where a binding is an assignment of colors to the transition variables. For instance a possible binding for the variables of pass is $c = IT, n_1 = 3, n_2 = 1$; a binding is valid only if it satisfies the transition guard. Hence a transition color domain corresponds to the set of all possible valid bindings for that transition.

The arcs connecting transition t to its input, output and inhibitor places $({}^\bullet t,\ t^\bullet,\ {}^\circ t)$ are respectively annotated with functions denoted by $W^-(p,t)$, $W^+(p,t)$ and $W^h(p,t)$. The formal definition of the arc functions is:

Definition 2 (Arc functions syntax). *A SN function W labeling an arc connecting transition t and place p, is a mapping $W : cd(t) \to Bag(cd(p))$*

$$W = \sum_i \lambda_i . T_i[p_i], \quad \lambda_i \in I\!N \tag{1}$$

where $cd(t)$ is the color domain of t, the sum is a multiset sum and λ_i are scalars, $T_i = \langle f_1, \ldots, f_k \rangle$ are tuples of class functions, and p_i is a guard. A class function (referring to class C, without loss of generality) is a mapping from $cd(t)$ to $Bag(C)$ whose syntax is:

$$f_i = \sum_{k=1}^{m} \alpha_k . c_k + \sum_{q=1}^{||C||} \beta_q . S_{C_q} + \sum_{k=1}^{m} \gamma_k .! c_k; \ \alpha_k, \beta_k, \gamma_k \in \mathbb{Z} \tag{2}$$

c_i are t's variables and denote 'projection' functions. $||C||$ is the number of static subclasses of C. S_{C_q} is the 'synchronization/diffusion' constant function which maps any color in $cd(t)$ to the set C_q. Symbol ! denotes the successor function, it is defined only for ordered classes and is followed by a variable name. Scalars in (2) must be such that no negative coefficients result from the evaluation of f_i for any legal variables binding.

An arc function is evaluated on a given binding of the transition: the value of a tuple is the Cartesian product of the value of its elements. A guarded tuple is evaluated as follows: if the guard is false (for a given binding) it evaluates to the empty multiset, otherwise its value corresponds to its standard evaluation without the guard. Function $\langle S_C, n \rangle$ on the outputs arc of start when applied to transition binding $n = 2$ evaluates in multiset $\sum_{c \in C} \langle c, 2 \rangle$. On the output arcs of pass to place Finished and Running there are two guarded functions which allow for a conditional behaviour, i.e., when the runner is the last of his team (i.e., the predecessor of the runner who started the race, stored in place First) then function $W^+(\text{pass},\text{Running}) = \langle c, ! n_1 \rangle [! n_1 \neq n_2]$ evaluates to *empty*, while $W^+(\text{pass},\text{Finished})$ evaluates to a multiset containing only one occurrence of color $\langle c \rangle$.

The model evolution in time can be simulated by starting from an initial marking (in the example all colors from class C in place Ready), and firing one of the enabled *transition instances*. A transition instance (t, b) is enabled in marking m if for each input place p of t ($p \in {}^{\bullet}t$) the multiset $W^-(p, t)(b)$ is included in $m(p)$, and for each inhibitor place p ($p \in {}^{\circ}t$), the multiplicity of each color in $W^h(p, t)(b)$ is greater than the multiplicity of the corresponding color in $m(p)$. An enabled transition instance may occur, withdrawing from each input place p the multiset $W^-(p, t)(b)$ and adding into each output place p the multiset $W^+(p, t)(b)$. In the example model, the inhibitor arc going from place Winner to transition win ensures that only the first team arriving at the end of the race is recorded as winner: the function $W^h(\text{win}, \text{Winner}) = \langle S_C \rangle$ means no tokens must be present in Winner, in order for win to be enabled. Finally the synchronization modelled by transition raceEnd makes use of function $\langle S_C - c \rangle$ that represents all the teams that have not won the race, i.e., the set of all elements in class C except for the one bound to variable c (the winner).

2.2   The Language to Express Structural Properties

Definition 3 (Language $\mathcal{L}$). *Let $\Sigma = \{A, B, \ldots, Z\}$ be the set of (finite and disjoint) basic color classes, and let $\mathcal{D}$ be any color domain built as Cartesian product of classes in Σ, ($\mathcal{D} = A^{e_A} \times B^{e_B} \times \ldots \times Z^{e_Z}, e_* \in \mathbb{N}$). Let $T_i : \mathcal{D} \to Bag(\mathcal{D}')$ and $[g_i']$ and $[g_i]$ SN standard predicates on $\mathcal{D}'$ and $\mathcal{D}$, respectively. The set of expressions:*

$$\mathcal{L} = \left\{ F : F = \sum_i \lambda_i.[g_i']T_i[g_i], \quad \lambda_i \in \mathbb{N}^+ \right\}$$

is the language used to express SN structural relations, where $T_i = \langle f_1, \ldots, f_l \rangle$ are function-tuples formed by class functions f_j, defined in turn as intersections of language elementary functions $\{a, !^k a, S_A, S - a, S - !^k a, \emptyset_A\}$ (projection, k^{th} successor, constant function corresponding to all elements of basic class A, projection/successor complement and the empty function; where A represents any basic class and a any variable of type A).

Language $\mathcal{L}$ actually extends the set of functions used in SN: indeed, predicate $[g_i']$, called *filter*, is not allowed in SN and permits the elements satisfying the boolean condition g_i' to be selected from the result of the application of $T_i[g_i]$. On the other hand, SN arc functions $W^-(p, t), W^+(p, t), W^h(p, t)$ can be written as elements of $\mathcal{L}$. The calculus we provide defines the following functional operators on $\mathcal{L}$:

F^t	Transpose	$F \cap F'$	Intersection	$\overline{F}$	Support
$F - F'$	Difference	$F + F'$	Sum	$\overline{F} \circ \overline{F'}$	Composition

All operators but composition apply to elements of $\mathcal{L}$ that map to multisets, and whose definition is consistent with the operator semantics. The composition is currently defined on a subset of $\mathcal{L}$ consisting of functions mapping to sets.

In the sequel the term *expression* will be used to indicate formulae that contain language functions and functional operators from the table above. The symbolic calculus is able to *solve* all the considered operators, that means $\mathcal{L}$ is closed w.r.t. them. Appropriate rewriting rules have been defined that simplify an arbitrary expression with operators until an expression in $\mathcal{L}$ is obtained. In some cases we are interested in obtaining an expression where terms are pairwise disjoint (i.e., when the expression is evaluated for any color in its domain, the multisets obtained by evaluating each term are disjoint). Each rewriting rule is based on the algebraic properties of functions appearing as operands.

A detailed description of these rules can be found in [3], where the difference, intersection, transpose operators rewriting rules have been first introduced. The details of composition have been recently completed.

3 Structural Properties Computation

In this section a number of structural properties will be defined and formalized through expressions in the language introduced in Sect. 2. Examples of computation of these structural properties follow each formal definition; the computation of the properties is performed by means of SNexpression [9], a tool implementing the calculus introduced in the previous section: it provides also direct computation of a few structural properties directly on SN models. With respect to the version presented in [9] the tool now has some new features, in particular the extension of all operators, except composition, to multisets. The tool can be downloaded from: http://www.di.unito.it/~depierro/SNex/.

Let us first define a symbolic relation between nodes in a Symmetric Net:

Definition 4 (Symbolic relation). *Given a binary relation $\mathcal{R}$ between the node instances of a SN such that (s,c) $\mathcal{R}$ (s',c') if and only if "(s,c) is in relation $\mathcal{R}$ with (s',c')" then its symbolic representation denoted $\mathcal{R}(s,s')$ is a mapping from $cd(s')$ to $2^{cd(s)}$ such that:*

$$\mathcal{R}(s,s')(c') = \{c \in cd(s) : (s,c) \ \mathcal{R} \ (s',c')\} \quad \text{for each } c' \in cd(s')$$

We are interested in deriving symbolic relations between instances of SN node pairs (place-transition, transition-place, transition-transition) in the form of an expression of $\mathcal{L}$. Such relations are derived by properly combining the SN functions surrounding the arcs connecting the nodes.

Table 1 provides the formulae showing how each symbolic structural relation depends on the arc functions. The calculus partially defined in [3] and recently completed allows us to apply transformations defined as rewriting rules that are repeatedly applied to the formulae according to the semantics of the operators appearing in them, until one obtains as a result an expression of language $\mathcal{L}$.

Some Auxiliary Relations: SbT, SfP, AbT, AtP. The first relations that are introduced here will be used to characterize more complex ones. They involve a pair of nodes, place and transition, directly connected through one or more arcs: $SbT(p,t) : cd(t) \rightarrow 2^{cd(p)}$, *Subtracted by Transition*: provides the set of colored tokens that a given instance of t withdraws from p; it is simply defined as

Table 1. Structural relations are obtained by properly combining the arc functions through intersection, transpose, sum, difference, support, and composition operators

$$
\begin{aligned}
SbT(p,t) &= \overline{W^-(t,p) - W^+(t,p)} \\
SfP(t,p) &= \overline{W^-(t,p) - W^+(t,p)}^{\,t} = SbT(p,t)^t \\
AbT(p,t) &= \overline{W^+(t,p) - W^-(t,p)} \\
AtP(t,p) &= \overline{W^+(t,p) - W^-(t,p)}^{\,t} = AbT(p,t)^t \\
SC(t,t') &= \bigcup_{p\in {}^\bullet t\cap {}^\bullet t'} SfP(t,p) \circ \overline{W^-(t',p)} \;\cup\; \bigcup_{p\in t^\bullet \cap \circ t'} AtP(t,p) \circ \overline{W^h(t',p)} \\
SC(t,t) &= \bigcup_{p\in {}^\bullet t} SfP(t,p) \circ \overline{W^-(t,p)} - Id \;\cup\; \bigcup_{p\in t^\bullet \cap \circ t} AtP(t,p) \circ \overline{W^h(t,p)} - Id \\
SCC(t,t') &= \bigcup_{p\in t^\bullet \cap {}^\bullet t'} AtP(t,p) \circ \overline{W^-(t',p)} \;\cup\; \bigcup_{p\in {}^\bullet t\cap \circ t'} SfP(t,p) \circ \overline{W^h(t',p)} \\
SME^s(t,t') &= \bigcup_{p\in {}^\bullet t\cap \circ t'} \overline{W^-(t,p)}^{\,t} \circ \overline{W^h(t',p)} \;\cup\; \bigcup_{p\in \circ t\cap {}^\bullet t'} \overline{W^h(t,p)}^{\,t} \circ \overline{W^-(t',p)}
\end{aligned}
$$

(the support of) the *multiset* difference of the function appearing on the input arc and the function appearing on the output arc connecting t and p;

$SfP(t,p) = SbT(p,t)^t : cd(p) \to 2^{cd(t)}$, *Subtracts from Place* (transpose of SbT): given a color of p it provides the set of instances of t that withdraw it;

$AbT(p,t) : cd(t) \to 2^{cd(p)}$, *Added by Transition*: provides the set of colored tokens an instance of t adds into p when it is fired; it is simply defined as (the support of) the *multiset* difference of the function appearing on the output arc and the function appearing on the input arc connecting t and p;

$AtP(t,p) = AbT(p,t)^t : cd(p) \to 2^{cd(t)}$, *Adds to Place* (transpose of AbT): given a color of p it provides the color instances of t that add it into p.

Structural Conflict: Two transition instances (t,c) and (t',c') are in conflict in a given marking M if the firing of the former produces a change in state that modifies the enabling condition of the latter, possibly disabling it. The structural conflict relation defines some conditions in the model structure and its annotations, that may lead to an actual conflict in some marking. The symbolic relation $SC(t,t')$ has color domain $cd(t')$ and co-domain $2^{cd(t)}$, so that when applied to a color c' in $cd(t')$ provides the subset of $cd(t)$ identifying the instances (t,c) of t that may disable (t',c'). An instance (t,c) may disable (t',c') either because it withdraws a token from an input place which is shared by the two transitions, or because it adds a token into an output place which is connected to t' through an inhibitor arc. Let us consider the two cases separately: let $p \in {}^\bullet t' \cap {}^\bullet t$, function $\overline{W^-(t',p)}$ gives the set of colored tokens in p required for the enabling of the instances of t'. Since $SfP(t,p)$ gives the instances of t that withdraw a given colored token from p, then the composition $SfP(t,p) \circ \overline{W^-(t',p)}$ provides the instances of t that may disable a given instance of t' because they require non-disjoint sets of colored tokens in the shared input place p. Similarly for the case of $p \in t^\bullet \cap \circ t'$ function $\overline{W^h(t',p)}$ gives the set of colored tokens in p that may disable t', while $AtP(t,p)$ gives the instances of t that add a given colored token in p, so that $AtP(t,p) \circ \overline{W^h(t',p)}$ provides the instances of t that may disable a given instance of t' because they add in p colored tokens that may disable t'. Finally $SC(t,t')$ is obtained by summing up over all common input places and common output-inhibitor places. The complete definition is shown in Table 1. Observe that it may be the case that different instances *of the same transition* are in conflict. The same expression can be used in this case, but at

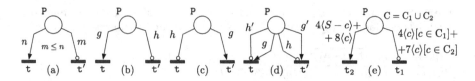

Fig. 2. Structural mutual exclusion patterns

the end one must subtract from the set of conflicting instances the instance to which SC is applied: this explains why Id is subtracted. Observe that the SC relation is not symmetric.

Structural Causal Connection: Two transition instances (t, c) and (t', c') are in causal connection in a given marking M if the firing of the former produces a change in state that modifies the enabling condition of the latter, possibly causing its enabling. The structural causal connection relation defines some conditions in the model structure and its annotations, that may lead to an actual causal connection in some marking. The symbolic relation $SCC(t, t')$ has color domain $cd(t')$ and co-domain $2^{cd(t)}$, so that when applied to a color c' in $cd(t')$ provides the subset of $cd(t)$ identifying the instances (t, c) of t that may cause the enabling of (t', c'). In this case we should concentrate on output places of t that are input places for t' and on input places of t that are inhibitor places for t', and in the former case the expression $AtP(t, p) \circ W^-(t', p)$ is used, while in the latter case the expression $SfP(t, p) \circ \overline{W^h(t', p)}$ is used. The complete definition is shown in Table 1.

Application to the Relay Race Example. Let us check some structural properties of the relay race model of Fig. 1 introduced in Sect. 2: for example let us consider the structural causal connection between transitions start and run; there is only one place that connects the two transitions, i.e. Running. According to Table 1 the formula for computing this property is $AtP(\text{start}, \text{Running}) \circ \overline{W^-(\text{run}, \text{Running})}$ (below denoted f'.g), where $AtP(\text{start}, \text{Running})$ is $\overline{W^+(\text{start}, \text{Running})}^t$. The tool can be exploited for the calculation of the property by submitting the following commands:

```
f := @N <S_C,n>    g := @C,N <c,n>
sf(f') => <n>    this expression corresponds to AtP(start,Running)
sf(f'.g) => <n>
```

here the symbol => is used to indicate the result returned by the tool, the symbol := instead allows to assign expressions to symbols, the syntax sf(*expression*) corresponds to a request to apply the operators. It is also possible to read the net (prepared in an appropriate textual format) and submit the command for SCC computation:

```
load"relayrace.sn"; SCC(start,run,Running) => <n>
```

The meaning of this result is: given an instance of transition run, e.g., with binding $c = IT, n = 1$, the instance of start that may enable it is that with

$n = 1$. Indeed, only when the race starts it happens that run becomes enabled due to the firing of start, since the following instances, until the end of the race, are instead activated by the firing of transition pass:

```
SCC(pass,run,Running) = <c_1,!-1n_1,S-n_1>
```

The instances of pass (whose variables are c, n_1 and n_2) that can enable a given instance of run (variables c and n) can only be those involving the same team (c has the same value in pass and run), and with n_1 and n_2 denoting the predecessor of n and any element of N but n, respectively.

Let us now evaluate a situation of potential *confusion* when the relay race model is interpreted as a Stochastic SN, with stochastically timed transitions, the white rectangles, and immediate transitions, the black ones. The presence of confusion in this kind of model is a symptom of an underspecified behaviour (from the point of view of a probabilistic characterization of conflicts resolution). In this case the potential *confusion* involves transitions pass and win: the folded structure of the high level model hides the structural conflict existing among the instances of win due to the presence of both an output arc and an inhibitor arc connecting this transition and place Winner. We want to compute the auto-conflicts of transition win: $SC(\text{win}, \text{win}, \text{Winner}) = AtP(\text{win}, \text{Winner}) \circ \overline{W^h(\text{win}, \text{Winner})} - Id$:

```
SC(win,win,Winner) =>   <S-c>
```

Indeed the firing of any instance of win, e.g. with $c = IT$, is in conflict with any other instance of the same transition (any $c \in C \backslash \{IT\}$) since only one team can win the race. Composing the function on the arc from place Finished to transition win with the outcome of the SC relation, we obtain the colored tokens the conflicting instances withdraw from place Finished; finally composing the transpose of the function on the arc from pass to Finished with the result of the last operation provides the instances of pass that may enable some instance of win in conflict with the instance we started with (e.g. the instance with $c = IT$). Summarizing $f := \overline{W^+(\text{pass}, \text{Finished})^t} \circ \overline{W^-(\text{win}, \text{Finished})} \circ \langle S - c \rangle$ is a function from $cd(\text{win})$ to $2^{cd(\text{pass})}$ indicating the possible presence of stochastic confusion in markings where both pass and win are concurrently enabled:

```
f1 := @C <c>   f2 := @C <S-c>   f3 := @C,N^2 <c>[!n_1=n_2]
sf(f1.f2) => <S-c>; sf(f3'.f1.f2) => [n_1 = !-1n_2]<S-c,S_N,S_N>
```

The interpretation of the result, a language expression with filter, is that given an instance of win, the instances of pass that may produce confusion, if enabled together with the former, are those involving a different team ($S - c$) and such that n_1 is the predecessor of n_2. In other words, n_1 is the identifier of the athlete running the last section of the race (otherwise the arc function from pass to place Finish would represent an empty set, and the instance of pass would not enable any instance of win).

3.1   Structural Mutual Exclusion

Two transition instances (t, c) and (t', c') are in (structural) mutual exclusion if the enabling of (t', c') in any M implies the fact that (t, c) is not enabled in M, and viceversa. This situation arises when in the net structure a place p exists that is input place for t and inhibitor place for t', and the number of tokens (of any color) required in the input place p for the enabling of t is greater than or equal to the upper bound on the number of tokens (of the same color) in p imposed by the inhibitor arc connecting p and t'.

In a (uncolored) Petri Net (possibly obtained by unfolding an SN) the necessary structural condition for two transitions to be in SME relation is the one depicted in Fig. 2.(a) where t and t' are in SME relation because, with respect to place P, the condition for the enabling of t in marking M is $M(P) \geq n$ while the condition for the enabling of t' in M is $M(P) < m$; since $((M(P) \geq n) \wedge (m \leq n)) \Rightarrow not(M(P) < m)$ and $((M(P) < m) \wedge (m \leq n)) \Rightarrow not(M(P) \geq n)$ the two transitions are indeed in SME relation, and the relation is symmetric.

When turning to SN models, we are looking for SME conditions on transition instances. The patterns that may lead to mutual exclusion of t and t' instances are depicted in Figs. 2.(b), (c) and (d).

Let us define a symbolic relation $SME(t, t') : cd(t') \rightarrow 2^{cd(t)}$ defined as follows: $SME(t, t')(c') = \{c \in cd(t) : (t, c)SME(t', c')\}$ i.e. a function giving the set of instances of t that are surely disabled in any marking where instance (t', c') is enabled. If all functions on input and inhibitor arcs were all mappings onto sets, then the SME relation could be computed by means of the expression shown in Table 1 (in the table it is called SME^s, because of the restriction on the functions W^- and W^h involved in the expression). The expression accounts for any possible structural pattern, including the general situation depicted in Fig. 2.(d). The expression for $SME^s(t, t')$ is the union of two parts, the former considers the case in which t' is connected to P through an inhibitor arc, while t is connected to P via an input arc; the latter considers the case in which t' is connected to P through an input arc and t instead through an inhibitor arc. Of course the two situations may overlap (as in Fig. 2.(d)), moreover this applies also when t and t' are the same transition (some instances of t may well be in mutual exclusion with other instances of the same transition, as shown in next section).

The expression $\overline{W^h(t, p)}^t \circ \overline{W^-(t', p)}$ (with $p \in {}^\circ t \cap {}^\bullet t'$) applied to a given color $c \in cd(t')$ first derives the set of colored tokens withdrawn from p by t', then by applying the transpose of $W^h(t, p)$ to this set, one obtains all t instances that would be inhibited by any color in such set. The other expression $\overline{W^-(t, p)}^t \circ \overline{W^h(t', p)}$ instead (with $p \in {}^\bullet t \cap {}^\circ t'$) applied to a given color $c \in cd(t')$ gives the set of colors that *should not appear* in p for t' to be enabled; by applying the transpose of $W^-(t, p)$ to this set one obtains all instances of t that require those colors in p to be enabled.

Let us now consider a more general case, where the input and inhibitor arcs are labelled with functions that map onto multisets. We first need to introduce an operator useful to define the SME relation in the general setting. Let $g : D_g \rightarrow Bag(D)$ and $h : D_h \rightarrow Bag(D)$ be two arc functions with same codomain, the comparison function $g \unrhd h$ is defined as:

$$g \unrhd h(c) = \{c' \in D_g : \exists d \in D, g(c')(d) \geq h(c)(d) > 0\} \ \forall c \in D_h$$

For any color $c \in D_h$ this function gives the set of all colors $c' \in D_g$ such that $g(c') \in Bag(D)$ contains at least one element d, also contained in $h(c)$, whose multiplicity is greater in $g(c')$ than in $h(c)$.

If we consider now a situation where g is the arc function $W^-(t, p) : cd(t) \to cd(p)$ associated with the input arc from p to t and h is the arc function $W^h(t', p) : cd(t') \to cd(p)$ associated with the inhibitor arc from p to t', then $SME_H(t, t', p) = g \trianglerighteq h$. In words: given an instance (t', c) of t' it corresponds to the set of instances of t which are surely disabled when (t', c) is enabled, because of place p, which is inhibitor for t' and input for t. If we are in the situation of Fig. 2.(d), where there is another pair of input-inhibitor arcs departing from p and directed towards t' and t, respectively, we can take the transpose of function $SME_H(t', t, p)$ to obtain another type of SME relation $SME_I(t, t', p) = SME_H(t', t, p)^t$ which, given an instance (t', c) of t', returns the set of instances of t which are surely disabled if (t', c) is enabled, because of place p which is input for t' and inhibitor for t. Finally:

$$SME(t, t') = \bigcup_p SME_H(t, t', p) + SME_I(t, t', p) = \bigcup_p SME_H(t, t', p) + (SME_H(t', t, p))^t$$

Let us define an algorithm implementing the computation of $SME_H(t, t', p)$: it is based on the representation of functions $W^h(t', p), W^-(t, p), W^-(t', p)$, $W^h(t, p)$ in the form of weighted sums of pairwise disjoint terms such that each term is in the form $[b'_i]\langle f_1, \dots, f_l \rangle [b_i]$, where functions f_i are intersections of language elementary functions (see Definition 3), and b'_i, b_i are standard predicates. In the sequel, let g be the function labelling the input arc (of t or t') and h the function labelling the inhibitor arc (of t or t'). They are in the form:

$$g^t = \sum_{i=1}^{K} m_i.G_i^t, \quad h = \sum_{i=1}^{K'} n_i.H_i$$

Since the terms G_i are disjoint (and hence so are the terms G_i^t) and the terms H_j are disjoint we can compare directly the weights of pairs G_i^t, H_j without instantiating the functions on a specific color. We have mutual exclusion when $m_i \geq n_j$, hence $SME_H(t, t') = \bigcup_{i,j:m_i \geq n_j} G_i^t \circ H_j$. The SME algorithm is:

procedure $SME_H(t, t', p)$:

Let: $g = W^-(t, p)$ and $h = W^h(t', p)$
$g^t = \sum_{i=1}^{K} m_i.G_i^t$, $G_i^t \cap G_j^t = \emptyset$, $\forall i \neq j$; $\quad h = \sum_{i=1}^{K'} n_i.H_i$, $H_i \cap H_j = \emptyset$, $\forall i \neq j$
$R = \emptyset$
for each $i = 1, \dots, K$ **do**

    for each $j = 1, \dots, K'$ **do**

        if $m_i \geq n_j$ **then** $R = R \cup G_i^t \circ H_j$
    return R

Let us apply the algorithm to the example in Fig. 2.(e) which corresponds to the pattern in Fig. 2.(b) with t2 corresponding to t and t1 corresponding to t'. The color class C has two static subclasses, C_1 and C_2. The functions on the arcs, both with domain and codomain C, are $g = 4\langle S - c \rangle + 8\langle c \rangle$ and $h = 4\langle c \rangle [c \in C_1] + 7\langle c \rangle [c \in C_2]$; observe that in both functions the two terms

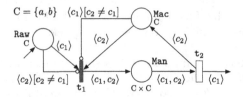

Fig. 3. A subnet from a FMS model

of the sum are disjoint. In this case the (multiset) transpose of g is equal to g itself $g^t = 4\langle S - c\rangle + 8\langle c\rangle$. In order to compute $SME_H(\mathbf{t2}, \mathbf{t1}, \mathbf{P})$ we compare the coefficients of the two terms in g^t (g_1^t and g_2^t) and those of h (h_1 and h_2): g_1^t has coefficient 4, which is equal to that of h_1, while g_2^t has coefficient 8, greater than those of both h_1 and h_2. Hence $SME_H(\mathbf{t2}, \mathbf{t1}, \mathbf{P}) = g_1^t \circ h_1 + g_2^t \circ h_1 + g_2^t \circ h_2 = \langle S - c\rangle[c \in C_1] + \langle c\rangle[c \in C_1] + \langle c\rangle[c \in C_2]$, after some simplifications we obtain: $SME_H(\mathbf{t2}, \mathbf{t1}, \mathbf{P}) = \langle S - c\rangle[c \in C_1] + \langle c\rangle$. Indeed, if $\mathbf{t1}$ is enabled for a given binding $c = a$ in C_1 there are less than 4 tokens of that color in P, hence all instances of $\mathbf{t2}$ with binding $c \neq a$ are not enabled because they need at least 4 tokens of color a in P; if $\mathbf{t1}$ is enabled for a given binding $c = b$ in C_2 there must be less than 7 tokens of that color in P, hence all instances of $\mathbf{t2}$ with same binding are not enabled since they require 8 tokens of that color. In this simple example $SME(\mathbf{t2}, \mathbf{t1}) = SME_H(\mathbf{t2}, \mathbf{t1}, \mathbf{P})$. Observe that $SME(\mathbf{t1}, \mathbf{t2}) = SME(\mathbf{t2}, \mathbf{t1})^t = \langle S - c \cap S_{C_1}\rangle[c \in C_1] + \langle S_{C_1}\rangle[c \in C_2] + \langle c\rangle$.

Machines Scheduling Policy in a Flexible Manufacturing System. Let us consider the model in Fig. 3, which is a small portion of a model representing a Flexible Manufacturing System (FMS) producing two types of parts. In [1] (Chap. 8) a GSPN representing such system is presented and studied. Here we concentrate only on the part of the model representing the scheduling policy for two machines that can process both part types. Place Raw represents a buffer of raw parts: the colors in class C allow one to distinguish between the parts of type a and b. There are two machines $M2$ and $M3$ that can process both part types, however machine $M2$ processes parts of type a more efficiently than $M3$, on the other hand machine $M3$ processes parts of type b more efficiently than $M2$, for this reason the scheduling policy for parts waiting in place Raw tries to allocate as much as possible parts of type a on $M2$ and parts of type b on $M3$ (but without leaving a machine idle if there is at least one waiting part in place Raw). Place Mac represents the idle machines: colors a and b are used to identify also the machines, since there is a natural association of each machine with a part type for efficiency reasons. The scheduling policy is hidden in transition $\mathbf{t_1}$, since its instances correspond to the possible scheduling choices. Using the calculus it is possible to discover the structural relations existing among the possible instances of $\mathbf{t_1}$. Structural conflict among $\mathbf{t_1}$ instances is computed through the following formula: $SfP(\mathbf{t_1}, \text{Raw}) \circ \overline{W^-(\mathbf{t_1}, \text{Raw})} - Id \cup SfP(\mathbf{t_1}, \text{Mac}) \circ \overline{W^-(\mathbf{t_1}, \text{Mac})}$. The two terms can be computed by SNexpression through the following commands:

```
load "FMS.sn"
```

```
SC(t1,t1,Raw) => <c_1,S-c_2>          SC(t1,t1,Mac) => <S-c_1,c_2>
sf(@C^2 <c_1,S-c_2> + <S-c_1,c_2>) => <c_1,S-c_2> + <S-c_1,c_2>
```

Hence the instances of t_1 potentially in conflict with a given instance of the same transition are those with variable c_1 bound to the same value as the first instance, and variable c_2 bound to a different value, or viceversa, different value for variable c_1 and same value for c_2. Since there are also inhibitor arcs connecting places Raw and Mac to transition t_1 we should also check the structural mutual exclusion relation. By applying the algorithm presented earlier in this section the following results are computed (here we show the commands to obtain them through SNexpression):

```
SME(t1,t1,Raw) => <S-c_1,c_1>[c_1 = c_2] + <c_2,S_C>[c_1 != c_2]
SME(t1,t1,Mac) => <S-c_2,c_1>[c_1 != c_2] + <c_2,S-c_2>
sf( <S-c_1,c_1>[c_1 = c_2] + <c_2,S_C>[c_1 != c_2] +
 <S-c_2,c_1>[c_1 != c_2] + <c_2,S-c_2>) => <S-c_1,c_1>[c_1 = c_2] +
 <S-c_2,c_1>[c_1 != c_2] + <c_1,S-c_1>[c_1 = c_2] + <c_2,S_C>[c_1 != c_2]
```

From this outcome we can infer that the scheduling policy is deterministic, hence the weights associated with the instances of t1 are irrelevant for the characterization of the stochastic process associated with the model. Let us consider the two cases: c_1=c_2 and c_1!=c_2. In the former case any other instance with the two variables bound to different values (represented by the two terms <S-c_1,c_1>[c_1=c_2] + <c_1,S-c_1>[c_1=c_2]) are in mutual exclusion with the considered instance; the only instance which is not mutually exclusive is the one with c_1=c_2 and both variables bound to a different value w.r.t. the reference instance. However we can verify that the two instances with c_1 and c_2 bound to the same value are not in conflict, so they are independent (if both enabled they can fire in any order without interfering with each other). In the case c_1!=c_2, the t1 reference instance is in mutual exclusion with all other instances, hence again there is no conflict to solve.

3.2 Symmetric SC Transitive Closure and Extended Conflict Sets

Extended Conflict Sets (ECS): In Stochastic SN (SSN) models, an extension of SNs comprising timed transitions (with exponentially distributed delays) and immediate transitions (firing in 0 time), structural conflict relations are used to identify *at the net level* subsets of immediate transitions whose firing order may influence the relative probability of alternative immediate transition firing sequences.[2] Immediate transitions that are in different ECSs are instead independent and can be fired in any order. ECS computation requires to introduce the reflexive and transitive closure of the symmetric SC relation: this new relation is denoted SSC^*. The first step to compute the desired relation consists in making the SC relation Symmetric: $SSC(t, t') = SC(t, t') \cup SC(t', t)^t$. The transitive closure is computed iteratively as follows: let us consider matrix $\mathbf{M}^0$ whose rows and columns are indexed on the (immediate) transitions $t_i \in I$,

---

[2] This is due to the way conflicts among enabled immediate transitions are probabilistically solved, by normalization of their weights to obtain the probabilities.

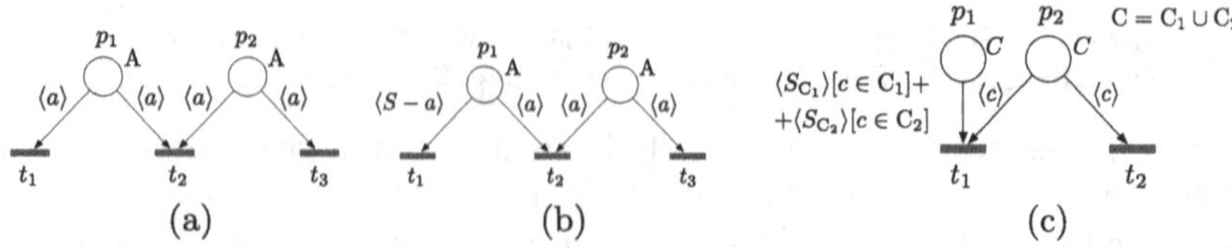

Fig. 4. ECS computation examples

and such that $\mathbf{M}^0(t_i, t_j) = SSC(t_i, t_j)$. A family $\{\mathbf{M}^1, \mathbf{M}^2, ..., \mathbf{M}^n\}$ of matrices can be derived by applying the following transformation: $\mathbf{M}_{i+1}(t_l, t_j) = \mathbf{M}^i \cup \bigcup_{t_k \in I} \mathbf{M}^i(t_l, t_k) \circ \mathbf{M}^i(t_k, t_l)$. Intuitively each iteration adds into element (t_l, t_j) of the matrix new (farther) indirect connections between t_l and t_j established by transitivity through an intermediate transition t_k. The iterative process eventually reaches a fixed point: $\mathbf{M}^{n+1} = \mathbf{M}^n$, and the elements of $\mathbf{M}^n$ contain the information needed to symbolically express all the ECS of the model (the upper bound where the iterations necessarily stop if it did not stop earlier, is a matrix full of functions in the form $\langle S_{C_j}, S_{C_i}, ..., S_{C_n} \rangle$).

Example of ECS Computation. The ECS computation technique is now illustrated through three examples, depicted in Fig. 4. The first example in Fig. 4(a) is very simple: the SC and SSC relations computation gives the following results: $SSC(t_1, t_2) = SC(t_1, t_2) = \langle a \rangle$, $SSC(t_2, t_3) = SC(t_2, t_3) = \langle a \rangle$ (there is no structural conflict between t_1 and t_3, due to the net structure, nor among the instances of the same transition, due to the arc functions). Hence matrix $\mathbf{M}^0$ and $\mathbf{M}^1$ are as follows:

$$\mathbf{M}^0 = \begin{pmatrix} 0 & a & 0 \\ a & 0 & a \\ 0 & a & 0 \end{pmatrix} \qquad \mathbf{M}^1 = \mathbf{M}^2 = \begin{pmatrix} 0 & a & a \\ a & 0 & a \\ a & a & 0 \end{pmatrix}$$

The elements of matrix $\mathbf{M}^1$ can be computed by applying the formulae presented above; so for example $\mathbf{M}^1(t_3, t_1)$ is derived by computing the result of the following expression: $ID_A + ID_A \circ 0_A + ID_A \circ ID_A + 0_A \circ ID_A = ID_A$, where $ID_A = \langle a \rangle$ and 0_A is the empty function (both ID_A and 0_A have domain A). In fact, only instances with same color $a \in A$ of t_1 and t_2 withdraw the same colored tokens from p_1, and only instances with same color $a \in A$ of t_2 and t_3 withdraw the same colored tokens from p_2. By transitive closure the instances with same color $a \in A$ of t_1 and t_3 are in relation SSC^* (through t_2). Hence there is one ECS for each distinct color $a \in A$, including the instances of t_1, t_2 and t_3 with color a. This can be derived from any column of matrix $\mathbf{M}^1$: in fact all functions in the column of transition t have a common domain, which is $cd(t)$. Given an element $a \in cd(t)$, the expression appearing in the row corresponding to transition t' provide the instances of t' that are in the same ECS as (t, a).

Now, let us consider the second example depicted in Fig. 4(b). In this case we can observe the ability of the tool to compute results that are parametric in the size of the classes. In particular, when computing the structural conflict relation between the instances of transition t_1 we obtain the following result:

```
SC(t1,t1,p1) = <O_A> : |A| = 2           SC(t1,t2,p1) = <S-a_1> : 2<=|A|<=n
               <S-a> : 3 <= |A|<= n       SC(t2,t2,p2) = <O_A> : 2<=|A|<=n
SC(t2,t3,p2) = <a> : 2 <= |A|<= n         SC(t3,t3,p2) = <O_A> : 2<=|A|<=n
```

The above SC relations are already symmetric (i.e. $SC(t_i, t_j) = SC(t_j, t_i)^t$), hence they lead directly to the elements of $\mathbf{M}^0$. The transitive closure becomes stable after three steps leading to the following result:

$$\mathbf{M}^3 = \mathbf{M}^4 = \left\{ \begin{pmatrix} 0 & S-a & S-a \\ S-a & 0 & a \\ S-a & a & 0 \end{pmatrix} |A| = 2, \quad \begin{pmatrix} S-a & S & S \\ S & S-a & S \\ S & S & S-a \end{pmatrix} |A| \geq 3 \right.$$

In this case if $|A| \geq 3$ all instances of the three transitions end up in a unique ECS. If instead $|A| = 2$ there is one ECS for each element $a \in A$ including instance (t_1, a) and $(t_2, A - a), (t_3, A - a)$, in other words there are two ECSs, comprising the instances of t_2 and t_3 with same color, and the instance of t_1 with the other color in A.

Finally let us consider the example in Fig. 4(c). In this case class C is partitioned into two static subclasses denoted C_1, C_2 (that for technical reasons and to simplify the discussion are both assumed to have cardinality > 2). The starting point is again the computation of the SC relation:

```
SC(t1,t1)= <S-c * S_C{1}>[c in C{1}] + <S-c * S_C{2}>[c in C{2}] =
         = <S_C{1}-c>[c in C{1}] + <S_C{2}-c>[c in C{2}]
SC(t2,t2) = <O_C>, SC(t1,t2,p2) = SC(t2,t1,p2) = <c_1>
```

Also in this case the relation is already symmetric (so that $SSC(t_i, t_j) = SC(t_i, t_j)$) and the above relations define the entries of matrix $\mathbf{M}^0$. The final result is the following ($\mathbf{M}^2 = \mathbf{M}^3$):

$$\mathbf{M}^3 = \begin{pmatrix} \langle S_{C_1} - c \rangle [c \in C_1] + \langle S_{C_2} - c \rangle [c \in C_2] & \langle S_{C_1} \rangle [c \in C_1] + \langle S_{C_2} \rangle [c \in C_2] \\ \langle S_{C_1} \rangle [c \in C_1] + \langle S_{C_2} \rangle [c \in C_2] & \langle S_{C_1} - c \rangle [c \in C_1] + \langle S_{C_2} - c \rangle [c \in C_2] \end{pmatrix}$$

That leads to the conclusion that there are two ECS in the model: the first one contains all the instances of t_1 and t_2 whose color belongs to static subclass C_1 while the second contains all the instances of t_1 and t_2 whose color belongs to static subclass C_2. Indeed if we consider a generic instance (t_1, c) of t_1 the functions in the first column show us that if $c \in C_i$ all instances of t_1 with a color in the same static subclass (see the expression in $\mathbf{M}^2(t_1, t_1)$) belong to the same ECS, and the same is true for all instances of t_2 with color in the same static subclass (see the expression in $\mathbf{M}^2(t_2, t_1)$). The same information can be derived from the second column (due to the symmetry of the involved relations).

4 Conclusions and Future Work

In this paper we have proposed an approach for the derivation of symbolic structural relations in SNs. The work extends that in [3], both from the point of view of the formalization of the set of structural relations together with their computation algorithms, and from the point of view of the underlying calculus, that has been generalized to support the relations computation. An implementation of the calculus is available in the SNexpression tool. There are several possible applications, including those discussed in the literature, namely ECS computation [5], Stubborn Sets computation [8], property preserving model reduction [6]: in these cases the results presented in this paper could extend the applicability to a wider class of models, or improve the effectiveness of the proposed method (in particular when either very simple or freely user defined arc function are considered). Another interesting application is the automatic generation of a reduced set of Ordinary Differential Equations from SSN models [2] with huge state spaces, to estimate the average colored marking at different points in time.

In most considered cases the symbolic formulae computed by the tool are intended as an information to be passed on to other analysis tools (e.g. simulators, state space generators) for increased efficiency or to perform some property check preliminar to other type of analysis, hence the readability of the symbolic formulae is not an issue. However for those situations where the user wants to interpret the results, we plan to develop in the future some graphical support (e.g. showing partial unfoldings of the model based on the computed structural relation). Finally work is in progress to extend the composition algorithm to functions mapping into multisets: this feature could be useful for example to check structure-based invariance properties.

References

1. Ajmone-Marsan, M., Balbo, G., Conte, G., Donatelli, S., Franceschinis, G.: Modelling with Generalized Stochastic Petri Nets. Wiley, Chichester, UK (1995)
2. Beccuti, M., Fornari, C., Franceschinis, G., Halawani, S., Ba-Rukab, O., Ahmad, A., Balbo, G.: From symmetric nets to differential equations exploiting model symmetries. Comput. J. **58**(1), 23–39 (2015)
3. Capra, L., De Pierro, M., Franceschinis, G.: A high level language for structural relations in well-formed nets. In: Ciardo, G., Darondeau, P. (eds.) ICATPN 2005. LNCS, vol. 3536, pp. 168–187. Springer, Heidelberg (2005)
4. Chiola, G., Dutheillet, C., Franceschinis, G., Haddad, S.: Stochastic Well-formed Coloured nets for symmetric modelling applications. IEEE Trans. Comput. **42**(11), 1343–1360 (1993)
5. Dutheillet, C., Haddad, S.: Conflict sets in colored petri nets. In: Proceedings of Petri Nets and Performance Models, pp. 76–85 (1993)
6. Evangelista, S., Haddad, S., Pradat-Peyre, J.-F.: Syntactical colored petri nets reductions. In: Peled, D.A., Tsay, Y.-K. (eds.) ATVA 2005. LNCS, vol. 3707, pp. 202–216. Springer, Heidelberg (2005)
7. Evangelista, S.: Syntactical rules for colored petri nets manipulation. Technical report CEDRIC-04-641, CEDRIC-CNAM Paris, January 2004
8. Evangelista, S., Pradat-Peyre, J.-F.: On the computation of stubborn sets of colored petri nets. In: Donatelli, S., Thiagarajan, P.S. (eds.) ICATPN 2006. LNCS, vol. 4024, pp. 146–165. Springer, Heidelberg (2006)
9. Franceschinis, G., Capra, L., De Pierro, M.: A tool for symbolic manipulation of arc functions in symmetric net models. In: 7th International Conference on Performance Evaluation Methodologies and Tools, VALUETOOLS 2013, ICST, January 2014

Optimizing Performance of Continuous-Time Stochastic Systems Using Timeout Synthesis

Tomáš Brázdil[1], Ľuboš Korenčiak[1(✉)], Jan Krčál[2],
Petr Novotný[3], and Vojtěch Řehák[1]

[1] Faculty of Informatics, Masaryk University, Brno, Czech Republic
{brazdil,korenciak,rehak}@fi.muni.cz
[2] Saarland University – Computer Science, Saarbrücken, Germany
krcal@cs.uni-saarland.de
[3] IST Austria, Klosterneuburg, Austria
petr.novotny@ist.ac.at

Abstract. We consider parametric version of fixed-delay continuous-time Markov chains (or equivalently deterministic and stochastic Petri nets, DSPN) where fixed-delay transitions are specified by parameters, rather than concrete values. Our goal is to synthesize values of these parameters that, for a given cost function, minimise expected total cost incurred before reaching a given set of target states. We show that under mild assumptions, optimal values of parameters can be effectively approximated using translation to a Markov decision process (MDP) whose actions correspond to discretized values of these parameters. To this end we identify and overcome several interesting phenomena arising in systems with fixed delays.

1 Introduction

Continuous-time Markov chains (CTMC) are a fundamental model of stochastic systems with discrete state-spaces that evolve in continuous-time. Several higher level modelling formalisms, such as stochastic Petri nets and stochastic process algebras, use CTMC as their semantics. As such, CTMC have been applied in performance and dependability analysis in various contexts ranging from aircraft communication protocols (see, e.g. [33]) to models of biochemical systems (see, e.g. [21]).

There are several equivalent definitions of CTMC (see, e.g. [15,29]). We may define a (uniformized, finite-state) CTMC to consist of a finite set of states S

The research leading to these results has received funding from the People Programme (Marie Curie Actions) of the European Union's Seventh Framework Programme (FP7/2007-2013) under REA grant agreement n° 291734]. This work is partly supported by the German Research Council (DFG) as part of the Transregional Collaborative Research Center AVACS (SFB/TR 14), by the EU 7th Framework Programme under grant agreement no. 295261 (MEALS) and 318490 (SENSATION), by the Czech Science Foundation, grant No. 15-17564S, and by the CAS/SAFEA International Partnership Program for Creative Research Teams.

© Springer International Publishing Switzerland 2015
J. Campos and B.R. Haverkort (Eds.): QEST 2015, LNCS 9259, pp. 141–159, 2015.
DOI: 10.1007/978-3-319-22264-6_10

coupled with a common rate λ and a stochastic matrix $P \in \mathbb{R}_{\geq 0}^{S \times S}$ specifying probabilities of transitions between states. An execution starts in a given initial state. In every step, the CTMC waits for a duration that is selected randomly according to the exponential distribution with the rate λ, and then moves to a state s' randomly chosen with probability $P(s, s')$.

The practical interpretation of the above semantics is that in every state the system waits for an event to occur and then reacts by changing its state. A typical example is a model of a simple queue to which new customers come in random intervals and are also served in random intervals. However, in practice, events are usually not exponentially distributed, and, in fact, their distributions may be quite far from being exponential. To deal with such events, phase-type approximation technique [28] is usually applied. Unfortunately, as already noted in [28], some distributions cannot be efficiently fit with phase-type approximation. In particular, degenerate distributions of events with fixed delays, i.e., events that occur after a fixed amount of time with probability 1, form a distinguished example of this phenomenon (for more details see [23]). However, as events with fixed delays play a crucial role in many systems, especially in communication protocols [30], time-driven real-time scheduling [32], etc., they should be handled faithfully in modelling and analysis.

Inspired by deterministic and stochastic Petri nets [26] and delayed CTMC [16] with at most one non-exponential transition enabled in any time, we study fixed-delay CTMC (fdCTMC), the CTMC extended with *fixed-delay transitions*. More concretely, we specify a set of states $S_{\mathrm{fd}} \subseteq S$ where fixed-delay transitions are enabled and add a stochastic matrix $F \in \mathbb{R}_{\geq 0}^{S_{\mathrm{fd}} \times S}$ specifying probabilities of fixed-delay transitions between states. In addition, we consider a *delay function* $\mathbf{d} : S_{\mathrm{fd}} \to \mathbb{R}_{>0}$. The semantics can be intuitively described as follows. Imagine a CTMC extended with an alarm clock. At the beginning of an execution, the alarm clock is turned off and the process behaves as the original CTMC. Whenever a state s of S_{fd} is visited and the alarm clock is off at the time, it is turned on and set to ring after $\mathbf{d}(s)$ time units. Subsequently, the process keeps behaving as the original CTMC until either a state of $S \setminus S_{\mathrm{fd}}$ is visited (in which case the alarm clock is turned off), or the alarm clock rings in a state s' of S_{fd}. In the latter case, a fixed-delay transition takes place, which means that the process changes the state randomly according to the distribution $F(s', \cdot)$, and the alarm clock is either turned off or newly set (when entering a state of S_{fd}).

In most practical applications mentioned above, fixed-delay transitions are determined by the design of the system and often strongly influence performance of the system. Indeed, both timeouts in network protocols as well as scheduling intervals in real-time systems directly influence performance of the respective systems and their manual setting usually requires considerable effort and expertise. This motivates us to consider the fixed-time delays $\mathbf{d}(s)$ as *free parameters* of the model, and develop techniques for their optimization with respect to a given performance measure.

Example 1. We demonstrate the concept on two different models of sending *one* segment of data in the *alternating bit protocol*. In the protocol, each segment of data is retransmitted until an acknowledgement is received. The delay between retransmissions has impact on throughput of the protocol as well as on network congestion. In the simpler model below on the left, the data is sent in state *init*. The exp-delay transitions, drawn as solid arrows, model message loss (with probability 0.2) and delivery (with probability 0.8). For simplicity we use rate 1 and omit self loops of exponential transitions in all examples. The fixed-delay transitions, drawn as dashed arrows, cause the data to be retransmitted. Note that whenever the data is retransmitted, the previous message with the data is canceled in this model.

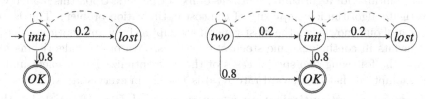

The more faithful model on the right models up to two messages with the data segment being delivered concurrently. For choosing an optimal delay between retransmissions, we need to formalize how to express performance of the protocol.

To express performance properties, we use standard cost (or reward) structures (see, e.g. [31]) that assign numerical rewards to states and transitions. More precisely, we consider the following three cost functions: $\mathcal{R} : S \to \mathbb{R}_{\geq 0}$, which assigns a cost rate $\mathcal{R}(s)$ to every state s so that the cost $\mathcal{R}(s)$ is paid for every unit of time spent in the state s, and functions $\mathcal{I}_P, \mathcal{I}_F : S \times S \to \mathbb{R}_{\geq 0}$ that assign to each exp-delay and fixed-delay transition, respectively, the cost that is immediately paid when the transition is taken. Note that $\mathcal{R}$ is usually used to express time spent in individual states, while the other two cost functions are used to quantify the difficulty of dealing with events corresponding to transitions. The performance measure itself is the *expected total cost incurred before reaching a given set of states G starting in a given initial state s_{in}*. For this moment, let us denote this measure by $E_\mathbf{d}$, stressing the fact that it depends on the delay function $\mathbf{d}$ which is the only variable quantity in our optimization task:

Problem 1 (Cost optimization). For a subspace of delay functions $D \subseteq (\mathbb{R}_{>0})^{S_\mathrm{fd}}$ and a given approximation error $\varepsilon > 0$, compute a delay function $\mathbf{d} \in D$ that is *ε-optimal within D*, i.e.

$$\left| \inf_{\mathbf{d}' \in D} E_{\mathbf{d}'} - E_\mathbf{d} \right| < \varepsilon.$$

Example 1 (cont.). We can model the expected cost of sending one data segment in our examples as follows. To take into account the expected time of data delivery, we set the cost rate of each state to, e.g., 1. To take into account the expected number of retransmissions, we set the cost of each fixed-delay transition, e.g., to 3. The cost of each exp-delay transition is set to 0. Now the goal for the left model is to find a delay $\mathbf{d}(init)$ optimizing the expected total cost incurred before reaching the state OK. Note that $\mathbf{d}$ is never set in the state *lost*. The goal is the same for the model on the right where $\mathbf{d}$ is set also in the state *two*. Note that it makes no sense to synthesize different delays $\mathbf{d}(init)$ and $\mathbf{d}(two)$ as the states *init* and *two* are indistinguishable in the implementation of the protocol. Therefore, we need to require that the synthesised delay function satisfies $\mathbf{d}(init) = \mathbf{d}(two)$.

Our Contribution: We consider fixed-delay CTMC as a natural extension of CTMC suitable for algorithmic synthesis of fixed timeouts. Upon this model, we investigate algorithmic complexity of the cost optimization problem. This is, to the best of our knowledge, the most general attempt at fully automatic synthesis of timeouts in continuous-time stochastic systems. We provide algorithms for solving the following two special cases of the cost optimization problem under the assumption that the reward rate $\mathcal{R}(s)$ is *positive* in every state s:

1. **Unconstrained optimization** where we demand $D = (\mathbb{R}_{>0})^{S_{\mathrm{fd}}}$, i.e. the set of all delay functions. We solve this problem by reduction to a finite Markov decision process (MDP) whose actions correspond to *discretized* (i.e. rounded onto a finite mesh) values of delays in the individual states, and then apply standard polynomial time algorithms for synthesis of the delays (note that a brute force search through a "discretized" subset of D would be exponentially worse). The most non-trivial part is to prove that the delays may be discretized. We show that a naïve rounding of a near-optimal delay function may cause arbitrarily high *absolute* error. Our solution, based on rather non-trivial insights into the structure of fdCTMCs, avoids this obstacle by identifying "safe" delay functions that may be rounded with an error bounded (exponentially) in the size of the system. This leads to an exponential time algorithm for solving the cost optimization problem.

2. **Bounded optimization under partial observation** where we introduce bounds $\underline{d}, \overline{d} > 0$ together with an equivalence relation $\equiv$ on S_{fd} and demand D to be the set of all delay functions $\mathbf{d}$ satisfying the following conditions:
   - $\underline{d} \leq \mathbf{d}(s) \leq \overline{d}$ for all $s \in S_{\mathrm{fd}}$,
   - $\mathbf{d}(s) = \mathbf{d}(s')$ whenever $s \equiv s'$.

   Like in the Example 1, the equivalence $\equiv$ can be used to hide information about detailed internal structure of states which is often needed in practical applications. In this paper, we show that the bounded optimization under partial observation can be solved in time doubly exponential in $\overline{d}$ and exponential in all other parameters.

   We also consider the corresponding approximate threshold variant: For a given x decide whether $\inf_{\mathbf{d}' \in D} E_{\mathbf{d}'} > x + \varepsilon$, or $\inf_{\mathbf{d}' \in D} E_{\mathbf{d}'} < x - \varepsilon$ (for $\inf_{\mathbf{d}' \in D} E_{\mathbf{d}'} \in [x - \varepsilon, x + \varepsilon]$ an arbitrary answer may be given). We show that this bounded optimization problem is NP-hard, thus a polynomial time solution of the bounded optimization under partial observation is unlikely.

   The assumption that all delays are between fixed thresholds $\underline{d}$ and $\overline{d}$ is crucial in our approach. As we discuss in Sect. 4, without this assumption the optimization under partial observation becomes much trickier and we leave its solution for future work.

Related Work. Various forms of continuous-time stochastic processes with fixed-delay transitions have already been studied, see e.g. [1,4,9,13,26]. In particular, as noted above, our definition of fdCTMC is closely related to the original definition of deterministic and stochastic Petri nets [26]. Papers on verification of continuous-time systems with timed automata (TA) specifications [4,5,12] are also related to our work as the constraints in timed automata resemble fixed-delay transitions. None of these works, however, considers synthesis of fixed-delays (or other parameters).

Parameter synthesis techniques have been developed for several models, such as parametric timed automata [2], parametric one-counter automata [17], parametric Markov models [18], etc. In continuous-time stochastic systems, [19,21] study synthesis of rates in CTMC which is a problem orthogonal to timeouts. Furthermore, optimal control of continuous-time (Semi)-Markov decision processes [6,8,10,27] can be viewed as synthesis of *discrete* parameters in continuous-time systems.

The problem of synthesizing timeouts as *continuous* parameters has been studied in variety of engineering contexts such as vehicle communication systems [22] and avionic subsystems [3,33]. To the best of our knowledge, no generic framework for synthesis of timeouts in stochastic continuous-time systems has been developed so far. In theoretical literature, only simpler cases have been addressed. For instance [11,34] consider a finite test case, a sequence of input and output actions, and synthesize times for input actions that maximise the probability of executing this acyclic sequence. Allowing cycles in fdCTMC makes the timeout synthesis problem much more demanding, e.g., due to potentially unbounded number of stochastic events between timeouts. Instead of static timeouts, [20,24] consider synthesis of "dynamic" timeouts where the delay is chosen based on the history of the execution so far. Consequently, the delay can be changed *while it is elapsing* whenever stochastic events occur. This makes it much simpler to solve and also adequate for a different application domain.

Section 2 introduces fixed-delay CTMC and cost structures. Sections 3 and 4 are devoted to unconstrained optimization and bounded optimization under partial observation, respectively. Due to space constraints, full proofs are in [7].

2   Preliminaries

We use $\mathbb{N}_0$, $\mathbb{R}_{\geq 0}$, and $\mathbb{R}_{>0}$ to denote the set of all non-negative integers, non-negative real numbers, and positive real numbers, respectively. Furthermore, for a countable set A, we denote by $\mathcal{D}(A)$ the set of discrete probability distributions over A, i.e. functions $\mu : A \to \mathbb{R}_{\geq 0}$ such that $\sum_{a \in A} \mu(a) = 1$. Encoding size of an object O is denoted by $||O||$.

Definition 1. *A* fixed-delay CTMC structure *(fdCTMC structure) C is a tuple $(S, \lambda, \mathrm{P}, S_{\mathrm{fd}}, \mathrm{F}, s_{in})$ where*

- *S is a finite set of states,*
- *$\lambda \in \mathbb{R}_{>0}$ is a (common) rate of exp-delay transitions,*

- P : $S \times S \to \mathbb{R}_{\geq 0}$ *is a stochastic matrix specifying probabilities of exp-delay transitions,*
- $S_{fd} \subseteq S$ *is a set of states where fixed-delay transitions are enabled,*
- F : $S_{fd} \times S \to \mathbb{R}_{\geq 0}$ *is a stochastic matrix specifying probabilities of fixed-delay transitions, and*
- $s_{in} \in S$ *is an initial state.*

A fixed-delay CTMC *(fdCTMC) is a pair* $C(\mathbf{d}) = (C, \mathbf{d})$ *where* C *is a fdCTMC structure and* $\mathbf{d} : S_{fd} \to \mathbb{R}_{>0}$ *is a delay function which to every state where fixed-delay transitions are enabled assigns a positive delay.*

A *configuration* of a fdCTMC is a pair (s, d) where $s \in S$ is the current state and $d \in \mathbb{R}_{>0} \cup \{\infty\}$ is the remaining time before a fixed-delay transition takes place. We assume that $d = \infty$ iff $s \notin S_{fd}$. To simplify notation, we similarly extend any delay function $\mathbf{d}$ to all states S by assuming $\mathbf{d}(s) = \infty$ iff $s \notin S_{fd}$.

An execution of $C(\mathbf{d})$ starts in the configuration (s_0, d_0) with $s_0 = s_{in}$ and $d_0 = \mathbf{d}(s_{in})$. In every step, assuming that the current configuration is (s_i, d_i), the fdCTMC waits for some time t_i and then moves to a next cofiguration (s_{i+1}, d_{i+1}) determined as follows:

- First, a waiting time t_{exp} for exp-delay transitions from s_i is chosen randomly according to the exponential distribution with the rate λ.
- Then
  - If $t_{exp} < d_i$, then an exp-delay transition occurs, which means that $t_i = t_{exp}$, s_{i+1} is chosen randomly with probability $P(s_i, s_{i+1})$, and d_{i+1} is determined by

$$d_{i+1} = \begin{cases} d_i - t_{exp} & \text{if} s_{i+1} \in S_{fd} \text{and} s_i \in S_{fd} \text{(previous delay remains)}, \\ \mathbf{d}(s_{i+1}) & \text{if} s_{i+1} \notin S_{fd} \text{or} s_i \notin S_{fd} \text{(delay is newly set or disabled)}. \end{cases}$$

  - If $t_{exp} \geq d_i$, then a fixed-delay transition occurs, which means that $t_i = d_i$, s_{i+1} is chosen randomly with probability $F(s_i, s_{i+1})$, and $d_{i+1} = \mathbf{d}(s_{i+1})$.

This way, the execution of a fdCTMC forms a *run*, an alternating sequence of configurations and times $(s_0, d_0)t_0(s_1, d_1)t_1 \cdots$. The probability measure $\mathrm{Pr}_{C(\mathbf{d})}$ over runs of $C(\mathbf{d})$ is formally defined in [7].

Total Cost Before Reaching a Goal. To allow formalization of performance properties, we enrich the model in a standard way (see, e.g. [31]) with costs (or rewards). A *cost structure* over a fdCTMC structure C with state space S is a tuple $Cost = (G, \mathcal{R}, \mathcal{I}_P, \mathcal{I}_F)$ where $G \subseteq S$ is a set of goal states, $\mathcal{R} : S \to \mathbb{R}_{\geq 0}$ assigns a cost rate to every state, and $\mathcal{I}_P, \mathcal{I}_F : S \times S \to \mathbb{R}_{\geq 0}$ assign an impulse cost to every exp-delay and fixed-delay transition, respectively. Slightly abusing the notation, we denote by *Cost* also the random variable assigning to each run $\omega = (s_0, d_0)t_0 \cdots$ the *total cost before reaching* G (in at least one transition), given by

$$Cost(\omega) = \begin{cases} \sum_{i=0}^{n-1} (t_i \cdot \mathcal{R}(s_i) + \mathcal{I}_i(\omega)) & \text{for minimal } n > 0 \text{ such that} s_n \in G, \\ \infty & \text{if there is no such } n, \end{cases}$$

where $\mathcal{I}_i(\omega)$ equals $\mathcal{I}_P(s_i, s_{i+1})$ for an exp-delay transition, i.e. when $t_i < d_i$, and equals $\mathcal{I}_F(s_i, s_{i+1})$ for a fixed-delay transition, i.e. when $t_i = d_i$.

We denote the expectation of *Cost* with respect to $\Pr_{C(\mathbf{d})}$ simply by $E_{C(\mathbf{d})}$, or by $E_{C(\mathbf{d})}^{Cost}$ when *Cost* is not clear from context. Our aim is to (approximatively) minimize the expected cost, i.e. to find a delay function $\mathbf{d}$ such that $E_{C(\mathbf{d})} \leq$ *Val* $[C] + \varepsilon$ where *Val* $[C]$ denotes the optimal cost $\inf_{\mathbf{d'}} E_{C(\mathbf{d'})}$.

Non-Parametric Analysis. Due to [25], we can easily analyze a fdCTMC where the delay function is *fixed*. Hence, both the expected total cost before reaching G and the reaching probabilities of states in G can be efficiently approximated.

Proposition 1. *There is a polynomial-time algorithm that for a given fdCTMC $C(\mathbf{d})$, cost structure Cost with goal states G, and an approximation error $\varepsilon > 0$ computes $x \in \mathbb{R}_{>0} \cup \{\infty\}$ and $p_s \in \mathbb{R}_{>0}$, for each $s \in G$, such that*

$$\left| E_{C(\mathbf{d})} - x \right| < \varepsilon \quad and \quad \left| \Pr_{C(\mathbf{d})}(\Diamond_G^s) - p_s \right| < \varepsilon$$

where $\Diamond_G^s$ is the set of runs that reach s as the first state of G (after at least one transition).

Markov Decision Processes. In Sect. 3 we use a reduction of fdCTMC to discrete-time Markov decision processes (DTMDP, see e.g. [31]) with uncountable space of actions.

Definition 2. *A DTMDP is a tuple $\mathcal{M} = (V, Act, T, v_{in}, V')$, where V is a finite set of vertices, Act is a (possibly uncountable) set of actions, $T: V \times Act \to \mathcal{D}(V) \cup \{\bot\}$ is a transition function, $v_{in} \in V$ is an initial vertex, and $V' \subseteq V$ is a set of goal vertices.*

An action a is *enabled* in a vertex v if $T(v, a) \neq \bot$. A *strategy* is a function $\sigma: V \to Act$ which assigns to every vertex v an action enabled in v. The behaviour of $\mathcal{M}$ with a fixed strategy σ can be intuitively described as follows: A run starts in the vertex v_{in}. In every step, assuming that the current vertex is v, the process moves to a new vertex v' with probability $T(v, \sigma(v))(v')$. Every strategy σ uniquely determines a probability measure $\Pr_{\mathcal{M}(\sigma)}$ on the set of *runs*, i.e. infinite alternating sequences of vertices and actions $v_0 a_1 v_1 a_2 v_2 \cdots \in (V \cdot Act)^\omega$; see [7] for details.

Analogously to fdCTMC, we can endow a DTMDP with a *cost function* $A: V \times Act \to \mathbb{R}_{\geq 0}$. We then define for each run $v_0 a_1 v_1 a_2 \ldots$, the *total cost incurred before reaching V'* as $\sum_{i=0}^{n-1} A(v_i, a_{i+1})$ if there is a minimal $n > 0$ such that $v_n \in V'$, and as ∞ otherwise. The expectation of this cost w.r.t. $\Pr_{\mathcal{M}(\sigma)}$ is similarly denoted by $E_{\mathcal{M}(\sigma)}$ or by $E_{\mathcal{M}(\sigma)}^A$ if the cost function is not clear from context.

Given $\varepsilon \geq 0$, we say that a strategy σ is *ε-optimal* in $\mathcal{M}$ if $E_{\mathcal{M}(\sigma)} \leq$ *Val* $[\mathcal{M}] + \varepsilon$ where *Val* $[\mathcal{M}] = \inf_{\sigma'} E_{\mathcal{M}(\sigma')}$; we call it *optimal* if it is 0-optimal. For any $s \in S$, let us denote by $\mathcal{M}[s]$ the DTMDP obtained from $\mathcal{M}$ by replacing the initial state by s. We call a strategy *globally (ε-)optimal* if it is (ε-)optimal in $\mathcal{M}[s]$ for every $s \in S$. Sometimes, we restrict to a subset D of all strategies and denote by *Val* $[\mathcal{M}, D]$ the restricted infimum $\inf_{\sigma' \in D} E_{\mathcal{M}(\sigma')}$.

3   Unconstrained Optimization

Theorem 1. *There is an algorithm that given a fdCTMC structure C, a cost structure Cost with $\mathcal{R}(s) > 0$ for all $s \in S$, and $\varepsilon > 0$ computes in exponential time a delay function $\mathbf{d}$ with*

$$\left| E_{C(\mathbf{d})} - \inf_{\mathbf{d}'} E_{C(\mathbf{d}')} \right| < \varepsilon.$$

The rest of this section is devoted to a proof of Theorem 1, which consists of two parts. First, we reduce the optimization problem in the fdCTMC to an optimization problem in a *discrete time* Markov decision process (DTMDP) with uncountably many actions. Second, we present the actual approximation algorithm based on a straightforward discretization of the space of actions of the DTMDP. However, the proof of its error bound is actually quite intricate. The time complexity is exponential because the discretized DTMDP needs exponential size in the worst case. We provide more detailed complexity analysis with respect to various parameters in [7].

For the rest of this section we fix a fdCTMC structure $C = (S, \lambda, \mathrm{P}, S_{\mathrm{fd}}, \mathrm{F}, s_{in})$, a cost structure $Cost = (G, \mathcal{R}, \mathcal{I}_{\mathrm{P}}, \mathcal{I}_{\mathrm{F}})$, and $\varepsilon > 0$. We assume that $Val\,[C] < \infty$. The opposite case can be easily detected by fixing an arbitrary delay function $\mathbf{d}$ and finding out whether $E_{C(\mathbf{d})} = \infty$ by Proposition 1. This is equivalent to $Val\,[C] = \infty$ by the following observation.

Lemma 1. *For any delay functions $\mathbf{d}, \mathbf{d}'$ we have $E_{C(\mathbf{d})} = \infty$ if and only if $E_{C(\mathbf{d}')} = \infty$.*

To further simplify our presentation, we assume that each state s of S_{fd} directly encodes whether the delay needs to be reset upon entering s. Formally, we assume $S_{\mathrm{fd}} = S^{reset} \uplus S^{keep}$ where $s' \in S^{reset}$ if $P(s, s') > 0$ for some $s \in S \setminus S_{\mathrm{fd}}$ or if $F(s, s') > 0$ for some $s \in S_{\mathrm{fd}}$; and $s' \in S^{keep}$ if $P(s, s') > 0$ for some $s \in S_{\mathrm{fd}}$. We furthermore assume that $s_{in} \in S^{reset}$ if $s_{in} \in S_{\mathrm{fd}}$. Note that each fdCTMC structure can be easily transformed to satisfy this assumption in polynomial time by duplication of states S_{fd}, see, e.g., Example 2.

3.1   Reduction to DTMDP $\mathcal{M}$ with Uncountable Space of Actions

We reduce the problem into a *discrete-time* problem by capturing the evolution of the fdCTMC only at *discrete moments* when a transition is taken after which the fixed-delay is (a) newly set, or (b) switched off, or (c) irrelevant as the goal set is reached. This happens exactly when one of the states of $S' = S^{reset} \cup (S \setminus S_{\mathrm{fd}}) \cup G$ is reached. We define a DTMDP $\mathcal{M} = (S', Act, T, s_{in}, G)$ with a cost function $\mathfrak{C}$:

- $Act := \mathbb{R}_{>0} \cup \{\infty\}$; where actions $\mathbb{R}_{>0}$ are enabled in $s \in S^{reset}$ and action ∞ is enabled in $s \in S \setminus S_{\mathrm{fd}}$.

– Let $s \in S'$ and d be an action of $\mathcal{M}$. Intuitively, we define $T(s,d)$ and $\text{\euro}(s,d)$ to summarize the behaviour of the fdCTMC starting in the configuration (s,d) until the first moment when S' is reached again.

Formally, let $C[s](d)$ denote a fdCTMC obtained from C by changing initial state to s and fixing a delay function that assigns d to s (and arbitrary values elsewhere). We define $\text{\euro}(s,d)$ as the cost accumulated before reaching another state of S' and $T(s,d)(s')$ as the probability that s' is the first such a reached state of S'. That is,

$$\text{\euro}(s,d) = E_{C[s](d)}^{Cost[S']} \qquad \text{and} \qquad T(s,d)(s') = \Pr\nolimits_{C[s](d)}(\Diamond_{S'}^{s'})$$

where $Cost[S']$ is obtained from $Cost$ by changing the set of goal states to S'. Note that the definition is correct as it does not depend on the delay function apart from its value d in the initial state s.

Example 2. Let us illustrate the construction on the fdCTMC from Sect. 1. The model, depicted on the left is modified to satisfy the assumption $S_{\text{fd}} = S^{reset} \uplus S^{keep}$: we duplicate the state *init* into another state *one* $\in S^{keep}$; the states from S^{reset} are then depicted in the top row. As in Sect. 1, we assign cost rate 1 to all states and impulse cost 3 to every fixed-delay transition (and zero to exp-delay transitions).

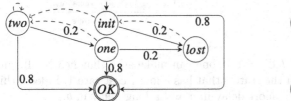 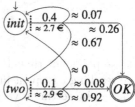

On the right, there is an excerpt of the DTMDP $\mathcal{M}$ and of the cost function $\text{\euro}$. For each non-goal state, we depict only one action out of uncountably many: for state *two* it is action 0.1 with cost ≈ 2.9, for state *init* it is action 0.4 with cost ≈ 2.7. The costs are computed in PRISM.   □

Note that there is a one-to-one correspondence between the delay functions in C and strategies in $\mathcal{M}$. Thus we use $\mathbf{d}, \mathbf{d}', \ldots$ to denote strategies in $\mathcal{M}$. Finally, let us state correctness of the reduction.

Proposition 2. *For any delay function* $\mathbf{d}$ *it holds* $E_{C(\mathbf{d})} = E_{\mathcal{M}(\mathbf{d})}$. *Hence,*

$$Val\,[C] = Val\,[\mathcal{M}].$$

In particular, in order to solve the optimization problem for C it suffices to find an ε-optimal strategy (i.e., a delay function) $\mathbf{d}$ in $\mathcal{M}$.

3.2   Discretization of the Uncountable MDP $\mathcal{M}$

Since the MDP $\mathcal{M}$ has uncountably many actions, it is not directly suitable for algorithmic solutions. We proceed in two steps. In the first and technically demanding step, we show that we can restrict to actions on a finite mesh. Second, we argue that we can also approximate all transition probabilities and costs by rational numbers of small bit length.

Restricting to a Finite Mesh. For positive reals $\delta > 0$ and $\bar{d} > 0$, we define a subset of delay functions $D(\delta, \bar{d}) = \{\mathbf{d} \mid \forall s \in S'\ \exists k \in \mathbb{N} : \mathbf{d}(s) = k\delta \leq \bar{d}\}$. Here, all delays are multiples of δ bounded by $\bar{d}$.

We need to argue that for the fixed $\varepsilon > 0$ there are some appropriate values δ and $\bar{d}$ such that $D(\delta, \bar{d})$ contains an ε-optimal delay function. A naïve approach would be to take any, say $\varepsilon/2$-optimal delay function $\mathbf{d}$, round it to closest delay function $\mathbf{d}^\star \in D(\delta, \bar{d})$ on the mesh, and show that the expected costs of these two functions do not differ by more than $\varepsilon/2$. However, this approach does not work as shown by the following example.

Example 3. Let us fix the fdCTMC structure C on the left (with cost rates in small boxes and zero impulse costs). An excerpt of $\mathcal{M}$ and $\mathfrak{C}$ is shown on the right (where only a few actions are depicted).

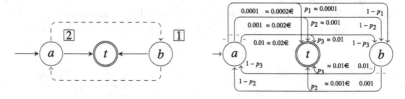

First, we point out that $Val\,[C] = 1$ as one can make sure that nearly all time before reaching t is spent in the state b that has a lower cost rate 1. Indeed, this is achieved by setting a very short delay in a and a long delay in b.

We claim that for any $\delta > 0$ there is a near-optimal delay function $\mathbf{d}$ such that rounding its components to the nearest integer multiples of δ yields a large error independent of δ. Indeed, it suffices to take a function $\mathbf{d}$ with $\mathbf{d}(b) = \delta$ and $\mathbf{d}(a)$ an arbitrary number significantly smaller than $\mathbf{d}(b)$, say $\mathbf{d}(a) = 0.01 \cdot \mathbf{d}(b)$. The error produced by the rounding can then be close to 0.5. For instance, given $\delta = 0.01$ we take a function with $\mathbf{d}(a) = 0.0001$ and $\mathbf{d}(b) = 0.01$, whose rounding to the closest delay function on the finite mesh yields a constant function $\mathbf{d}^\star = (0.01, 0.01)$. Then $E_{C(\mathbf{d})} \approx 1.01$ and $E_{C(\mathbf{d}^\star)} \approx 1.5$, even though the rounding does not change any transition probability or cost by more than 0.02!

The reason why the delay function $\mathbf{d}$ is so sensitive to small perturbations is that it makes a very large number of steps before reaching t (around 200 on average) and thus the small one-step errors caused by a perturbation accumulate into a large global error. The number of steps of an ε-optimal delay functions is not bounded, in general. By multiplying both $\mathbf{d}(a)$ and $\mathbf{d}(b)$ by the same infinitesimally small factors we obtain an ε-optimal delay functions that make an arbitrarily high expected number of steps before reaching t. □

A crucial observation is that we do not have to show that the "naïve" rounding works for *every* near-optimal delay function. To prove that $D(\delta, \bar{d})$ contains an ε-optimal function, it suffices to show that there is *some* $\varepsilon/2$-optimal function whose rounding yields error at most $\varepsilon/2$. Proving the existence of such well-behaved functions forms the technical core of our discretization process which is formalized below.

We start by formalizing the concept of "small perturbations". We say that a delay function $\mathbf{d}^\star$ is α-*bounded* by a delay function $\mathbf{d}$ if for all states $s, t \in S'$ we have:

1. $|T(s, \mathbf{d}^\star(s))(t) - T(s, \mathbf{d}(s))(t)| \leq \alpha$ and
2. $\mathbf{\in}(s, \mathbf{d}^\star(s)) - \mathbf{\in}(s, \mathbf{d}(s)) \leq \alpha$;

and furthermore, $T(s, \mathbf{d}(s))(t) = 0$ iff $T(s, \mathbf{d}^\star(s))(t) = 0$, i.e. the qualitative transition structure is preserved. (Note that $\mathbf{d}^\star$ may incur much smaller one-step costs than $\mathbf{d}$, but not significantly higher).

Using standard techniques of numerical analysis, we express the increase in accumulated cost caused by a bounded perturbation as a function of the worst-case (among all possible initial states) expected cost and expected number of steps before reaching the target. The number of steps is essential as discussed in Example 3 and can be easily expressed by a cost function $\#$ that assigns 1 to every action in every state. To express the worst-case expectation of some cost function \$, we denote $Bound\,[\$, \mathbf{d}] := \max_{s \in S'} E^{\$}_{\mathcal{M}[s](\mathbf{d})}$.

Lemma 2. *Let $\alpha \in [0, 1]$ and let, $\mathbf{d}'$ be a delay function that is α-bounded by another delay function $\mathbf{d}$. If $\alpha \leq \frac{1}{2 \cdot Bound\,[\in, \mathbf{d}] \cdot |S'|}$, then*

$$E_{\mathcal{M}(\mathbf{d}')} \leq E_{\mathcal{M}(\mathbf{d})} + 2 \cdot \alpha \cdot Bound\,[\#, \mathbf{d}] \cdot (1 + Bound\,[\in, \mathbf{d}] \cdot |S'|).$$

The next lemma shows how to set the parameters δ and $\overline{d}$ to make the finite mesh $D(\delta, \overline{d})$ "dense" enough, i.e. to ensure that for any $\mathbf{d}$, $D(\delta, \overline{d})$ contains a delay function that is α-bounded by $\mathbf{d}$.

Lemma 3. *There are positive numbers $D_1, D_2 \in \exp(||C||^{\mathcal{O}(1)})$ computable in time polynomial in $||C||$ such that the following holds for any $\alpha \in [0, 1]$ and any delay function $\mathbf{d}$: If we put*

$$\delta := \alpha/D_1 \quad and \quad \overline{d} := |\log(\alpha)| \cdot D_2 \cdot Bound\,[\in, \mathbf{d}],$$

then $D(\delta, \overline{d})$ contains a delay function which is α-bounded by $\mathbf{d}$.

Proof (Sketch). Computing the value of δ is easy as the derivatives of the probabilities and costs are bounded from above by the rate λ and the maximal cost rate, respectively. For $\overline{d}$ we need additional technical observations, see [7] for further details.

Unfortunately, as shown in Example 3, the value $Bound\,[\#, \mathbf{d}]$ can be arbitrarily high, even for near-optimal functions $\mathbf{d}$. Hence, we cannot use Lemma 2 right away to show that a delay function in $D(\delta, \overline{d})$ that is α-bounded by some near-optimal $\mathbf{d}$ is also near-optimal. The crucial insight is that for any $\varepsilon' > 0$ there are (globally) ε'-optimal delay functions that use number of steps that is *proportional* to their expected cost.

Lemma 4. *There is a positive number $N \in \exp(\|C\|^{\mathcal{O}(1)})$ computable in time polynomial in $\|C\|$ such that the following holds: for any $\varepsilon' > 0$, there is a globally $\varepsilon'/2$-optimal delay function $\mathbf{d}'$ with*

$$Bound\,[\#, \mathbf{d}'] \ \leq \ \frac{Bound\,[\text{€}, \mathbf{d}']}{\varepsilon'} \cdot N. \tag{1}$$

Proof (Sketch). After proving the existence of globally near-optimal strategies, we suitably define the number N and take an arbitrary globally ε''-optimal delay function $\mathbf{d}''$, where $\varepsilon'' \ll \varepsilon'$. If this function *does not* satisfy (1), we conclude that it must induce the following pathological behaviour in C: the system stays for a long time in a component of its state space such that (a) fixed-delay transitions are active in each state of the component, each such transition within the component having zero impulse cost; and (b) function $\mathbf{d}''$ assigns very small (in a well-defined sense) delays to all states of the component. We call such a component a *bad sink*. Intuitively, inside a bad sink the system rapidly performs one fixed-delay transition after another, incurring only a tiny cost between two successive transitions. This allows the delay function to perform many steps while staying ε''-optimal. (In Example 3, $\{a, b\}$ would be a bad sink for $\mathbf{d}$, as with high probability the cycle on these two states is completed every 0.0101 units of time, with cost 0.0102 incurred per cycle.)

To obtain a globally ε'-optimal delay function satisfying (1), we carefully modify $\mathbf{d}''$ so as to remove all bad sinks. This is done by selecting a suitable state in each bad sink and "inflating" its delay to a sufficiently high threshold. Choosing the right state and threshold is a rather delicate process, since an improper choice might significantly increase the incurred cost. Also note that Lemma 2 cannot be used to bound the increase in cost caused by the modification, as we do not know the value of $Bound\,[\#, \mathbf{d}'']$. Instead, we utilize non-trivial insights into the structure of C and $\mathcal{M}$. $\qquad\square$

By using these proportional delay functions, we reduce the perturbation error of Lemma 2 only to a function of $Bound\,[\text{€}, \mathbf{d}]$. Combining this with Lemma 3, we obtain that the delay functions in $D(\delta, \overline{d})$ approximate all the proportional delay functions $\mathbf{d}$ of Lemma 4, and thus $Val\,[\mathcal{M}, D(\delta, \overline{d})]$ approximates $Val\,[\mathcal{M}]$. The parameters $\delta, \overline{d}$ depend on ε and $Bound\,[\text{€}, \mathbf{d}]$ of any such $\mathbf{d}$ from Lemma 4. As these delay functions are *globally* ε-optimal, all such $Bound\,[\text{€}, \mathbf{d}]$ can be ε-approximated by $\overline{Val}\,[\mathcal{M}] := \max_{s \in S'} Val\,[\mathcal{M}[s]]$.

Proposition 3. *For N from Lemma 4, D_1 and D_2 from Lemma 3, it holds that*

$$\left|\, Val\,[\mathcal{M}] - Val\,[\mathcal{M}, D(\delta, \overline{d})]\,\right| \ \leq \ \frac{\varepsilon}{2}$$

where $\delta := \dfrac{\alpha}{D_1}, \quad \overline{d} := |\log(\alpha)| \cdot D_2 \cdot (\overline{Val}\,[\mathcal{M}] + \varepsilon), \quad \alpha := \dfrac{\varepsilon^2}{64N \cdot |S'| \cdot (1 + \overline{Val}\,[\mathcal{M}])^2}.$

Bounding $\overline{Val}\,[\mathcal{M}]$. In Proposition 3, the allowed perturbation α and hence the fineness of the mesh δ needed to obtain the required precision depend on the bound $\overline{Val}\,[\mathcal{M}]$. We first provide the following theoretical worst-case bound.

Lemma 5. *There is a number $M \in \exp(\|C\|^{\mathcal{O}(1)})$ computable in time polynomial in $\|C\|$ such that $\overline{Val}\,[\mathcal{M}] \leq M$.*

In practice, one can obtain better bounds by computing $\max_{s \in S} E_{C[s]}(\mathbf{d})$ for an arbitrary $\mathbf{d}$ as $\max_{s \in S} E_{C[s](\mathbf{d})} \geq \max_{s \in S} \inf_{\mathbf{d}'} E_{C[s](\mathbf{d}')} = \overline{Val}\,[\mathcal{M}]$. One can set $\mathbf{d}$ by some heuristics (e.g. to the constant function $1/\lambda$) or randomly. One can even use the minimum from a series of such computations. We believe that in most cases, this yields a significant improvement. For instance, for the 3-state model from Sect. 1, we get a bound $\max_{s \in S} E_{C[s](1/\lambda)} \approx 4.3$ instead of the theoretical bound $\overline{Val}\,[\mathcal{M}] \approx 55000$.

Representing the Finite Mesh. Since one-step costs and probabilities produced by delay functions in $D(\delta, \overline{d})$ may be irrational, we need to approximate them by rational numbers. So let us fix δ and $\overline{d}$ from Proposition 3. For any $\kappa > 0$ we define DTMDP $\mathcal{M}_\kappa = (S', Act_\kappa, T_\kappa, s_{in}, G)$ with a cost function $\mathfrak{C}_\kappa$ where

- the strategies are exactly delay functions from $D(\delta, \overline{d})$, i.e. $Act_\kappa = \{k\delta \mid k \in \mathbb{N}, \delta \leq k\delta \leq \overline{d}\} \cup \{\infty\}$ where again ∞ is enabled in $s \in S' \setminus S_{\mathrm{fd}}$ and the rest is enabled in $s \in S_{\mathrm{fd}}$; and
- for all $(s, \mathbf{d}) \in S' \times Act_\kappa$ the transition probabilities in $T_\kappa(s, \mathbf{d})$ and costs in $\mathfrak{C}_\kappa(s, \mathbf{d})$ are obtained by rounding the corresponding numbers in $T(s, \mathbf{d})$ and $\mathfrak{C}(s, \mathbf{d})$ up (using the algorithm of Proposition 1) to the closest multiple of κ.[1]

Proposition 4. *Let $\varepsilon > 0$ and fix $\kappa = (\varepsilon \cdot \delta \cdot minR)/(2 \cdot |S'| \cdot (1 + \overline{Val}\,[\mathcal{M}])^2)$, where $minR$ is a minimal cost rate in C. Then it holds*

$$\left| \; Val\,[\mathcal{M}, D(\delta, \overline{d})] - Val\,[\mathcal{M}_\kappa] \; \right| \; \leq \; \frac{\varepsilon}{2}.$$

Proof (Sketch). We use similar technique as in Lemma 2, taking advantage of the fact that probabilities and costs of each action are changed by at most κ by the rounding. $\square$

The Algorithm for Theorem 1. First the discretization step δ, maximal delay $\overline{d}$, and rounding error κ are computed. Then the discretized DTMDP $\mathcal{M}_\kappa$ is constructed according to the above-mentioned finite mesh representation. Finally the globally optimal delay function from $\mathcal{M}_\kappa$ is chosen using standard polynomial algorithms for finite MDPs [14,31]. From Propositions 3 and 4 it follows that this delay function is ε-optimal in $\mathcal{M}$, and thus also in C (Proposition 2).

The size of $\mathcal{M}_\kappa$ (and its construction time) can be stated in terms of a polynomial in $\|C\|$, $\overline{Val}\,[\mathcal{M}]$, $1/\delta$, $\overline{d}$, and $1/\kappa$. Examining the definitions of these parameters in Propositions 3 and 4, as well as the bound on $\overline{Val}\,[\mathcal{M}]$ from Lemma 5, we conclude that the size of $\mathcal{M}_\kappa$ and the overall running time of our algorithm are exponential in $\|C\|$ and polynomial in $1/\varepsilon$. The pseudo-code of the whole algorithm is given in [7].

---

[1] More precisely, all but the largest probability in $T(s, \mathbf{d})$ are rounded up, the largest probability is suitably rounded down so that the resulting vector adds up to 1.

4   Bounded Optimization Under Partial Observation

In this section, we address the cost optimization problem for delay functions chosen under partial observation. For an equivalence relation $\equiv$ on S_{fd} specifying observations, and $\underline{d}, \overline{d} > 0$, we define $D(\underline{d}, \overline{d}, \equiv) = \{\mathbf{d} \mid \forall s, s' : \underline{d} \le \mathbf{d}(s) \le \overline{d}, s \equiv s' \Rightarrow \mathbf{d}(s) = \mathbf{d}(s')\}$.

Theorem 2. *There is an algorithm that for a fdCTMC structure C, a cost structure Cost with $\mathcal{R}(s) > 0$ for all $s \in S$, an equivalence relation $\equiv$ on S_{fd}, $\underline{d}, \overline{d} > 0$, and $\varepsilon > 0$ computes in time exponential in $\|C\|$, $\|\underline{d}\|$, and $\overline{d}$ a delay function $\mathbf{d}$ such that*

$$\left| \inf_{\mathbf{d}' \in D(\underline{d}, \overline{d}, \equiv)} E_{C(\mathbf{d}')} - E_{C(\mathbf{d})} \right| < \varepsilon.$$

Also, one cannot hope for polynomial complexity as the corresponding threshold problem is NP-hard, even if we restrict to instances where $\overline{d}$ is of magnitude polynomial in $\|C\|$.

Theorem 3. *For a fdCTMC structure C, a cost structure Cost with $\mathcal{R}(s) > 0$ for all $s \in S$, an equivalence relation $\equiv$ on S_{fd}, $\underline{d}, \overline{d} > 0$, $\varepsilon > 0$, and $x \in \mathbb{R}_{\ge 0}$, it is NP-hard to decide*

$$\text{whether} \qquad \inf_{\mathbf{d} \in D(\underline{d}, \overline{d}, \equiv)} E_{C(\mathbf{d})} > x + \varepsilon \qquad or \qquad \inf_{\mathbf{d} \in D(\underline{d}, \overline{d}, \equiv)} E_{C(\mathbf{d})} < x - \varepsilon$$

(if the optimal cost lies in the interval $[x - \varepsilon, x + \varepsilon]$, an arbitrary answer may be given). The problem remains NP-hard even if $\overline{d}$ is given in unary encoding.

For $\overline{d}$ given in unary we get a matching upper bound.

Theorem 4. *The approximate threshold problem of Theorem 3 is in NP, provided that $\overline{d}$ is given in unary.*

We leave the task of settling the exact complexity of the general problem (where $\overline{d}$ is given in binary) to future work.

For the rest of this section we fix a fdCTMC structure $C = (S, \lambda, \mathrm{P}, S_{\mathrm{fd}}, \mathrm{F}, s_{in})$, a cost structure $Cost = (G, \mathcal{R}, \mathcal{I}_{\mathrm{P}}, \mathcal{I}_{\mathrm{F}})$, $\varepsilon > 0$, and an equivalence relation $\equiv$ on S_{fd}, $\underline{d}, \overline{d} > 0$. We simply write D instead of $D(\underline{d}, \overline{d}, \equiv)$ and again assume that $Val\,[C, D] < \infty$.

4.1   Approximation Algorithm

In this Section, we address Theorem 2. First observe, that the MDP $\mathcal{M}$ introduced in Sect. 3 can be due to Proposition 2 also applied in the bounded partial-observation setting. Indeed, $E_{C(\mathbf{d})} = E_{\mathcal{M}(\mathbf{d})}$ for each $\mathbf{d} \in D$ and thus, $Val\,[C, D] = Val\,[\mathcal{M}, D]$ (where analogously $Val\,[C, D]$ denotes $\inf_{\mathbf{d} \in D} E_{C(\mathbf{d})}$). Furthermore, by fixing a mesh δ and a round-off error κ, we define a finite DTMDP $\mathcal{M}_D^*$ where

- actions are restricted to a finite mesh of multiples of δ within the bounds $\underline{d}$ and $\overline{d}$; and
- probabilities and costs are rounded to multiples of κ as in Sect. 3.

To show that $\mathcal{M}_D^\star$ suitably approximates $\mathcal{M}$ we use similar techniques as in Sect. 3. However, thanks to the constraints $\underline{d}$ and $\overline{d}$ we can show that for *every* delay function $\mathbf{d} \in D$ the values *Bound* $[\#, \mathbf{d}]$ and *Bound* $[\mathord{\in}, \mathbf{d}]$, which feature in Lemma 2, are bounded by a function of $\|C\|$, $\underline{d}$ and $\overline{d}$ (in particular, the bound is independent of $\mathbf{d}$). This substantially simplifies the analysis. We state just the final result.

Proposition 5. *There is a number* $B \in \exp((\|C\| \cdot \|\underline{d}\| \cdot \overline{d})^{\mathcal{O}(1)})$ *such that for* $\delta = \varepsilon/B$ *and* $\kappa = (\varepsilon \cdot \delta)/B$ *it holds* $|\,Val\,[\mathcal{M}, D] - Val\,[\mathcal{M}_D^\star]| < \varepsilon$.

The proof of Theorem 2 is finished by the following algorithm.

- For δ and κ from Proposition 5, the algorithm first constructs (in the same fashion as in Sect. 3) in 2-exponential time the MDP $\mathcal{M}_D^\star$.
- Then it finds an optimal strategy $\mathbf{d}$ (which also satisfies $|E_{C(\mathbf{d})} - \inf_{\mathbf{d}'} E_{C(\mathbf{d}')}| < \varepsilon$) by computing $E_{\mathcal{M}'_\varepsilon(\mathbf{d})}$ for every (MD) strategy $\mathbf{d}$ of $\mathcal{M}'_\varepsilon$ in the set D.

The algorithm runs in 2-EXPTIME because there are $\leq |Act_\varepsilon|^{|S|}$ strategies which is exponential in $\|C\|$, $\|\underline{d}\|$, and $\overline{d}$ as $|Act_\varepsilon|$ is exponential in these parameters. The correctness follows from Propositions 2 and 5, proving Theorem 2.

Challenges of Unbounded Optimization. The proof of Proposition 5 is simpler than the techniques from Sect. 3 because we work with the compact space bounded by $\underline{d}$ and $\overline{d}$. This restriction is not easy to lift; the techniques from Sect. 3 cannot be easily adapted to *unbounded* optimization under partial observation.

The reason is that local adaptation of the delay function (heavily applied in the proof of Lemma 4) is not possible as the delays are not independent. Consider on the right a variant of Example 3 with components a and b being switched by fixed-delay transitions. All states have cost rate 1 and all transitions have cost 0; furthermore, all states are in one class of equivalence of $\equiv$. If in state a or b more than one exp-delay transition is taken before a

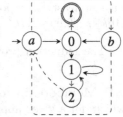

fixed-delay transition, a long detour via state 1 is taken. In order to avoid it and to optimize the cost, one needs to set the one common delay as close as possible to 0. Contrarily, in order to decrease the expected number of visits from a to b from a before reaching t which is crucial for the error bound, one needs to increase the delay.

4.2   Complexity of the Threshold Problem

We now turn our attention to Theorem 3. We show the hardness by reduction from SAT. Let $\varphi = \varphi_1 \wedge \cdots \wedge \varphi_n$ be a propositional formula in conjunctive normal form (CNF) with $\varphi_i = (l_{i,1} \vee \cdots \vee l_{i,k_i})$ for each $1 \le i \le n$ and with the total number of literals $k = \sum_{i=1}^{n} k_i$. As depicted in the following figure, the fdCTMC structure C_φ is composed of n components (one per clause), depicted by rectangles. The component of each clause is formed by a cycle of sub-components (one per literal) connected by fixed-delay transitions. Positive literals are modelled differently from negative literals.

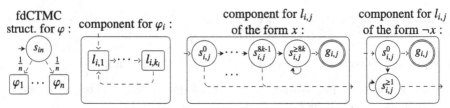

The cost structure $Cost_\varphi$ assigns rate cost 1 to every state, and impulse cost 0 to every transition; the goal states are depicted by double circles and exp-delay transitions are depicted with heavier heads to distinguish from the dashed fixed-delay transitions. We require $s_{i,j}^0 \equiv s_{i',j'}^0$ iff the literals $l_{i,j}$ and $l_{i',j'}$ have the same variable. Furthermore, let D denote $D(0.01, 16k, \equiv)$. Note that $\overline{d} = 16k$ is linear in $\|\varphi\|$ and thus it can be encoded in unary. We obtain the following:

Proposition 6. *For a formula φ in CNF with k literals, C_φ and $Cost_\varphi$ are constructed in time polynomial in k and, furthermore,*

$$Val\,[C_\varphi, D] < 17k^2 \quad \text{if } \varphi \text{ is satisfiable} \quad \text{and} \quad Val\,[C_\varphi, D] > 17k^2 + 1, \text{ otherwise.}$$

Proof (Sketch). The proof is based on the facts that the probability to take no exponential transition within time 0.01 is > 0.99 and the probability to take at least $8k$ exponential transitions within time $16k$ is > 0.99 and that $(16k \cdot k)/0.99 < 17k^2$. □

The reduction proves NP-hardness as it remains to set $x := 2k^2 + \frac{1}{2}$ and $\varepsilon := \frac{1}{2}$.

NP Membership for Unary $\overline{d}$. To prove Theorem 4 we give an algorithm which, for a given approximate threshold $x > 0$, consists of

- *first* guessing the delay function $\mathbf{d}$ of $\mathcal{M}_D^\star$ that is in the set D such that $E_{\mathcal{M}_D^\star(\mathbf{d})} < x$,
- *then* constructing just the fragment $\mathcal{M}_\mathbf{d}$ of $\mathcal{M}_D^\star$ used by the guessed function $\mathbf{d}$. Here $\mathcal{M}_\mathbf{d} = (S', \{\infty\}, T_\mathbf{d}, G, \text{€}_\mathbf{d})$ where the transition probabilities and costs coincide with $\mathcal{M}$ for states in $S' \setminus S_{\mathrm{fd}}$ and in any state $s \in S_{\mathrm{fd}}$ are defined by $T_\mathbf{d}(s, \infty) = T^\star(s, \mathbf{d}(s))$ and $\text{€}_\mathbf{d}(s, \infty) = \text{€}^\star(s, \mathbf{d}(s))$ (here $T^\star(s, \mathbf{d}(s))$ and $\text{€}^\star(s, \mathbf{d}(s))$ are as in $\mathcal{M}_D^\star$).

- Last, for $\sigma : s \mapsto \infty$, the algorithm computes $y = E_{\mathcal{M}_d(\sigma)}$ by standard methods and *accepts* iff $y < x$.

Note that when $\overline{d}$ is encoded in unary, both d and $\mathcal{M}_d$ are of bit size that is polynomial in the size of the input. Hence, d and $\mathcal{M}_d$ can be constructed in non-deterministic polynomial time (although the whole $\mathcal{M}_D^*$ is of exponential size in this unary case). The expected total cost x in $\mathcal{M}_d(\sigma)$ that has polynomial size can be also computed in polynomial time. The correctness of the algorithm easily follows from Proposition 5; for an explicit proof see [7].

5 Conclusions

In this paper, we introduced the problem of synthesising timeouts for fixed-delay CTMC. We study two variants of this problem, show that they are effectively solvable, and obtain provable worst-case complexity bounds. First, for *unconstrained optimization*, we present an approximation algorithm based on a reduction to a discrete-time Markov decision process and a standard optimization algorithm for this model. Second, we approximate the case of *bounded optimization under partial observation* also by a MDP. However, a restriction of the class of strategies twists it basically into a partial-observation MDP (where only memoryless deterministic strategies are considered). We give a 2-exponential approximation algorithm (which becomes exponential if one of the constraints is given in unary) and show that the corresponding decision problem is NP-hard.

The correctness of our algorithms stems from non-trivial insights into the behaviour of fdCTMC that we deem to be interesting in their own right. Hence, we believe that techniques presented in this paper lay the ground for further development of performance optimization via timeout synthesis.

References

1. Alur, R., Courcoubetis, C., Dill, D.: Model-checking for probabilistic real-time systems. In: Albert, J.L., Monien, B., Artalejo, M.R. (eds.) Automata, Languages and Programming. LNCS, vol. 510, pp. 115–136. Springer, Heidelberg (1991)
2. Alur, R., Henzinger, T.A., Vardi, M.Y.: Parametric real-time reasoning. In: STOC, pp. 592–601. ACM (1993)
3. Audsley, N.C., Grigg, A.: Timing analysis of the ARINC 629 databus for real-time applications. Microprocess. Microsyst. **21**(1), 55–61 (1997)
4. Baier, C., Bertrand, N., Bouyer, P., Brihaye, T., Größer, M.: Probabilistic and topological semantics for timed automata. In: Arvind, V., Prasad, S. (eds.) FSTTCS 2007. LNCS, vol. 4855, pp. 179–191. Springer, Heidelberg (2007)
5. Baier, C., Bertrand, N., Bouyer, P., Brihaye, T., Größer, M.: Almost-sure model checking of infinite paths in one-clock timed automata. In: LICS, pp. 217–226. IEEE (2008)
6. Brázdil, T., Forejt, V., Krčál, J., Křetínský, J., Kučera, A.: Continuous-time stochastic games with time-bounded reachability. Inf. Comput. **224**, 46–70 (2013)

7. Brázdil, T., Korenčiak, Ľ., Krčál, J., Novotný, P., Řehák, V.: Optimizing performance of continuous-time stochastic systems using timeout synthesis. CoRR, abs/1407.4777 (2014)

8. Brázdil, T., Krčál, J., Křetínský, J., Kučera, A., Řehák, V.: Stochastic real-time games with qualitative timed automata objectives. In: Gastin, P., Laroussinie, F. (eds.) CONCUR 2010. LNCS, vol. 6269, pp. 207–221. Springer, Heidelberg (2010)

9. Brázdil, T., Krčál, J., Křetínský, J., Řehák, V.: Fixed-delay events in generalized semi-Markov processes revisited. In: Katoen, J.-P., König, B. (eds.) CONCUR 2011. LNCS, vol. 6901, pp. 140–155. Springer, Heidelberg (2011)

10. Buchholz, P., Hahn, E.M., Hermanns, H., Zhang, L.: Model checking algorithms for CTMDPs. In: Gopalakrishnan, G., Qadeer, S. (eds.) CAV 2011. LNCS, vol. 6806, pp. 225–242. Springer, Heidelberg (2011)

11. Carnevali, L., Ridi, L., Vicario, E.: A quantitative approach to input generation in real-time testing of stochastic systems. IEEE Trans. Softw. Eng. 39(3), 292–304 (2013)

12. Chen, T., Han, T., Katoen, J.-P., Mereacre, A.: Quantitative model checking of continuous-time Markov chains against timed automata specifications. In: LICS, pp. 309–318. IEEE (2009)

13. Choi, H., Kulkarni, V.G., Trivedi, K.S.: Transient analysis of deterministic and stochastic Petri nets. In: Marsan, M.A. (ed.) Application and Theory of Petri Nets, vol. 691, pp. 166–185. Springer, Heidelberg (1993)

14. Etessami, K., Wojtczak, D., Yannakakis, M.: Recursive stochastic games with positive rewards. In: Aceto, L., Damgård, I., Goldberg, L.A., Halldórsson, M.M., Ingólfsdóttir, A., Walukiewicz, I. (eds.) ICALP 2008, Part I. LNCS, vol. 5125, pp. 711–723. Springer, Heidelberg (2008)

15. Feller, W.: An Introduction to Probability Theory and Its Applications, vol. 1. Wiley, New York (1968)

16. Guet, C.C., Gupta, A., Henzinger, T.A., Mateescu, M., Sezgin, A.: Delayed continuous-time Markov chains for genetic regulatory circuits. In: Madhusudan, P., Seshia, S.A. (eds.) CAV 2012. LNCS, vol. 7358, pp. 294–309. Springer, Heidelberg (2012)

17. Haase, C., Kreutzer, S., Ouaknine, J., Worrell, J.: Reachability in succinct and parametric one-counter automata. In: Bravetti, M., Zavattaro, G. (eds.) CONCUR 2009. LNCS, vol. 5710, pp. 369–383. Springer, Heidelberg (2009)

18. Hahn, E.M., Hermanns, H., Zhang, L.: Probabilistic reachability for parametric Markov models. STTT 13(1), 3–19 (2011)

19. Han, T., Katoen, J.P., Mereacre, A.: Approximate parameter synthesis for probabilistic time-bounded reachability. In: Real-Time Systems Symposium, pp. 173–182. IEEE (2008)

20. Jensen, P.G., Taankvist, J.H.: Learning optimal scheduling for time uncertain settings. Aalborg University, Student project (2014)

21. Jha, S.K., Langmead, C.J.: Synthesis and infeasibility analysis for stochastic models of biochemical systems using statistical model checking and abstraction refinement. TCS 412(21), 2162–2187 (2011)

22. Khaksari, M., Fischione, C.: Performance analysis and optimization of the joining protocol for a platoon of vehicles. In: ISCCSP, pp. 1–6. IEEE (2012)

23. Korenčiak, Ľ., Krčál, J., Řehák, V.: Dealing with zero density using piecewise phase-type approximation. In: Horváth, A., Wolter, K. (eds.) EPEW 2014. LNCS, vol. 8721, pp. 119–134. Springer, Heidelberg (2014)

24. Kwiatkowska, M., Norman, G., Segala, R., Sproston, J.: Verifying quantitative properties of continuous probabilistic timed automata. In: Palamidessi, C. (ed.) CONCUR 2000. LNCS, vol. 1877, pp. 123–137. Springer, Heidelberg (2000)
25. Lindemann, C.: An improved numerical algorithm for calculating steady-state solutions of deterministic and stochastic Petri net models. Perform. Eval. **18**(1), 79–95 (1993)
26. Marsan, M.A., Chiola, G.: On Petri nets with deterministic and exponentially distributed firing times. In: Rozenberg, G. (ed.) Advances in Petri Nets, pp. 132–145. Springer, Heidelberg (1987)
27. Neuhäusser, M.R., Zhang, L.: Time-bounded reachability probabilities in continuous-time Markov decision processes. In: QEST, pp. 209–218. IEEE (2010)
28. Neuts, M.F.: Matrix-geometric Solutions in Stochastic Models: An Algorithmic Approach. Courier Dover Publications, Mineola (1981)
29. Norris, J.R.: Markov Chains. Cambridge University Press, Cambridge (1998)
30. Obermaisser, R.: Time-Triggered Communication. CRC Press, Boca Raton (2011)
31. Puterman, M.L.: Markov Decision Processes. Wiley, Hoboken (1994)
32. Ramamritham, K., Stankovic, J.A.: Scheduling algorithms and operating systems support for real-time systems. Proc. IEEE **82**(1), 55–67 (1994)
33. Tiassou, K.B.: Aircraft operational reliability - a model-based approach and case studies. Ph.D. thesis, Universié de Toulouse (2013)
34. Wolovick, N., D'Argenio, P.R., Qu, V: Optimizing probabilities of real-time test case execution. In: ICST, pp. 446–455. IEEE (2009)

Lumping-Based Equivalences in Markovian Automata and Applications to Product-Form Analyses

Andrea Marin and Sabina Rossi[✉]

DAIS - Università Ca' Foscari, Venice, Italy
{marin,srossi}@dais.unive.it

Abstract. The analysis of models specified with formalisms like Markovian process algebras or stochastic automata can be based on equivalence relations among the states. In this paper we introduce a relation called *exact equivalence* that, differently from most aggegation approaches, induces an exact lumping on the underlying Markov chain instead of a strong lumping. We prove that this relation is a congruence for Markovian process algebras and stochastic automata whose synchronisation semantics can be seen as the master/slave synchronisation of the Stochastic Automata Networks (SAN). We show the usefulness of this relation by proving that the class of quasi-reversible models is closed under exact equivalence. Quasi-reversibility is a pivotal property to study product-form models, i.e., models whose equilibrium behaviour can be computed very efficiently without the problem of the state space explosion. Hence, exact equivalence turns out to be a theoretical tool to prove the product-form of models by showing that they are exactly equivalent to models which are known to be quasi-reversible.

1 Introduction

Stochastic modelling plays an important role in computer science since it is widely used for performance evaluation and reliability analysis of software and hardware architectures, including telecommunication systems. In this context, Continuous Time Markov Chains (CTMCs) are the underlying stochastic processes of the models specified with many formalisms such as Stochastic Petri nets [22], Stochastic Automata Networks (SAN) [24], queueing networks [3] and a class of Markovian process algebras (MPAs), e.g., [12,14]. The aim of these formalisms is to provide a high-level description language for complex models and automatic analysis methods. Modularity in the model specification is an important feature of both MPAs and SANs that allows for describing large systems in terms of cooperations of simpler components. Nevertheless, one should notice that a modular specification does not lead to a modular analysis, in general. Thus, although the intrinsic compositional properties of such formalisms are extremely helpful in the specification of complex systems, in many cases carrying out an exact analysis for those models (e.g., those required by quantitative model checking) may be extremely expensive from a computational point of view.

© Springer International Publishing Switzerland 2015
J. Campos and B.R. Haverkort (Eds.): QEST 2015, LNCS 9259, pp. 160–175, 2015.
DOI: 10.1007/978-3-319-22264-6_11

The introduction of equivalence relations among quantitative models is an important formal approach to comparing different systems and also improving the efficiency of some analysis. Indeed, if we can prove that a model P is in some sense equivalent to Q and Q is much simpler than P, then we can carry out an analysis of the simplest component to derive the properties of the original one.

Bisimulation based relations on stochastic systems inducing the notions of ordinary (or strong) and exact lumpability for the underlying Markov chains have been studied in [2,7,10,14,26]. In this paper, we apply this idea by introducing the notion of *exact equivalence* on the states of synchronising stochastic automata and study its compositionality properties. Exact equivalence is a congruence for the synchronisation semantics that we consider, i.e., it is preserved by the synchronising operator. Moreover, we prove that an exact equivalence relation among the states of a non-synchronising automaton induces an exact lumping on its underlying CTMC. This is opposed to the usual notions of bisimulation-based equivalences previously introduced in the literature that induce a strong lumping [18] on the underlying CTMC [2,6,8,14,20]. Interestingly, we show that an exact equivalence over a non-synchronising stochastic automata induces a *strong lumping* on the time-reversed Markov chain underlying the model. This important observation, allows us to prove that exact equivalence preserves the quasi-reversibility property [17] defined for stochastic networks. Quasi-reversibility is one of the most important and widely used characterisation of product-form models, i.e., models whose equilibrium distribution can be expressed as the product of functions depending only on the local state of each component. Informally, we can say that product-forms project the modularity in the model definition to the model analysis, thus drastically reducing the computational costs of the derivation of the quantitative indices. Basically, a synchronisation of quasi-reversible components whose underlying chain is ergodic has a product-form solution, meaning that one can check the quasi-reversibility modularly for each component, without generating the whole state space.

In this paper we provide a new methodology to prove (disprove) that a stochastic automaton is quasi-reversible by simply showing that it is exactly equivalent to another model which is known to be (to be not) quasi-reversible. In practice, this approach can be useful because proving the quasi-reversibility of a model may be a hard task since it requires one to reverse the underlying CTMC and check some conditions on the reverse process, see, e.g., [11,17]. Conversely, by using exact equivalence, one can prove or disprove the quasi-reversibility property by considering only the forward model, provided that it is exactly equivalent to another (simpler) quasi-reversible model known in the wide literature of product-forms. Moreover, while automatically proving quasi-reversibility is in general unfeasible, checking the exact equivalence between two automata can be done algoritmically by exploiting a partition refinement strategy, similar to that of Paige and Tarjan's algorithm for bisimulation [23].

The paper is structured as follows. Section 2 introduces the notation and recalls the basic definitions on Markov chains. In Sect. 3 we give the definition of stochastic automata and specify their synchronisation semantics. Section 4 presents the definition of quasi-reversibility for stochastic automata. Exact equivalence is introduced in Sect. 5 and the fact that it preserves quasi-reversibility is proved. Finally, Sect. 6 concludes the paper.

2 Preliminaries

Let $X(t)$ be a stochastic process taking values into a state space S for $t \in \mathbb{R}^+$. $X(t)$ is said *stationary* if $(X(t_1), X(t_2), \ldots, X(t_n))$ has the same distribution as $(X(t_1+\tau), X(t_2+\tau), \ldots, X(t_n+\tau))$ for all $t_1, t_2, \ldots, t_n, \tau \in \mathbb{R}^+$. Moreover, $X(t)$ satisfies the *Markov property*, and it is called *Markov process*, if the conditional (on both past and present states) probability distribution of its future behaviour is independent of its past evolution until the present state. A Continuous-Time Markov Chain (CTMC) is a Markov process with a discrete state space S.

A CTMC $X(t)$ is said to be *time-homogeneous* if the conditional probability $P(X(t + \tau) = s \mid X(t) = s')$ does not depend upon t, and is *irreducible* if every state in S can be reached from every other state. A state in a Markov process is called *recurrent* if the probability that the process will eventually return to the same state is one. A recurrent state is called *positive-recurrent* if the expected number of steps until the process returns to it is finite. A CTMC is *ergodic* if it is irreducible and all its states are positive-recurrent. For finite Markov chains, irreducibility is sufficient for ergodicity.

An ergodic CTMC possesses an *equilibrium* (or *steady-state*) *distribution*, that is the *unique* collection of positive numbers $\pi(s)$ with $s \in S$ such that

$$\lim_{t \to \infty} P(X(t) = s \mid X(0) = s') = \pi(s).$$

The transition rate between two states s and s' is denoted by $q(s, s')$, with $s \neq s'$. The infinitesimal generator matrix $\mathbf{Q}$ of a Markov process is such that the $q(s, s')$'s are the off-diagonal elements while the diagonal elements are formed as the negative sum of the extra diagonal elements of each row. Any non-trivial vector of real numbers $\boldsymbol{\mu}$ satisfying the system of global balance equations (GBEs)

$$\boldsymbol{\mu}\mathbf{Q} = \mathbf{0} \tag{1}$$

is called *invariant measure* of the CTMC. For irreducible CTMCs, if $\boldsymbol{\mu}_1$ and $\boldsymbol{\mu}_2$ are both invariant measures of the same chain, then there exists a constant $k > 0$ such that $\boldsymbol{\mu}_1 = k\boldsymbol{\mu}_2$. If the CTMC is ergodic, then there exists a unique invariant measure $\boldsymbol{\pi}$ whose components sum to unity, i.e., $\sum_{s \in S} \pi(s) = 1$. In this case $\boldsymbol{\pi}$ is the equilibrium or steady-state distribution of the CTMC.

It is well-known that the solution of system (1) is often unfeasible due to the large number of states of the CTMC underlying the model of a real system. The analysis of an ergodic CTMC in equilibrium can be greatly simplified if it satisfies the property that when the direction of time is reversed the stochastic behaviour of the process remains the same.

Given a stationary CTMC $X(t)$ with $t \in \mathbb{R}^+$, we call $X(\tau - t)$ its reversed process. In the following we denote by $X^R(t)$ the reversed process of $X(t)$. It can be shown that $X^R(t)$ is also a stationary CTMC [17]. We say that

$X(t)$ is *reversible* if it is stochastically identical to $X^R(t)$, i.e., the process $(X(t_1), \ldots, X(t_n))$ has the same distribution as $(X(\tau - t_1), \ldots, X(\tau - t_n))$ for all $t_1, \ldots, t_n, \tau \in \mathbb{R}^+$ [17].

For a stationary Markov process there exists a necessary and sufficient condition for reversibility expressed in terms of the equilibrium distribution π and the transition rates.

Proposition 1 (Transition rates and probabilities of reversible processes [17]). *A stationary CTMC with state space S and infinitesimal generator $\mathbf{Q}$ is reversible if there exists a vector of positive real numbers π summing to unity, such that for all $s, s' \in S$ with $s \neq s'$,*

$$\pi(s)q(s,s') = \pi(s')q(s',s).$$

In this case π is the equilibrium distribution of the chain.

The *reversed process* $X^R(t)$ of a Markov process $X(t)$ can always be defined even when $X(t)$ is not reversible. In [11,17] the authors show that $X^R(t)$ is a CTMC and its transition rates are defined according to the following proposition.

Proposition 2 (Transition rates of reversed processes [11]). *Given the stationary CTMC $X(t)$ with state space S and infinitesimal generator $\mathbf{Q}$, the transition rates of the reversed process $X^R(t)$, forming its infinitesimal generator $\mathbf{Q}^R$, are defined as follows: for all $s, s' \in S$,*

$$q^R(s', s) = \frac{\mu(s)}{\mu(s')} q(s, s'), \tag{2}$$

where $q^R(s', s)$ denotes the transition rate from s' to s in the reversed process and μ is an invariant measure of $X(t)$.

The forward and the reversed processes share all the invariant measures and in particular they possess the same equilibrium distribution π.

In the following, for a given CTMC with state space S and for any state $s \in S$ we denote by $q(s)$ (resp., $q^R(s)$) the quantity $\sum_{s' \in S, s \neq s'} q(s, s')$ (resp., $\sum_{s' \in S, s \neq s'} q^R(s, s')$).

In the context of performance and reliability analysis, the notion of *lumpability* is used for generating an aggregated Markov process that is smaller than the original one but allows one to determine exact results for the original process. More precisely, the concept of lumpability can be formalized in terms of equivalence relations over the state space of the Markov chain. Any such equivalence induces a *partition* on the state space of the Markov chain and aggregation is achieved by clustering equivalent states into macro-states, thus reducing the overall state space. In general, when a CTMC is aggregated the resulting stochastic process will not have the Markov property. However, if the partition can be shown to satisfy the so called *strong* lumpability condition [1,18], the Markov property is preserved and the equilibrium solution of the aggregated process may be used to derive an exact solution of the original one.

Strong lumpability has been introduced in [18] and further studied in [9,27].

Definition 1 (Strong lumpability). *Let $X(t)$ be a CTMC with state space S and $\sim$ be an equivalence relation over S. We say that $X(t)$ is* strongly lumpable *with respect to $\sim$ (resp., $\sim$ is a* strong lumpability *for $X(t)$) if $\sim$ induces a partition on the state space of $X(t)$ such that for any equivalence class $S_i, S_j \in S/\sim$ with $i \neq j$ and $s, s' \in S_i$,*

$$\sum_{s'' \in S_j} q(s, s'') = \sum_{s'' \in S_j} q(s', s'').$$

Thus, an equivalence relation over the state space of a Markov process is a strong lumpability if it induces a partition into equivalence classes such that for any two states within an equivalence class their aggregated transition rates to any other class are the same. Notice that every Markov process is strongly lumpable with respect to the identity relation, and so it is the trivial relation having only one equivalence class.

A probability distribution π is *equiprobable* with respect to a partition of the state space S of an ergodic Markov process if for all the equivalence classes $S_i \in S/\sim$ and for all $s, s' \in S_i$, $\pi(s) = \pi(s')$.

In [25] the notion of exact lumpability is introduced as a sufficient condition for a distribution to be equiprobable with respect to a partition.

Definition 2 (Exact lumpability). *Let $X(t)$ be a CTMC with state space S and $\sim$ be an equivalence relation over S. We say that $X(t)$ is* exactly lumpable *with respect to $\sim$ (resp., $\sim$ is an* exact lumpability *for $X(t)$) if $\sim$ induces a partition on the state space of $X(t)$ such that for any $S_i, S_j \in S/\sim$ and $s, s' \in S_i$,*

$$\sum_{s'' \in S_j} q(s'', s) = \sum_{s'' \in S_j} q(s'', s').$$

An equivalence relation is an exact lumpability if it induces a partition on the state space such that for any two states within an equivalence class the aggregated transition rates into such states from any other class are the same.

The proof of next proposition is given in [25].

Proposition 3. *Let $X(t)$ be an ergodic CTMC with state space S and $\sim$ be an equivalence relation over S. If $X(t)$ is exactly lumpable with respect to $\sim$ (resp., $\sim$ is an exact lumpability for $X(t)$) then for all $s, s' \in S$ such that $s \sim s'$, $\mu(s) = \mu(s')$, where μ is an invariant measure for $X(t)$.*

3   Stochastic Automata

Many high-level specification languages for stochastic discrete-event systems are based on *Markovian process algebras* [7,13,14] that are characterized by powerful composition operators and timed actions whose delay is governed by independent random variables with a continuous-time exponential distribution. The expressivity of such languages allows the development of well-structured specifications and efficient analyses of both qualitative and quantitative properties in a single framework. Their semantics is given in terms of stochastic automata, an extension of labelled automata with clocks that often are exponentially distributed random variables. In this paper we consider stochastic concurrent automata with an underlying continuous time Markov chain as common denominator of a wide set of Markovian stochastic process algebra. Stochastic automata are equipped with a *composition operation* by which a complex automaton can be constructed from simpler components. Our model draws a distinction between *active* and *passive* action types, and in forming the composition of automata only active/passive synchronisations are permitted. An analogue semantics is proposed for Stochastic Automata Networks in [24].

Definition 3 (Stochastic Automaton (SA)). *A stochastic automaton P is a tuple $(S_P, A_P, P_P, \leadsto_P, q_P)$ where*

- S_P *is a denumerable set of states called* state space *of P,*
- A_P *is a denumerable set of* active *types,*
- P_P *is a denumerable set of* passive *types,*
- τ *denotes the* unknown *type,*
- $\leadsto_P \subseteq (S_P \times S_P \times T_P)$ *is a transition relation where $T_P = (A_P \cup P_P \cup \{\tau\})$ and for all $s \in S_P$, $(s, s, \tau) \notin \leadsto_P$,*[1]
- q_P *is a function from $\leadsto_P$ to $\mathbb{R}^+$ such that $\forall s_1 \in S_P$ and $\forall a \in P_P$, $\sum_{s_2 : (s_1, s_2, a) \in \leadsto_P} q_P(s_1, s_2, a) \le 1$.*

In the following we denote by $\to_P$ the relation containing all the tuples of the form (s_1, s_2, a, q) where $(s_1, s_2, a) \in \leadsto_P$ and $q = q_P(s_1, s_2, a)$. We say that $q_P(s, s', a) \in \mathbb{R}^+$ is the *rate* of the transition from state s to s' with type a if $a \in A_P \cup \{\tau\}$. Notice that this is indeed the apparent transition rate from s to s' relative to a. If a is passive then $q_P(s, s', a) \in (0, 1]$ denotes the *probability* that the automaton synchronises on type a with a transition from s to s'. Hereafter, we assume that $q_P(s, s', a) = 0$ whenever there are no transitions with type a from s to s'. If $s \in S_P$, then for all $a \in T_P$ we write $q_P(s, a) = \sum_{s' \in S} q_P(s, s', a)$. Moreover we denote by $q_P(s, s') = \sum_{a \in T_P} q_P(s, s', a)$ and $q_P(s) = \sum_{a \in T_P} q_P(s, a)$. We say that P is *closed* if $P_P = \emptyset$. We use the notation $s_1 \overset{a}{\leadsto}_P s_2$ to denote the tuple $(s_1, s_2, a) \in \leadsto_P$; we denote by $s_1 \xrightarrow{(a,r)}_P s_2$ (resp., $s_1 \xrightarrow{(a,p)}_P s_2$) the tuple $(s_1, s_2, a, r) \in \to_P$ (resp., $(s_1, s_2, a, p) \in \to_P$).

Definition 4 (CTMC underlying a closed SA). *The CTMC underlying a closed stochastic automaton P, denoted $X_P(t)$, is defined as the CTMC with state space S_P and infinitesimal generator matrix $\mathbf{Q}$ defined as: for all $s_1 \ne s_2 \in S_P$,*

$$q(s_1, s_2) = \sum_{a, r : (s_1, s_2, a, r) \in \to_P} r \,.$$

For ergodic chains, we denote an invariant measure and the equilibrium distribution of the CTMC underlying P by μ_P and π_P, respectively.

---

[1] Notice that τ self-loops do not affect the equilibrium distribution of the CTMC underlying the automaton. Moreover, the choice of excluding τ self-loops will simplify the definition of automata synchronisation.

Table 1. Operational rules for SA synchronisation

$$\frac{s_{p_1} \xrightarrow{(a,r)}_P s_{p_2} \quad s_{q_1} \xrightarrow{(a,p)}_Q s_{q_2}}{(s_{p_1}, s_{q_1}) \xrightarrow{(a,pr)}_{P \otimes Q} (s_{p_2}, s_{q_2})} \quad (a \in \mathcal{A}_P = \mathcal{P}_Q)$$

$$\frac{s_{p_1} \xrightarrow{(a,p)}_P s_{p_2} \quad s_{q_1} \xrightarrow{(a,r)}_Q s_{q_2}}{(s_{p_1}, s_{q_1}) \xrightarrow{(a,pr)}_{P \otimes Q} (s_{p_2}, s_{q_2})} \quad (a \in \mathcal{P}_P = \mathcal{A}_Q)$$

$$\frac{s_{p_1} \xrightarrow{(\tau,r)}_P s_{p_2}}{(s_{p_1}, s_{q_1}) \xrightarrow{(\tau,r)}_{P \otimes Q} (s_{p_2}, s_{q_1})} \qquad \frac{s_{q_1} \xrightarrow{(\tau,r)}_Q s_{q_2}}{(s_{p_1}, s_{q_1}) \xrightarrow{(\tau,r)}_{P \otimes Q} (s_{p_1}, s_{q_2})}$$

We say that an automaton is *irreducible* if each state can be reached by any other state after an arbitrary number of transitions. We say that a closed automaton P is *ergodic* if its underlying CTMC is ergodic.

The synchronisation operator between two stochastic automata P and Q is defined in the style of master/slave synchronisation of SANs [24] based on the Kronecker's algebra and the active/passive cooperation used in Markovian process algebra such as PEPA [14].

Definition 5 (SA synchronisation). *Given two stochastic automata P and Q such that $\mathcal{A}_P = \mathcal{P}_Q$ and $\mathcal{A}_Q = \mathcal{P}_P$ we define the automaton $P \otimes Q$ as follows:*

- $\mathcal{S}_{P \otimes Q} = \mathcal{S}_P \times \mathcal{S}_Q$,
- $\mathcal{A}_{P \otimes Q} = \mathcal{A}_P \cup \mathcal{A}_Q = \mathcal{P}_P \cup \mathcal{P}_Q$,
- $\mathcal{P}_{P \otimes Q} = \emptyset$,
- τ *is the unknown type,*
- $\leadsto_{P \otimes Q}$ *and $q_{P \otimes Q}$ are defined according to the rules for $\rightarrow_{P \otimes Q}$ depicted in Table 1: indeed, the relation $\rightarrow_{P \otimes Q}$ contains the tuples $((s_{p_1}, s_{q_1}), (s_{p_1}, s_{q_2}), a, q)$ with $((s_{p_1}, s_{q_1}), (s_{p_1}, s_{q_2}), a) \in \leadsto_{P \otimes Q}$ and $q = q_{P \otimes Q}((s_{p_1}, s_{q_1}), (s_{p_1}, s_{q_2}), a)$.*

Given a closed stochastic automaton P we can define its reversed P^R in the style of [4], that is a stochastic automaton whose underlying CTMC $X_{P^R}(t)$ is identical to $X_P^R(t)$.

Definition 6 (Reversed SA [4]). *Let P be a closed stochastic automaton with an underlying irreducible CTMC and let μ_P be an invariant measure. Then we define the stochastic automaton P^R reversed of P as follows:*

- $\mathcal{S}_{P^R} = \{s^R \mid s \in \mathcal{S}_P\}$
- $\mathcal{A}_{P^R} = \mathcal{A}_P$ *and* $\mathcal{P}_{P^R} = \mathcal{P}_P = \emptyset$
- $\leadsto_{P^R} = \{(s_1^R, s_2^R, a) : (s_2, s_1, a) \in \leadsto_P, a \in \mathcal{A}_P \cup \{\tau\}\}$
- $q_{P^R}(s_1^R, s_2^R, a) = \mu_P(s_2)/\mu_P(s_1)q_P(s_2, s_1, a)$

It can be easily proved that for any invariant measure (including the equilibrium distribution) μ_P for P there exists an invariant measure μ_{P^R} for P^R such that for all $s \in \mathcal{S}_P$ it holds $\mu_P(s) = \mu_{P^R}(s^R)$, and viceversa.

4   Quasi-Reversible Automata

In this section we review the definition of quasi-reversibility given by Kelly in [17] by using the notation of stochastic automata. In order to clarify the exposition, we introduce a closure operation over stochastic automata that allows us to assign to all the transitions with the same passive type the same rate λ.

Definition 7 (SA closure). *The closure of a stochastic automaton P with respect to a passive type $a \in \mathcal{P}_P$ and a rate $\lambda \in \mathbb{R}^+$, written $P^c = P\{a \leftarrow \lambda\}$, is the automaton defined as follows:*

- $\mathcal{S}_{P^c} = \{s^c \mid s \in \mathcal{S}_P\}$
- $\mathcal{A}_{P^c} = \mathcal{A}_P$ *and* $\mathcal{P}_{P^c} = \mathcal{P}_P \smallsetminus \{a\}$
- $\leadsto_{P^c} = \{(s_1^c, s_2^c, b) \mid (s_1, s_2, b) \in \leadsto_P, \ a \neq b\} \cup \{(s_1^c, s_2^c, \tau) \mid (s_1, s_2, a) \in \leadsto_P\}$
-

$$q_{P^c}(s_1^c, s_2^c, b) = \begin{cases} q_P(s_1, s_2, b) & \text{if } b \neq a, \tau \\ q_P(s_1, s_2, a)\lambda + q_P(s_1, s_2, \tau) & \text{if } b = \tau \end{cases}$$

where we assume that $q_P(s_1, s_2, b) = 0$ if $(s_1, s_2, b) \notin \leadsto_P$.

Notice that for a closure P^c of a stochastic automaton P with respect to all its passive types in $\mathcal{P}_P$ we can compute the equilibrium distribution, provided that the underlying CTMC is ergodic (see Definition 4).

Definition 8 (Quasi-reversible SA [17,21]). *An irreducible stochastic automaton P with $\mathcal{P}_P = \{a_1, \ldots, a_n\}$ and $\mathcal{A}_P = \{b_1, \ldots b_m\}$ is quasi-reversible if*

- *for all $a \in \mathcal{P}_P$ and for all $s \in \mathcal{S}_P$, $\sum_{s' \in \mathcal{S}_P} q_P(s, s', a) = 1$*
- *for each closure $P^c = P\{a_1 \leftarrow \lambda_1\} \ldots \{a_n \leftarrow \lambda_n\}$ with $\lambda_1, \ldots, \lambda_n \in \mathbb{R}^+$ there exists a set of positive real numbers $\{k_{b_1}, \ldots, k_{b_m}\}$ such that for each $s \in \mathcal{S}_{P^c}$ and $1 \leq i \leq m$*

$$k_{b_i} = \frac{\sum_{s' \in \mathcal{S}_{P^c}} \mu_{P^c}(s') q_{P^c}(s', s, b_i)}{\mu_{P^c}(s)}, \tag{3}$$

where μ_{P^c} denotes any non-trivial invariant measure of the CTMC underlying P^c.

Notice that in the definition of quasi-reversibility we do not require the closure of P with respect to all its passive types to originate a stochastic automaton with an ergodic underlying CTMC because we assume μ_{P^c} to be an invariant measure, i.e., we do not require that $\sum_{s \in \mathcal{S}_{P^c}} \mu_{P^c}(s) = 1$. However, the irreducibility of the CTMC underlying the automaton ensures that all the invariant measures differ for a multiplicative constant, hence Eq. (3) is independent of the choice of the invariant measure.

The next theorem states that a network of quasi-reversible stochastic automata exhibits a product-form invariant measure and, if the joint state space is ergodic, a product-form equilibrium distribution. For the sake of simplicity, we state the theorem for two synchronising stochastic automata although the result holds for any finite set of automata which synchronise pairwise [11,17,21].

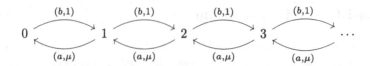

Fig. 1. Stochastic automaton underlying a Jackson's queue.

Theorem 1 (*Product-form solution based on quasi-reversibility*). *Let P and Q be two quasi-reversible automata such that $\mathcal{A}_P = \mathcal{P}_Q$ and $\mathcal{A}_Q = \mathcal{P}_P$ and let $S = P \otimes Q$. Assume that there exists a set of positive real numbers $\{k_a : a \in \mathcal{A}_P \cup \mathcal{A}_Q\}$ such that if we define the following automata $P^c = P\{a \leftarrow k_a\}$ for each $a \in \mathcal{P}_P$ and $Q^c = Q\{a \leftarrow k_a\}$ for each $a \in \mathcal{P}_Q$ it holds that:*

$$k_a = \frac{\sum_{s' \in S_{P^c}} \mu_{P^c}(s') q_{P^c}(s', s, a)}{\mu_{P'}(s)} \qquad \forall s \in \mathcal{S}_{P^c}, a \in \mathcal{A}_P$$

$$k_a = \frac{\sum_{s' \in S_{Q^c}} \mu_{Q'}(s') q_{Q^c}(s', s, a)}{\mu_{Q^c}(s)} \qquad \forall s \in \mathcal{S}_{Q^c}, a \in \mathcal{A}_Q$$

Then, given the invariant measures μ_{P^c} and μ_{Q^c} it holds that

$$\mu_S(s_1, s_2) = \mu_{P^c}(s_1^c) \mu_{Q^c}(s_2^c)$$

is an invariant measure for all the positive-recurrent states $(s_1, s_2) \in \mathcal{S}_S$ where s_1^c and s_2^c are the states in $\mathcal{S}_{P^c}$ and $\mathcal{S}_{Q^c}$ corresponding to $s_1 \in \mathcal{S}_P$ and $s_2 \in \mathcal{S}_Q$ according to Definition 7. In this case we say that P and Q have a quasi-reversibility based product-form.

Example 1 (*Product-form solution of Jackson networks*). Jackson networks provide an example of models having a product-form solution. A network consists of a collection of exponential queues with state-independent probabilistic routing. Jobs arrive from the outside at each queuing station in the network according to a homogeneous Poisson process. It is well-known that the queues of Jackson networks are quasi-reversible and hence the product-form is a consequence of Theorem 1. Figure 1 shows the automaton underlying a Jackson's queue where a is an active type while b is a passive one. It is worth of notice that also the queues considered in [5, 19] are quasi-reversible. □

5   Lumpable Bisimulations and Exact Equivalences

In this section we introduce two coinductive definitions, named *lumpable bisimulation* and *exact equivalence*, over stochastic automata which provide a sufficient condition for strong and exact lumpability of the underlying CTMCs.

The lumpable bisimulation is developed in the style of Larsen and Skou's bisimulation [16]. In this section we restrict ourself to the class of irreducible stochastic automata. Let $\mathcal{S}^*$ be the set of all states of all irreducible stochastic automata and $\mathcal{T}^*$ be the set of all action types of all stochastic automata. As expected, for any $s, s' \in \mathcal{S}^*$ and for any $a \in \mathcal{T}^*$, $q(s, s', a)$ denotes $q_P(s, s', a)$ if $s, s' \in \mathcal{S}_P$ for some stochastic automaton P, otherwise $q(s, s', a)$ is equal to 0. Analogously, we write $q(s, a)$ to denote $q_P(s, a)$ when $s \in \mathcal{S}_P$ for some stochastic automaton P.

Definition 9 (Lumpable bisimulation). *An equivalence relation $\mathcal{R} \subseteq \mathcal{S}^* \times \mathcal{S}^*$ is a* lumpable bisimulation *if whenever $(s, s') \in \mathcal{R}$ then for all $a \in \mathcal{T}^*$ and for all $C \in \mathcal{S}^*/\mathcal{R}$ such that*

- *either $a \neq \tau$,*
- *or $a = \tau$ and $s, s' \notin C$,*

it holds $\sum_{s'' \in C} q(s, s'', a) = \sum_{s'' \in C} q(s', s'', a)$.

It is clear that the identity relation is a lumpable bisimulation. In [15] we proved that the transitive closure of a union of lumpable bisimulations is still a lumpable bisimulation. Hence, the maximal lumpable bisimulation, denoted $\sim_s$, is defined as the union of all the lumpable bisimulations. We say that two stochastic automata P and Q are equivalent according to the lumpable bisimulation equivalence relation, denoted $P \sim_s Q$, if there exists $s_p \in \mathcal{S}_P$ and $s_q \in \mathcal{S}_Q$ such that $(s_p, s_q) \in \sim_s$. For any stochastic automaton P, $\sim_s$ induces a partition on the state space of the underlying Markov process that is a strong lumping (see Definition 1) [15].

We now introduce the notion of exact equivalence for stochastic automata. An equivalence relation over $\mathcal{S}^*$ is an *exact equivalence* if for any action type $a \in \mathcal{T}^*$, the total conditional transition rates from two equivalence classes to two equivalent states, via activities of this type, are the same. Moreover, for any type a, equivalent states have the same apparent conditional exit rate.

Definition 10 (Exact equivalence). *An equivalence relation $\mathcal{R} \subseteq \mathcal{S}^* \times \mathcal{S}^*$ is an* exact equivalence *if whenever $(s, s') \in \mathcal{R}$ then for all $a \in \mathcal{T}^*$ and for all $C \in \mathcal{S}^*/\mathcal{R}$ it holds*

- *$q(s, a) = q(s', a)$,*
- *$\sum_{s'' \in C} q(s'', s, a) = \sum_{s'' \in C} q(s'', s', a)$.*

The transitive closure of a union of exact equivalences is still an exact equivalence. Hence, the maximal exact equivalence, denoted $\sim_e$, is defined as the union of all exact equivalences. We say that two stochastic automata P and Q are *exactly equivalent*, denoted $P \sim_e Q$, if there exists $s_p \in \mathcal{S}_P$ and $s_q \in \mathcal{S}_Q$ such that $(s_p, s_q) \in \sim_e$.

Example 2. Let us consider a queueing model of a system with two identical processors, named κ_1 and κ_2. Each job is assigned to one of the processors which are assumed not to work in parallel. At each service completion event of processor κ_i, the next job is assigned to κ_j, for $i \neq j$, with probability p, and is assigned

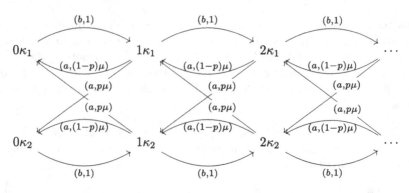

Fig. 2. Queue with alternating servers.

to processor κ_i with probability $1 - p$. Automaton P underlying this model is depicted in Fig. 2 where state $n\kappa_i$, for $n > 0$ and $i = 1, 2$, denotes the state in which processor κ_i is being used and there are n customers waiting to be served. State $0\kappa_i$ denotes the empty queue. It is easy to prove that the equivalence relation $\sim$ obtained by the reflexive closure of $\{(n\kappa_1, n\kappa_2), (n\kappa_2, n\kappa_1) : n \in \mathbb{N}\}$ is an exact equivalence over the state space of P. Let us consider the automaton Q depicted in Fig. 1, then it holds that the equivalence relation given by the symmetric and reflexive closure of $\sim' = \sim \cup \{(n\kappa_1, n), (n, n\kappa_2) : n \in \mathbb{N}\}$, where each n denotes a state of Q, is still an exact equivalence.   □

The next proposition states that, for any stochastic automaton P, $\sim_e$ induces an exactly lumpable partition on the state space of the Markov process underlying P.

Proposition 4 (Exact lumpability). *Let P be a closed stochastic automaton with state space $\mathcal{S}_P$ and $X_P(t)$ its underlying Markov chain with infinitesimal generator matrix $\mathbf{Q}$. Then for any equivalence class $S_i, S_j \in \mathcal{S}_P/\sim_e$ and $s, s' \in S_i$,*

$$\sum_{s'' \in S_j} q(s'', s) = \sum_{s'' \in S_j} q(s'', s')$$

i.e., $\sim_e$ is an exact lumpability for $X_P(t)$.

The next theorem plays an important role in studying the product-form of exactly equivalent automata. Informally, it states that the exact equivalence preserves the invariant measure of equivalent states.

Theorem 2. *Let P and Q be two closed stochastic automata such that $P \sim_e Q$ and let $\boldsymbol{\mu}_P$ and $\boldsymbol{\mu}_Q$ be two invariant measures of P and Q, respectively. Then, there exists a positive constant K such that for each $s_1 \in \mathcal{S}_P$ and $s_2 \in \mathcal{S}_Q$ with $s_1 \sim_e s_2$ it holds that $\mu_P(s_1)/\mu_Q(s_2) = K$.*

Corollary 1. *Let P and Q be two closed stochastic automata such that $P \sim_e Q$ and let $\boldsymbol{\pi}_P$ and $\boldsymbol{\pi}_Q$ be the stationary distributions of P and Q, respectively.*

Then, for all $s_1, s_2 \in \mathcal{S}_P$ and $s_1', s_2' \in \mathcal{S}_Q$ such that $s_i \sim_e s_i'$ for $i = 1, 2$, it holds that $\pi_P(s_1)/\pi_P(s_2) = \pi_Q(s_1')/\pi_Q(s_2')$.

We can prove that both lumpable bisimulation and exact equivalence are congruences for SA synchronisation.

Proposition 5. (Congruence) *Let* P, P', Q, Q' *be stochastic automata.*

- *If* $P \sim_s P'$ *and* $Q \sim_s Q'$ *then* $P \otimes Q \sim_s P' \otimes Q'$.
- *If* $P \sim_e P'$ *and* $Q \sim_e Q'$ *then* $P \otimes Q \sim_e P' \otimes Q$.

The proof is based on the fact that if $\sim_i$ is a lumpable bisimulation (resp. an exact equivalence) then the relation

$$\mathcal{R} = \{((s_{p_1}, s_{q_1}), (s_{p_2}, s_{q_2})) \mid s_{p_1} \sim_i s_{p_2} \text{ and } s_{q_1} \sim_i s_{q_2}\}$$

is also a lumpable bisimulation (resp., an exact equivalence) over $\mathcal{S}_P \times \mathcal{S}_Q$.

The next theorem proves that any exact equivalence between two stochastic automata induces a lumpabale bisimulation between the corresponding reversed automata.

Theorem 3 (Exact equivalence and lumpable bisimulation). *Let* P *and* Q *be two closed stochastic automata,* P^R *and* Q^R *be the corresponding reversed automata defined according to Definition 6 and* $\sim \subseteq \mathcal{S}_P \times \mathcal{S}_Q$ *be an exact equivalence. Then* $\sim' = \{(s_1^R, s_2^R) \in \mathcal{S}_{PR} \times \mathcal{S}_{QR} \mid (s_1, s_2) \in \sim\}$ *is a lumpable bisimulation.*

As a consequence any exact equivalence over the state space of a stochastic automaton P induces a lumpable bisimulation over the state space of the reversed automaton P^R.

Corollary 2. *Let* P *be a closed stochastic automaton and* $\sim \subseteq \mathcal{S}_P \times \mathcal{S}_P$ *be an exact equivalence. Then the relation* $\sim' = \{(s_1^R, s_2^R) \in \mathcal{S}_{PR} \times \mathcal{S}_{PR} \mid (s_1, s_2) \in \sim\}$ *is a lumpable bisimulation.*

The following lemma provides a characterization of quasi-reversibility in terms of lumpable bisimulation. Informally, it states that an automaton is quasi-reversible if and only if for each closure its reversed is lumpable bisimilar to an automaton with a single state.

Lemma 1. (Quasi-reversibility and lumpable bisimulation) *An irreducible stochastic automaton* P *is quasi-reversible if and only if the following properties hold for every closure* P^c *of* P *with reversed automaton* P^{cR}:

- *if* $s^R \in \mathcal{S}_{PcR}$, *then* $[s^R]_{\sim_s} = \mathcal{S}_{PcR}$
- *if* $a \in \mathcal{P}_P$ *then* $q_P(s, a) = 1$ *for all* $s \in \mathcal{S}_P$.

The following proposition states that both lumpable bisimulations and exact equivalences are invariant with respect to the closure of automata where any closure P^c of P is defined according to Definition 7.

Proposition 6. *Let P and Q be two stochastic automata with $\mathcal{A}_P = \mathcal{A}_Q$, $\mathcal{P}_P = \mathcal{P}_Q = \{a_1, \ldots, a_n\}$ and $\sim \subseteq \mathcal{S}_P \times \mathcal{S}_Q$ be an exact equivalence (resp., a lumpable bisimulation). Then for every closure $P^c = P\{a_1 \leftarrow \lambda_1\} \ldots \{a_n \leftarrow \lambda_n\}$ and $Q^c = Q\{a_1 \leftarrow \lambda_1\} \ldots \{a_n \leftarrow \lambda_n\}$ the relation $\sim' = \{(s_1^c, s_2^c) \in \mathcal{S}_{P^c} \times \mathcal{S}_{Q^c} | (s_1, s_2) \in \sim\}$ is an exact equivalence (resp., a lumpable bisimulation).*

The next theorem proves that the class of quasi-reversible stochastic automata is closed under exact equivalence.

Theorem 4. *Let P and Q be two stochastic automata such that $P \sim_e Q$. If Q is quasi-reversible then also P is quasi-reversible.*

Proof. We have to prove that:

1. The outgoing transitions for each passive type $a \in \mathcal{P}_P$ sums to unity.
2. For each closure $P^c = P\{a_1 \leftarrow \lambda_1\} \ldots \{a_n \leftarrow \lambda_n\}$ of P with $\lambda_1, \ldots, \lambda_n \in \mathbb{R}^+$ there exists a set of positive real numbers $\{k_1, \ldots, k_m\}$ such that for each $s \in \mathcal{S}_{P^c}$ and $1 \le i \le m$, Eq. (3) is satisfied.

The first claim follows immediately from the first item of Definition 10. Now observe that, by Definition 10, if $P \sim_e Q$ then $\mathcal{P}_P = \mathcal{P}_Q$ and $\mathcal{A}_P = \mathcal{A}_Q$. Let $\mathcal{P}_P = \mathcal{P}_Q = \{a_1, \ldots, a_n\}$ and $\mathcal{A}_P = \mathcal{A}_Q = \{b_1, \ldots, b_m\}$. By Proposition 6, for any closure $P^c = P\{a_1 \leftarrow \lambda_1\} \ldots \{a_n \leftarrow \lambda_n\}$ and $Q^c = Q\{a_1 \leftarrow \lambda_1\} \ldots \{a_n \leftarrow \lambda_n\}$ the relation $\sim' = \{(s_1^c, s_2^c) \in \mathcal{S}_{P^c} \times \mathcal{S}_{Q^c} | (s_1, s_2) \in \sim \text{ and} (s_1, s_2) \in \mathcal{S}_P \times \mathcal{S}_Q\}$ is an exact equivalence. By Theorem 3, the relation $\sim'' = \{(s_1^{cR}, s_2^{cR}) \in \mathcal{S}_{P^{cR}} \times \mathcal{S}_{Q^{cR}} | (s_1, s_2) \in \sim'\}$ is a lumpable bisimulation. By Lemma 1 since Q is quasi-reversible then for all $s^R \in \mathcal{S}_{Q^{cR}}$ it holds $[s^R]_{\sim_s} = \mathcal{S}_{Q^{cR}}$, i.e., there exists a set of positive real numbers $\{k_{b_1}, \ldots, k_{b_m}\}$ such that for each $s^R \in \mathcal{S}_{Q^{cR}}$ and $1 \le i \le m$

$$k_{b_i} = \sum_{s' \in \mathcal{S}_{Q^{cR}}} q_{Q^{cR}}(s^R, s', b_i) = \frac{\sum_{s' \in \mathcal{S}_{Q^c}} \mu_{Q^c}(s') q_{Q^c}(s', s, b_i)}{\mu_{Q^c}(s)},$$

which can be written as

$$k_{b_i} = \frac{\sum_{C \in \mathcal{S}_{Q^c}/\sim_e} \sum_{s' \in C} \mu_{Q^c}(s') q_{Q^c}(s', s, b_i)}{\mu_{Q^c}(s)}.$$

By Proposition 4, $\sim_e$ induces an exact lumping on the CTMC underlying Q^c and, by Proposition 3, for all s and s' in the same equivalence class $\mu_{Q^c}(s) = \mu_{Q^c}(s')$. Hence we can write

$$k_{b_i} = \frac{\sum_{C \in \mathcal{S}_{Q^c}/\sim_e} \mu_{Q^c}(C) \sum_{s' \in C} q_{Q^c}(s', s, b_i)}{\mu_{Q^c}(s)}.$$

where $\mu_{Q^c}(C)$ denotes $\mu_{Q^c}(s)$ for an arbitrary state $s \in C$. Now from the fact that $P^c \sim_e Q^c$, we have that for each class $C \in \mathcal{S}_{Q^c}/\sim_e$ there exists a class

$C' \in S_{P^c}/\sim_e$ such that all the states $s \in C$ are equivalent to the states in C'. Moreover, by Definition 10, we have $\sum_{s' \in C} q_{Q^c}(s', s_1, b_i) = \sum_{s' \in C'} q_{P^c}(s', s_2, b_i)$ for every state $s_1 \sim_e s_2$ with $s_1 \in Q^c$ and $s_2 \in P^c$. Therefore, we can write:

$$
\begin{aligned}
k_{b_i} &= \frac{\sum_{C \in S_{Q^c}/\sim_e} \mu_{Q^c}(C) \sum_{s' \in C} q_{Q^c}(s', s_1, b_i)}{\mu_{Q^c}(s_1)} \\
&= \frac{\sum_{C \in S_{Q^c}/\sim_e} \mu_{Q^c}(C) \sum_{s' \in C} q_{P^c}(s', s_2, b_i)}{\mu_{Q^c}(s_1)} \\
&= \frac{\sum_{C' \in S_{P^c}/\sim_e} K \mu_{P^c}(C') \sum_{s' \in C'} q_{P^c}(s', s_2, b_i)}{K \mu_{P^c}(s_2)} \\
&= \frac{\sum_{s' \in S_{P^c}} \mu_{P^c}(s') q_{P^c}(s', s_2, b_i)}{\mu_{P^c}(s_2)},
\end{aligned}
$$

where K is the positive constant given by Theorem 2. Summing up, since every closure Q^c of Q corresponds to a closure P^c for P and Q^c satisfies Eq. (3) for all states s and active types b_i, then the set of positive rates $\{k_{b_i}\}$ defined for Q^c are the same that satisfy Eq. (3) for P^c. Therefore, P is also quasi-reversible. $\square$

Example 3. Let us consider the automata Q and P depicted in Figs. 1 and 2, respectively. We already observed in Example 2 that there exists an exact equivalence $\sim'$ such that $n, n\kappa_1$ and $n\kappa_2$ belong to the same equivalence class, where n is a state of Q and $n\kappa_i$ belongs to the state space of P. Then, since Q is well-known to be quasi-reversible, by Theorem 2 also P is quasi-reversible. As a consequence, the queueing station modelled by P can be embdded in quasi-reversible product-form queueing networks maintining the property that the equilibrium distribution is separable. $\square$

The next example shows that, differently from exact equivalence, lumpable bisimulation does not preserve quasi-reversibility.

Example 4. Consider the automaton R depicted in Fig. 3. It is easy to prove that R is lumpable bisimilar to Jackson's queue Q depicted in Fig. 1. However, R is not quasi-reversible, i.e., the corresponding reversed automaton is not lumpable bisimilar to a single-state automaton. More precisely, one can observe that in the reversed automaton there is one type a transition exiting from state 0^R but there is no type a transition from state $1'^R$. This is sufficient to claim that states 0^R and $1'^R$ cannot belong to the same equivalence class. $\square$

The following final result is an immediate consequence of Theorems 1 and 4.

Corollary 3. *Let P, P', Q, Q' be stochastic automata such that $P \sim_e P'$ and $Q \sim_e Q'$. If P and Q have a quasi-reversibility based product-form then also P' and Q' are in product-form.*

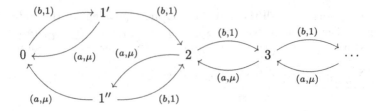

Fig. 3. Stochastic automaton strongly equivalent to a Jackson queue.

6   Conclusion

In this paper we have introduced the notion of exact equivalence, defined on the states of cooperating stochastic automata [24]. With respect to most stochastic equivalences defined for process algebras, exact equivalence induces an exact lumping in the underlying CTMC rather than a strong lumping. We show that this fact has important implications not only from a theoretical point of view but also in reducing the computational complexity of the analysis of cooperating models in equilibrium. Indeed, the class of quasi-reversible automata, whose composition is known to be in product-form and hence analysable efficiently, is closed under the exact equivalence. This leads to a new approach for proving the quasi-reversibility of a stochastic component which does not require to study the reverse-time underlying CTMC but to find a model exactly equivalent to the considered one that is already known to be (or to be not) quasi-reversible, or whose quasi-reversibility can be decided easier.

References

1. Baarir, S., Beccuti, M., Dutheillet, C., Franceschinis, G., Haddad, S.: Lumping partially symmetrical stochastic models. Perf. Eval. **68**(1), 21–44 (2011)
2. Baier, C., Katoen, J.-P., Hermanns, H.: Comparative branching-time semantics for markov chains. Inf. Comput. **200**(2), 149–214 (2005)
3. Balsamo, S., Marin, A.: Queueing networks. In: Bernardo, M., Hillston, J. (eds.) SFM 2007. LNCS, vol. 4486, pp. 34–82. Springer, Heidelberg (2007)
4. Balsamo, S., Dei Rossi, G.-L., Marin, A.: Lumping and reversed processes in cooperating automata. In: Al-Begain, K., Fiems, D., Vincent, J.-M. (eds.) ASMTA 2012. LNCS, vol. 7314, pp. 212–226. Springer, Heidelberg (2012)
5. Baskett, F., Chandy, K.M., Muntz, R.R., Palacios, F.G.: Open, closed, and mixed networks of queues with different classes of customers. J. ACM **22**(2), 248–260 (1975)
6. Bernardo, M.: Weak Markovian bisimulation congruences and exact CTMC-level aggregations for concurrent processes. In: Proceedings of the 10th Workshop on Quantitative Aspects of Programming Languages and Systems (QALP12), pp. 122–136 (2012)
7. Bernardo, M., Gorrieri, R.: A tutorial on empa: a theory of concurrent processes with nondeterminism, priorities, probabilities and time. Theo. Comput. Sci. **202**, 1–54 (1998)

8. Bravetti, M.: Revisiting interactive markov chains. Electr. Notes Theor. Comput. Sci. **68**(5), 65–84 (2003)
9. Buchholz, P.: Exact and ordinary lumpability in finite markov chains. J. Appl. Probab. **31**, 59–75 (1994)
10. Buchholz, P.: Exact performance equivalence: an equivalence relation for stochastic automata. Theor. Comput. Sci. **215**(1–2), 263–287 (1999)
11. Harrison, P.G.: Turning back time in markovian process algebra. Theo. Comput. Sci. **290**(3), 1947–1986 (2003)
12. Hermanns, H.: Interactive Markov Chains. Springer, Heidelberg (2002)
13. Hermanns, H., Herzog, U., Katoen, J.P.: Process algebra for performance evaluation. Theor. Comput. Sci. **274**(1–2), 43–87 (2002)
14. Hillston, J.: A Compositional Approach to Performance Modelling. Cambridge Press, Cambridge (1996)
15. Hillston, J., Marin, A., Piazza, C., Rossi, S.: Contextual lumpability. In: Proceedings of Valuetools 2013 Conference. ACM Press (2013)
16. Skou, A., Larsen, K.G.: Bisimulation through probabilistic testing. Inf. Comput. **94**(1), 1–28 (1991)
17. Kelly, F.: Reversibility and Stochastic Networks. Wiley, New York (1979)
18. Kemeny, J.G., Snell, J.L.: Finite Markov Chains. Springer, Heidelberg (1976)
19. Le Boudec, J.Y.: A BCMP extension to multiserver stations with concurrent classes of customers. In: SIGMETRICS 1986/PERFORMANCE 1986: Proceedings of the 1986 ACM SIGMETRICS International Conference on Computer Performance Modelling, Measurement and Evaluation, pp. 78–91. ACM Press, New York, NY (1986)
20. Marin, A., Rossi, S.: Autoreversibility: exploiting symmetries in Markov chains. In: Proceedings of MASCOTS 2013, pp. 151–160. IEEE Computer Society (2013)
21. Marin, A., Vigliotti, M.G.: A general result for deriving product-form solutions of Markovian models. In: Proceedings of First Joint WOSP/SIPEW International Conference on Performance Engineering, pp. 165–176. ACM, San Josè, CA, USA (2010)
22. Molloy, M.K.: Performance analysis using stochastic petri nets. IEEE Trans. on Comput. **31**(9), 913–917 (1982)
23. Paige, R., Tarjan, R.E.: Three partition refinement algorithms. SIAM J. Comput. **16**(6), 973–989 (1987)
24. Plateau, B.: On the stochastic structure of parallelism and synchronization models for distributed algorithms. SIGMETRICS Perf. Eval. Rev. **13**(2), 147–154 (1985)
25. Schweitzer, P.: Aggregation methods for large Markov chains. In: Mathematical Computer Performance and Reliability (1984)
26. Sproston, J., Donatelli, S.: Backward bisimulation in markov chain model checking. IEEE TSE **32**(8), 531–546 (2006)
27. Sumita, U., Reiders, M.: Lumpability and time-reversibility in the aggregation-disaggregation method for large markov chains. Commun. Stat. Stoch. Models **5**, 63–81 (1989)

A Numerical Analysis of Dynamic Fault Trees Based on Stochastic Bounds

J.M. Fourneau[1]([⊠]) and Nihal Pekergin[2]

[1] PRiSM, CNRS and University Versailles St Quentin, Versailles, France
jmf@prism.uvsq.fr
[2] LACL, UPEC, Créteil, France
Nihal.Pekergin@u-pec.fr

Abstract. We present a numerical method based on stochastic bounds and computations of discrete distributions to get the transient reliability of a system described by a Dynamic Fault Tree. We show that the gates of the tree are associated to simple operators on the probability distributions of the time to failure. We also prove that almost all operators are stochastically monotone, which allows us to simplify the computation complexity using stochastic bounds. We show that replicated events and functional dependency gates can be analyzed with conditional probabilities which can be handled with the same techniques but with a higher complexity. Finally we show the tradeoff between the precision of the approach and the complexity of the computations.

1 Introduction

Static Fault Tree (SFT) analysis is a standard technique for reliability modeling. Dynamic Fault Trees (DFT) have been proposed as an extension to model more complex systems where the duration and the sequences of transitions are taken into account (see [3,14] for a detailed description). DFTs are much more difficult to solve than SFTs. Fault Trees are composed of a set of leaves which model the components of the system and some gates whose inputs are connected to the leaves or to the outputs of other gates. The value of each leaf is a boolean which is True if the component is down and False otherwise. The whole topology of the connection must be a tree where there are no replicated gate inputs (shared components and/or subsystems) The root of the tree is a boolean value which must be True when the system has failed.

The SFTs contain 3 types of gates: OR, AND and K out of N (or voting) gates, which are all logical gates. DFTs have four new types of gate: PAND (priority AND), FDEP (functional dependency), SEQ (sequential failures) and SPARE gates. Typically the analysis of a DFT is based on a decomposition in dynamic or static subtrees using an efficient modularization algorithm. Static subtrees are analyzed using BDD or minimal cut-sets while dynamic subtree analysis are performed using Markov chains or simulations [8]. Indeed, when the basic components fail according to exponential distributions, one can build a Markov representation of the model directly or with the help of some high level

J. Campos and B.R. Haverkort (Eds.): QEST 2015, LNCS 9259, pp. 176–191, 2015.
DOI: 10.1007/978-3-319-22264-6_12

formalism like Petri Nets [7] or I/O Interactive Markov chains. But, as usual with component-based models, the curse of dimensionality occurs quickly and limits our ability to solve the models even by means of sophisticated techniques like lumpability [4]. Simulation techniques allow to deal with general distribution for component failure times but this implies extremely large simulation times and dealing with a large set of replicated simulations to obtain accurate results for transient processes [11]. Recently, the analysis of the structure function associated with continuous distributions for the inputs with direct analytical techniques has been investigated in [9] (see also [5] for a tool based on a mathematical software).

The aim of this paper is to present a new approach to find upper and lower stochastic bounds on the distribution of failure time for a system modeled by a DFT. This algorithm is based on exact numerical computations for discrete distributions and stochastic bounds to reduce the complexity of the numerical approach when the sizes of the distributions become too large. The methodology can be generalized to any problems in performance evaluation, stochastic model checking or reliability, involving discrete distributions and stochastically monotone operators.

As usual in reliability modeling, we assume that all the failure events are mutually independent. We only represent failures of the components, and the components cannot be repaired. Indeed, there is no clear description, in the literature, of the reparation scheduler associated with a DFT. We do not assume that the component failure time distributions follow an exponential or a Weibull process. We rather assume that these processes are described by discrete distributions which come from empirical data. The statistical analysis of these data to obtain empirical distributions is out of the scope of this paper. Note that a typical preprocessing consists in the fitting of the experimental data to obtain a continuous distribution that we can use in a simulation engine or an analytical model. We completely avoid this preprocessing here, as our inputs are the experimental measurements. Knowing the input discrete distributions of the failure times for all components of the DFT, we want an algorithm to compute the exact distribution or some stochastic bounds of the distribution for the time to failure of the system.

We illustrate the approach for a simple tree with one AND gate connected to two components for which the failure times are modeled by discrete distributions I_1 and I_2. Let $\mathcal{R}_1$ (resp. $\mathcal{R}_2$) be the support for I_1 (resp. I_2) of size n_1 (resp. n_2). $\mathcal{R}_1(i) \in \mathbb{R}^+$ denotes the ith failure instant for the distribution I_1, while $P_1(i)$ is the probability of failure at time $\mathcal{R}_1(i)$ for component 1. Similarly $P_2(i)$ is the probability of failure at time $\mathcal{R}_2(i)$ for component 2. If the system is modeled by an AND gate, it means that the system fails when both components fail. Therefore if component 1 fails at time $\mathcal{R}_1(i)$ and component 2 fails at time $\mathcal{R}_2(j)$, the system fails at time $max(\mathcal{R}_1(i), \mathcal{R}_2(j))$ with probability $P_1(i).P_2(j)$. This is a simple consequence of the independence of component failures and it seems possible to compute the distribution of the failure time of this simple system. However, if we look at the size of the output distribution, we see easily that the number of time instants may be as large as $n_1 + n_2 - 1$ (due to the *max* the smallest failure time is excluded). This is typically the case for many gates in a DFT (will be discussed in Sect. 3). And the computation complexity at any stage of the DFT analysis also depends on these sizes which increase after each computation stage. Therefore we need at each step of the analysis to simplify the distributions by reducing their sizes. And we have to find conditions to prove that such an approach provide bounds rather than approximations. We advocate that the stochastic bound theory applied on these discrete distributions and the stochastic monotonicity of some gates of a DFT will provide an efficient answer to the computation of the time to failure of a DFT with a trade-off between accuracy and computation time.

The technical part of the paper is organized as follows. In Sect. 2, we give a brief introduction to stochastic bounds for discrete distributions. Section 3 is devoted to DFTs without replicated events or FDEP gates. We analyze such trees with a bottom up algorithm on discrete distributions. We also show that all the gates except PAND gates are stochastically monotone thus bounds can be built during the numerical analysis. In Sect. 4, we present the analysis of the replicated events by conditioning, and study the complexity issues.

2   Stochastic Bounds and Calculus on Distributions

During the computation steps, the successively obtained distributions become larger and larger in size. This phenomenon cannot be avoided. Thus we advocate that one can obtain upper and lower stochastic bounds with a reduction of the size of the distributions making the computations of bounds easier than the exact analysis. Let us first define stochastic bounds on discrete distributions and show how we can change the size of the distribution while building a stochastic bound.

2.1   Stochastic Bounds

We refer to [10] for theoretical issues of the stochastic comparison method. We consider state space $\mathcal{G} = \{1, 2, \ldots, n\}$ endowed with a total order denoted as $\leq$. Let X and Y be two discrete random variables taking values on $\mathcal{G}$, with cumulative distribution functions (cdf) F_X and F_Y, and probability mass functions (pmf) $\boldsymbol{d}_X$ and $\boldsymbol{d}_Y$ ($\boldsymbol{d}_X(i) = \text{Prob}(X = i)$, and $\boldsymbol{d}_Y(i) = \text{Prob}(Y = i)$, for $i = 1, 2, \ldots, n$).

Definition 1. *X is said to be less than Y in the sense of the $\leq_{st}$ order, denoted by $X \leq_{st} Y$ or $\boldsymbol{d}_X \leq_{st} \boldsymbol{d}_Y$:*

generic definition : $X \leq_{st} Y \iff \mathbb{E}f(X) \leq \mathbb{E}f(Y)$, *for all non decreasing functions $f : \mathcal{G} \to \mathbb{R}^+$ whenever expectations exist.*

cumulative distribution functions : $X \leq_{st} Y \leftrightarrow \forall a \in \mathcal{G}, \ F_X(a) \geq F_Y(a)$.

probability mass functions : $X \leq_{st} Y \leftrightarrow \forall i, \ \sum_{k=i}^{n} \boldsymbol{d}_X(k) \leq \sum_{k=i}^{n} \boldsymbol{d}_Y(k)$.

Property 1. Let $(X_1, X_2, \cdots X_n)$ and $(Y_1, Y_2, \cdots Y_n)$ be mutually independent real-valued random variables such that $\forall i, \quad X_i \leq_{st} Y_i$ If $f : R^n \to R$ be a function which is non decreasing with respect to each entry i, then

$f(X_1, X_2, \cdots X_n) \leq_{st} f(Y_1, Y_2, \cdots Y_n)$.

The following theorem asserts the closure under mixture of the $\leq_{st}$ ordering (Theorem 1.2.15 in page 6 of [10]).

Theorem 1. *If X, Y and Θ are random variables such that $[X \mid \Theta = \theta] \leq_{st} [Y \mid \Theta = \theta]$ for all θ in the support of Θ, then $X \leq_{st} Y$.*

Example 1. Consider $d2 = [0.1, 0.2, 0.1, 0.2, 0.05, 0.1, 0.25]$, $d1 = [0.25, 0.05, 0.1, 0.15, 0.15, 0.3]$ taking values respectively on supports $\{1, \ldots, 7\}$ and $\{2, \ldots, 7\}$. The set $\mathcal{G}$ is the union of the supports and we can check that $d2 \leq_{st} d1$: the probability mass of $d1$ is concentrated to higher states such that the cdf of $d1$ is always below or equal to the cdf of $d2$ (Fig. 1).

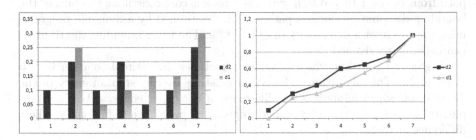

Fig. 1. $d2 \leq_{st} d1$: pmfs (left) and cdfs (right).

2.2 Algorithms to Construct Bounding Distributions

Several strategies have been proposed to stochastically bound a distribution of size n by a distribution of size k. For an arbitrary distribution of failure time, we assume that the support $\mathcal{R}(j), 1 \leq j \leq n$ is increasingly ordered and it consists of failure times such that the corresponding probabilities $P(j) > 0$.

Size Reduction of a Discrete Distribution. We propose Algorithm 1 to obtain an upper bounding distribution by dividing the support into almost equal groups of consecutive bins and by mapping each group into one single bin. The probability mass in a group is moved to the largest bin to provide an upper bounding distribution in the sense of the $\leq_{st}$ ordering. The lower bounding discrete distribution can be also built by moving the probability mass to the smallest bin. The time complexity of the algorithm is linear $(O(n))$.

Optimal Bounding Distributions with Respect to a Reward. In [2], we have proposed an algorithm to construct bounding distributions in the sense of the $\leq_{st}$ order which are optimal with respect to a given positive, increasing reward function, r. Formally, for a given distribution d defined with n bins, the upper bounding distribution with k bits, d^u is constructed such that :

Algorithm 1. Compute an Upper-Bounding distribution d^u with k bins on the original distribution d with n bins.

1: Obtain, by an Euclidean division of n by k, the quotient q and the remainder r.
2: **for** i=1 to r **do**
3:     $\mathcal{R}^u(i) = \mathcal{R}(i * (q+1))$ and $P^u(i) = \sum_{l=1}^{q+1} P((i-1) * (q+1) + l)$
4: **end for**
5: **for** i=r+1 to k **do**
6:     $\mathcal{R}^u(i) = \mathcal{R}(i * q + r)$ and $P^u(i) = \sum_{l=1}^{q} P((i-1) * q + r + l)$
7: **end for**

---

1. $d \leq_{st} d^u$,
2. $\sum_{i=1}^{k} r(\mathcal{R}^u(i)) P^u(i) - \sum_{i=1}^{n} r(\mathcal{R}(i)) P(i)$ is minimal among the set of distributions on k bins that are stochastically upper than d.

The problem can be reduced to a graph problem which consists in finding the path with minimal cost with a fixed number of hops. The vertices of the k-hop path from vertex 1 to vertex n with the smallest cost constitute the bins of the bounding distribution. Such a path can be determined by a dynamic programming approach with complexity of $O(n^2 k)$ [6]. The lower bounding distribution can be similarly defined.

A greedy algorithm which provides sometimes an optimal solution is also given in [2]. This algorithm is based on deleting step by step an entry of the distribution which minimizes the difference of the expected rewards. The complexity is $O(nlogn)$ due to the underlying sorting procedure to find the smallest difference at each step.

3   Dynamic Fault Trees Without Replicated Events

In this section, we first give the functionality of gates with independent inputs and the complexity in terms of the output distribution size and the computation complexity. The monotonicity properties of these gates are established in order to be able to apply some algorithms which compute upper and lower bounding distributions within a smaller support. Afterward, the algorithms to analyze DFTs are presented.

A DFT is not a tree when it contains replicated events. It is almost a tree in the following sense: when we remove the replicated leaves and the $FDEP$ gates, and we only consider the remaining gates, the structure is a tree. We define subtrees based on this structure. When we consider two subtrees of a DFT, if they intersect, one of them is included into the other one. This is the argument used in the following algorithm, the aim of which is to reduce the size of the number of subtrees we have to consider. For an arbitrary DFT T, let us denote by $\mathcal{S}$ a set of subtrees of T.

3.1   Functionality of Gates

We give the outputs of DFT gates with independent inputs. Note that we consider two inputs (when it is possible) but the multiple input cases can be easily

---

Algorithm 2. Deforestation Algorithm.

1: **while** there exist in S two subtrees $T1$ and $T2$ such that $T1 \cap T2 \neq \emptyset$ **do**
2:     **if** $T1 \subset T2$ **then**
3:         delete $T1$ from S
4:     **else if** $T2 \subset T1$ **then**
5:         delete $T2$ from S
6:     **end if**
7: **end while**

---

derived. Let V_i be the discrete random variable having distribution I_i representing the failure times for a leaf or the input of a gate. Similarly, Q will represent the failure time at the output of a gate. We denote by MT the mission time which means that we observe the system during time interval 0 and MT. Thus we compute the transient reliability which is also the availability of the DFT at any instant $t : 0 \leq t \leq MT$. Note that it may happen that some failures occur after MT, but they are not taken into account in the analysis. We also consider an abstract date (EoT, or End of Time) larger than any possible value of the mission time and which is used to define some gates.

Static Fault Tree Gates. These are AND, OR, K out of N gates. The output of an AND gate becomes True, when all its inputs are True, while the output of OR gate becomes true when at least one input becomes True. K out of N gate is a voting gate with multiple inputs: the output becomes True when at least $1 \leq K \leq N$ inputs become True. The random variable Q associated with the output of these gates are (remember that Q, V_1 and V_2 are failure times):

$$Q = \mathrm{AND}(V_1, V_2) = \max(V_1, V_2), \tag{1}$$

$$Q = \mathrm{OR}(V_1, V_2) = \min(V_1, V_2), \tag{2}$$

$$Q = \mathrm{K\,out\,of\,N}(V_1, V_2, \cdots V_N) = V_{[K]}, \tag{3}$$

where $V_{[K]}$ denotes the Kth order statistics of the input: if $(V_1, V_2, \cdots V_N)$ are ordered from the smallest to the greatest one such as $V_{[1]} \leq V_{[2]} \cdots \leq V_{[N]}$, the Kth order statistics is the Kth element of the list [1]. Remark that the OR gate can be specified as 1 *out of N* and the AND gate can be specified as N *out of N*.

Property 2. The distribution of the random variable Q which is the output for the gates $AND, OR, K\,out\,of\,N$ gates can be built by merging the distributions of the inputs (we assume that they are ordered with respect to values of their supports). Then the complexity of these operations is $O(\sum_i n_i)$, where n_i is the size of input i. The size of the output distribution is $O(\sum_i n_i)$.

PAND. The output of the PAND gate becomes True when all of its inputs have failed in a pre-assigned order (from left to right in graphical notation). When the sequence of failures is not respected, the output of the gate is False. Since we consider discrete time models, both inputs may fail at the same time.

Thus we distinguish two gates, PAND and SPAND whose outputs differ when both inputs fail at the same time:

$$Q = \text{PAND}(V_1, V_2) = \mathbb{1}_{(V_1 \leq V_2)} V_2 + \mathbb{1}_{(V_1 > V_2)} EoT, \qquad (4)$$

$$Q = \text{SPAND}(V_1, V_2) = \mathbb{1}_{(V_1 < V_2)} V_2 + \mathbb{1}_{(V_1 \geq V_2)} EoT. \qquad (5)$$

Property 3. The complexity of the output distribution construction for the PAND and SPAND gates is $O(n_1 + n_2)$. The size of the output distribution is $O(n_2)$.

FDEP. The FDEP gate has one main input connected to a component or to the output of another gate and it has several links connected to components. When the main input becomes True, all the components connected by the links must become True, irrespective of their current value. For the FDEP gate given in Fig. 2, when A becomes True, input B of gate1 and input C of gate2 become simultaneously True. $FDEP$ gate will be considered in the next section since an equivalence in terms of OR gates with replicated inputs can be given (Fig. 2).

SEQ. The output of the SEQ gate becomes True when all of its inputs have failed in a pre-assigned order but it is not possible that the failures occur in another order.

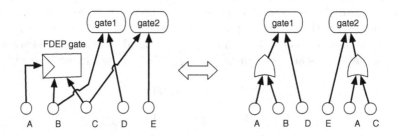

Fig. 2. Equivalence between an FDEP gate and multiple OR gates with replicated leaves (with A label).

Property 4. The output of SEQ gate is computed by the convolution of distributions, the complexity is $O(n \ log(n))$ with $n = n_1 + n_2$. The size of the output distribution is $O(n_1 n_2)$.

SPARE Gates. They are used to represent the replacement of a primary component by a spare with the same functionality. Spare components may fail even if they are dormant (standby). A spare component may be "cold" if it is not subject to failures when it is dormant, "hot" if the dormant has the same failure time distribution as an operating one. A spare is "warm" if it has different failure time distributions while it is dormant and active. Under the considered assumptions, a cold spare (CSPARE) gate is equivalent to a SEQ gate and a hot spare (HSPARE) gate is equivalent to an AND gate.

Let us now consider a warm spare (WSPARE) gate. We model a spare leaf (component) with two failure time distributions associated to two events,

$e1$ and $e2$. The spare is subject to failure (event e_1) according to the first distribution during its whole life (when it is dormant or active). When the spare becomes active, an extra failure event $e2$ can occur according to the second distribution. Let A be the random variable for the failure time of the primary component and B_1 and B_2 be the random variables for the spare component.

- When $B_1 < A$, the spare becomes faulty before the primary component. Therefore the failure time of the WSPARE is A.
- When $A < B_1$, the spare becomes active at time instant A. The spare becomes faulty due to event $e1$ or event $e2$ but event $e2$ can occur after the failure of the primary. Hence the failure time is $min(B_1, A + B_2)$
- When $A = B_1$, both formulas give a failure time equal to A.

After some algebraic manipulations, we get:

$$Q = \text{WSPARE}(A, B_1, B_2) = min(max(A, B_1), A + B_2). \tag{6}$$

Property 5. The time complexity for a warm spare gate is $O(n_a + n_{b1} + (n_a + n_{b2})log(n_a + n_{b2}))$ with an FFT. The size of the output is $O(n_a + n_{b1} + n_a \times n_{b2})$.

Clearly, for most of the gates the failure time distribution at the output has a much larger support than that of its inputs. Therefore, such a numerical analysis cannot be efficient for large DFTs. We now show how we can, thanks to stochastic monotonicity property, derive stochastic bounds of a distribution with a much smaller complexity.

3.2 Monotonicity of Gates

First we give the definition for the monotonicity of a gate g. Intuitively, if such a property holds, when the inputs increase in some sense, so does the outputs.

Definition 2. *Let g be a DFT gate with n independent inputs V_i, and output Q. Gate g is said to be stochastically monotone, if inputs V_is are upper-bounded by independent V_i^us, then the output Q^u with inputs V_i^u provides an upper bound of the output Q with inputs V_i (all comparisons are with $\leq_{st}$ order). More formally,*

$$\forall i, \ V_i \leq_{st} V_i^u \implies Q \leq_{st} Q^u.$$

Lemma 1. *The AND, OR, K out of N, HSPARE, WSPARE, CSPARE and SEQ gates are stochastically monotone.*

Proof. These are the direct results of Property 1 of the $\leq_{st}$ order. From Eqs. 1, 2, 3 and 6 it follows that the output of gates can be written as $Q = f(V_1, V_2, \cdots, V_n)$ where f is a non decreasing function with respect to each mutually independent V_i.

Lemma 2. *The PAND and SPAND gates are not stochastically monotone.*

Proof. It can be seen from Eqs. 4 and 5 that the output function is not non decreasing with respect to V_1 and V_2, thus Property 1 can not be applied. We give the following counter-example. The supports of the input are both equal to $\{1, 2, 3, 4\}$ and $EoT = 4$. For V_1, the probabilities are $[0.1, \ 0.4, \ 0.3, \ 0.2]$ while they are $[0.1, \ 0.1, \ 0.2, \ 0.6]$ for V_2. Thus $Q = \mathrm{PAND}(V_1, V_2) = [0.01, 0.05, 0.16, 0.78]$ on the same support. Consider the upper bound V_2^u $[0.1, \ 0.0, \ 0.3, \ 0.6]$. Clearly, $V_2 \leq_{st} V_2^u$. Then $Q^u = PAND(V_1, V_2^u) = [0.01, \ 0.0, \ 0.24, \ 0.75]$. Neither $Q \leq_{st} Q^u$ nor $Q^u \leq_{st} Q$ does not hold.

3.3   DFT Evaluation and Bounding Algorithm

We present first the exact evaluation of a DFT with G gates with independent component failures and then give the bounding algorithm if the DFT is composed of stochastically monotone gates.

Property 6 (Bottom-Up-Exact-Evaluation). Consider an arbitrary DFT without replication of events neither explicitly nor implicitly with an FDEP gate (see Fig. 2). Since such a the DFT is associated with a tree and we can use a bottom-up algorithm (Algorithm 3 below) to evaluate the exact distribution of the failure time of the system.

Algorithm 3. Bottom-Up-Exact Algorithm for a DFT without replicated events.

 1: Label the gates using the topological order from the bottom to the top, for the internal nodes of the tree.
 2: **for** g from 1 to G **do**
 3:     Evaluate the output distribution of gate g according to the gate type and the methods described in Sections 3.1 and 3.2.
 4: **end for**

Proof. This algorithm is based on the labeling of the gates such that for a given gate g, its inputs are known or already computed when the output of g is computed. Indeed, the inputs of g are either a distribution representing the component failure time, if the input is a leaf or an already computed output of a gate j with $j < g$ due to the labeling of the inner nodes. The key property here is the independence of the leaves.

We do not claim that such an algorithm is original (even if we do not find it in the literature). The main contribution of this paper is on the application of stochastic bounds. Indeed, this algorithm becomes too complex to be used on trees with many gates or distributions with large supports. When the underlying DFT is composed of monotone gates (see Lemma 1) with independent inputs, one can reduce the distribution sizes by using the algorithms given in Sect. 2.2. Thus, we exclude PAND gates because they are not monotone and we derive a bottom up bounding algorithm (Algorithm 4 below), where the maximum size of a distribution is limited to k.

Algorithm 4. Bottom-Up-Bounding Algorithm for a DFT without replicated events and without PAND gates.

1: Label the gates using the topological order from the bottom to the top, for the internal nodes of the tree.
2: **for** g from 1 to G **do**
3:     Evaluate the output distribution of gate g according to the gate type and the methods described in Sections 3.1 and 3.2
4:     If the size of the output distribution is larger than k, use one of the compression methods, given in Section 2.2, to reduce the size to k.
5: **end for**

In the sequel, T_g denotes a tree or a subtree rooted in gate g, and $D(T_g)$ is the exact distribution of the failure time of T_g. Let $D^a(T_g)$ (resp. $D^b(T_g)$) be the failure time distribution for the root of tree T_g after instruction 3 (resp. after instruction 4) of Algorithm 4.

Proposition 1. *Consider an arbitrary DFT T_G, rooted in gate G without replication of events and without PAND gates. Let k be the upper bound on the distribution size. The DFT is still associated with a tree so we use Algorithm 4 to build upper and lower bounds on the failure time distribution of the system. More formally, $D(T_G) \leq_{st} D^b(T_G)$.*

Proof. We proceed by induction on g.

For $g = 1$, $D(T_1) = D^a(T_1)$ because the inputs of the first gate are the leaves and the distribution of the leaves are supposed to be exact (in the algorithm, we modify the outputs when they are too large but we suppose that the size of the inputs are smaller than k). And by construction, we have: $D^a(T_1) \leq_{st} D^b(T_1)$.

For an arbitrary $g > 1$, $D^a(T_g)$ is computed with input distributions which may be an exact distribution for a leaf or an upper bound $D^b(T_j)$ with $j < g$. By induction we have $D(T_j) \leq_{st} D^b(T_j)$.

According to Lemma 1, all the gates g are stochastically monotone. Therefore: $D(T_g) \leq_{st} D^a(T_g)$. By construction, we still have $D^a(T_g) \leq_{st} D^b(T_g)$. Finally we get:

$$D(T_g) \leq_{st} D^b(T_g),$$

and the proof is complete.

Example. We consider a small example to report some numerical results. We analyze a DFT with two leaves connected to a CSPARE gate. We generate two couples of random discrete distributions as inputs. For the first couple the distribution size is 10 while it is 1000 for the second one. We compute the output distributions of the CSPARE gate for each set of inputs. We use the distributions with 10 bins to draw readable curves (Fig. 3) while the distributions of size 1000 are used to measure the execution times. The failure time distribution of the CSPARE gate has 10^6 bins with inputs of 1000 bins, and the results are obtained in half a second on a usual laptop. The results with input distributions of 10 bins is obtained instantaneously. For this case, the upper bound with 50 bins instead of 100 bins illustrates the accuracy of the method.

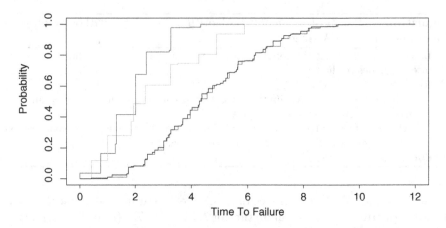

Fig. 3. The input distributions are blue and green; the output distribution of the cold spare gate is black (100 bins), and a stochastic upper bound with 50 bins is red (Color figure online).

4   Replicated Events and Conditional Probabilities

4.1   Evaluation of PAND Gates

We now explain how we accommodate PAND gates.

Proposition 2. *Consider an arbitrary DFT with PAND gates but without replication of events. Let k be the upper bound of the distribution size. Algorithm 5 computes a bound on the distribution after decomposition of the DFT into the subtrees rooted in the PAND gates and a remaining subtree.*

Proof. It is based on the following fact: we cannot use stochastic bounds inside a subtree rooted in a PAND gate because the PAND gate is not stochastically monotone (Lemma 2). Therefore we compute the exact solution for these subtrees. As the DFT is a tree (the replicated events are not allowed here), there is only one way for the intersection of subtrees: the inclusion. If a subtree is included by another one, it is removed from the set of subtrees for which exact distribution will be computed to avoid several computations of its output distribution.

We now consider DFT with replication of events. Note that the explicit replication only concerns leaves. But the components connected to an FDEP gate can be a leaf of a subtree. For the sake of readability, we assume that the FDEP gates are only connected to input leaves. If an FDEP gate is connected to subtrees, we use the same decomposition approach: we evaluate the distribution of the subtrees in a first step, then we replace the subtrees by the distribution with an equivalent leaf, and we can continue the analysis with a DFT which satisfies the assumptions. Thus, we assume without loss of generality that the FDEP gates

Algorithm 5. Compute a bounding distribution for a DFT without replicated events, with PAND gates.

1: Build all the subtrees rooted in the PAND gates.
2: Apply Algorithm 2 to this set of subtrees.
3: Evaluate the output distributions for each subtree of this set using Algorithm 3
4: If the size of an output distribution is larger than k, use one of the compression methods to reduce its size to k.
5: Evaluate the bounds on the failure time distribution of the system using Algorithm 4 on the remaining tree.

have been transformed to make appear explicitly the replicated leaves. We now have to deal with this replication for the computation of the output distributions for the gate and for the computation of the stochastic bounds on the outputs of the gates. However we do not consider DFTs with replicated leaves modeling spare components because their analysis require an ad-hoc algorithm. The technique also relies on conditioning but it needs more space to be presented.

The main assumption we need for the computation of the output is the independence of the inputs. This is not the case anymore as all the replicated leaves are obviously dependent. We have two strategies to deal with these replicated leaves: factorization and conditioning on the replicated events (see [13] for an application of conditioning for Static Fault Trees). The tree is associated with a formula whose terms are the leaves and whose operators are the internal gates. A replicated leaf appeared in two separate parts of the formula. In some limited cases, it is possible to factorize the formula to remove the replication. From this new formula, one can build a new DFT with less replications. But as the process is not general we do not develop further this approach here and we put more emphasis on the conditioning on the replicated leaves.

4.2 DFT with Replication but Without PAND Gates

We first build the set $\mathcal{S}$ of subtrees associated to replicated leaves as follows: for each leaf i, we consider the set of leaves $\mathcal{L}_i$ which are replications of leaf i. By construction, $i \in \mathcal{L}_i$. We build T_i as the subtree of the DFT, which contains $\mathcal{L}_i$, which contains all the leaves connected to the gates of T_i, which is a tree when we do not consider the replicated leaves in $\mathcal{L}_i$, and which is minimal for the inclusion.

We first assume that there is no PAND gate and that the subtrees of $\mathcal{S}$ do not intersect. Thus, it is not needed to apply deforestation procedure (Algorithm 2). This allows us to present a simpler algorithm. The general case will be presented in the next section. The main idea is to first compute the distribution of the failure time for all subtrees after conditioning on the replicated leaves. We obtain the distribution of the failure time of the subtree by means of the Total Probability law. As the trees do not intersect, we replace each of them by a new leaf with the computed discrete distribution. Finally, we evaluate the remaining tree as usual. Remember that I_i is the failure time distribution for leaf i with support $\mathcal{R}_i$ and probability P_i. Let x be a time instant in $\mathcal{R}_i$, $T_i^{(x)}$ will denote the subtree when leaf i and all the leaves replicated are equal to x. The algorithm is divided into two parts such that the internal part can be used elsewhere.

Algorithm 6. Compute a bounding distribution for a subtree with replicated events and without PAND gates.

1: **for** all time instant x in $\mathcal{R}_i$ **do**
2:    Make the failure time of all the leaves in $\mathcal{L}_i$ equal to x with probability 1.
3:    Evaluate the distribution of failure time of the subtree $T_i^{(x)}$ with Algorithm 4
4: **end for**
5: Using the Law of Total Probability, evaluate the distribution for the failure time of the output for T_i

$$D^u(T_i) = \sum_{x \in \mathcal{R}_i} D^u(T_i^{(x)}) P_i(x).$$

Algorithm 7. Compute a bounding distribution for a DFT with replicated events and without PAND gates.

1: **for** all replicated variable i, **do**
2:    Build set $\mathcal{L}_i$, and subtree T_i associated with $\mathcal{L}_i$.
3:    Compute $D^u(T_i)$ with Algorithm 6
4:    Replace subtree T_i by a leaf with probability distribution $D^u(T_i)$.
5: **end for**
6: Evaluate the failure time distribution of the system using Algorithm 4 on the resulting tree.

Proposition 3. *Consider an arbitrary DFT with replicated events, then Algorithm 7 computes upper and lower stochastic bounds on the failure time distribution of the DFT. Furthermore, the number of operations for Algorithm 6 is in $O(n_i\, U_i)$ where n_i is the size of $\mathcal{R}_i$ and U_i is the number of operations to obtain the failure distribution for tree T_i.*

Proof. We make the proof for the computation of upper bounds. The proof for lower bound is similar and it is omitted. We assume that there are no PAND gates in the DFT. Thus, it is possible to use Algorithm 4 to obtain bounds for the failure time in a subtree. In Proposition 1, we state that for all x: $D(T_i^{(x)}) \leq_{st} D^u(T_i^{(x)})$. Indeed we have used Algorithm 4 with all the leaves in $\mathcal{L}_i$ being associated with the Dirac distribution δ_x. And the proof of the algorithm does not take into account the distribution of the failure time for the leaves. Finally, Theorem 1 shows the compatibility of conditional probabilities with stochastic bounds. Therefore: $D(T_i) \leq_{st} D^u(T_i)$.

Proposition 4. *We can compute the exact distribution as well, by using Algorithm 3 rather than Algorithm 4 in instruction 3 of Algorithm 6 (see below Algorithm 8) and then the same thing in instruction 2 of Algorithm 7.*

Algorithm 8. Compute the exact distribution for a subtree with replicated events, and without PAND gates.

1: **for** all time instant x in $\mathcal{R}_i$ **do**
2:     Make the failure time of all the leaves in $\mathcal{L}_i$ equal to x with probability 1.
3:     Evaluate the distribution of failure time of the subtree $T_i^{(x)}$ with Algorithm 3.
4: **end for**
5: Using the Law of Total Probability, evaluate the distribution for the failure time of the output for T_i

$$D(T_i) = \sum_{x \in \mathcal{R}_i} D(T_i^{(x)}) P_i(x).$$

4.3 DFT with Replicated Events and PAND Gates

We now consider the case of a DFT with some PAND gates and with some replicated leaves. We begin with building a new set of subtrees which contains all the subtrees rooted in the PAND gates and all the subtrees T_i constraining subset $\mathcal{L}_i$ of leaves to be minimal for inclusion (as in the beginning of this section). As we consider two types of subtrees, it is now possible that they intersect. Thus, we use deforestation algorithm (Algorithm 2) to obtain a minimal set of subtrees in $\mathcal{S}$. We now label these subtrees. If a subtree contains a PAND gate, it receives label "p". If a subtree contains replicated leaves, it is labeled with a "d". If it contains several replicated leaves, we only use one "d" label. Note that a tree may receive both labels "d" and "p" because it results of a merge operation during the deforestation (Algorithm 2). We compute the exact distribution or a bound depending on the labels of the subtree using the algorithms presented previously. Finally, as usual, we remove all the subtrees in $\mathcal{S}$ and replace each subtree by a leaf with the computed distribution.

The main problem arises when, in Algorithm 2, we merge two subtrees associated with replicated events. Suppose that tree T_i and T_j intersect. Suppose that T_i is built on replicated leaves in $\mathcal{L}_j$ and T_i in $\mathcal{L}_j$. Assume without loss of generality that $T_j \subset T_i$. To analyze T_i, we have to condition the distributions on both the values of Leaf i and Leaf j. Therefore, in Algorithms 7 and 8, we have to consider the product of the distributions and the support is now $\mathcal{R}_i \times \mathcal{R}_j$, where $\times$ is the Cartesian product. Clearly, the intersection of subtrees does not change the used probabilistic technique, but increases the time complexity of the analysis.

Proposition 5. *Consider an arbitrary DFT with replicated events and some PAND gates, then Algorithm 9 computes upper and lower stochastic bounds of the distribution of the failure time of the DFT.*

Proof. As the proof is based on the same arguments than the previous algorithms, it is omitted. The time complexity of this last algorithm could be much larger if the number of subtrees with d or dp labels is large or if the conditioning is made on a support with a large size (for instance a product space, due to the intersection of d trees).

Algorithm 9. Compute a bounding distribution for a complex DFT

1: Build set S of subtrees
2: Apply Algorithm 2 to reduce the number of subtrees
3: Label the remaining subtrees
4: **for** all subtrees T_i **do**
5:     **if** T_i is only labelled with "p" **then**
6:         compute $D(T_i)$ with Algorithm 3
7:     **else if** T_i is only labelled with "d" **then**
8:         compute $D^u(T_i)$ with Algorithm 6
9:     **else if** T_i is labelled with "dp" **then**
10:         compute $D(T_i)$ with Algorithm 8
11:     **end if**
12:     Replace subtree T_i by a leaf with probability distribution $D^u(T_i)$ or $D(T_i)$
13: **end for**
14: Evaluate the failure time distribution for the system using Algorithm 4 on the resulting tree.

5   Open Questions and Conclusions

Using an approach similar to the one presented in Sect. 4, we can analyze a DFT with correlated inputs. We also investigate how we can add in the model a process to represent reparation or maintenance. Such an approach will greatly enlarge the modeling power of DFTs and it will be developed in the future. We also plan to complete the development of a software tool like Galileo [12] or the tool presented in [5]. A preliminary version is already available but we need to add a GUI. It will show the trade-off between accuracy and complexity with respect to the sizes of distributions, and it will let to deal with correlated failure time distributions of the leaves. Finally note that the methodology we have presented here is not limited to DFTs. It can be improved to any formalism with monotone operators on distributions applied for stochastic model checking.

Acknowledgements. The authors are supported by grant ANR-12-MONU-00019.

References

1. Arnold, B., Balakrishnan, N., Nagaraja, H.: A First Course in Order Statistics. Society for Industrial and Applied Mathematics, Philadelphia (2008)
2. Salaht, F.A., Cohen, J., Taleb, H.C., Fourneau, J. M., Pekergin, N.: Accuracy vs. complexity: the stochastic bound approach. In: 11th International Workshop on Discrete Event Systems (WODES 2012) (2012)
3. Bechta Dugan, J., Bavuso, S.J., Boyd, M.A.: Dynamic fault-tree models for fault-tolerant computer systems. IEEE Trans. Reliab. **41**(3), 363–377 (1992). Sep
4. Boudali, H., Crouzen, P., Stoelinga, M.: Dynamic fault tree analysis using input/output interactive Markov chains. In: The 37th IEEE/IFIP International Conference on Dependable Systems and Networks, pp. 708–717 (2007)

5. Bucci, G., Carnevali, L., Vicario, E.: A tool supporting evaluation of non-markovian fault trees. In: Fifth International Conference on the Quantitative Evaluation of Systems, pp. 115–116 (2008)
6. Guérin, R., Orda, A.: Computing shortest paths for any number of hops. IEEE/ACM Trans. Networking **10**(5), 613–620 (2002)
7. Malhotra, M., Trivedi, K.S.: Dependability modeling using Petri-nets. IEEE Trans. Reliab. **44**(3), 428–440 (1995). Sep
8. Manian, R., Dugan, B.J., Coppit, D., Sullivan, K.J.: Combining various solution techniques for dynamic fault tree analysis of computer systems. In: High-Assurance Systems Eng. Symposium, pp. 21–28, November 1998
9. Merle, G., Roussel, J.-M., Lesage, J.-J., Bobbio, A.: Probabilistic algebraic analysis of fault trees with priority dynamic gates and repeated events. IEEE Trans. Reliab. **59**(1), 250–261 (2010)
10. Muller, A., Stoyan, D.: Comparison Methods for Stochastic Models and Risks. Wiley, New York (2002)
11. Rao, K.D., Gopika, V., Sanyasi Rao, V.V.S., Kushwaha, H.S., Verma, A.K., Srividya, A.: Dynamic fault tree analysis using monte carlo simulation in probabilistic safety assessment. Reliab. Eng. Syst. Safety **94**(4), 872–883 (2009)
12. Sullivan, K.J., Dugan, B.J., Coppit, D.: The galileo fault tree analysis tool. In: Digest of Papers: FTCS-29, The Twenty-Ninth International Symposium on Fault-Tolerant Computing, USA, pp. 232–235 (1999)
13. Trivedi, K.S.: Probability and Statistic with Reliability, Queueing and Computer Science Applications, 2nd edn. Wiley, New York (2002)
14. Vesely, W.E., Goldberg, F.F., Roberts, N.H., Hassl, D.F.: Fault Tree Handbook, NUREG-0492, T.R., U.S. nuclear regulatory commission (1981)

Applications

Quantitative Placement of Services in Hierarchical Clouds

Asser N. Tantawi[✉]

IBM T.J. Watson Research Center, Yorktown Heights, NY, USA
tantawi@us.ibm.com

Abstract. In this paper, we consider a hierarchical cloud topology and address the problem of optimally placing a group of logical entities according to some policy constraining the allocation of the members of the group at the various levels of the hierarchy. We introduce a simple group hierarchical placement policy, parametrized by lower and upper bounds, that is generic enough to include several existing policies such as collocation and anti-collocation, among others, as special cases. We present an efficient placement algorithm for this group hierarchical placement policy and demonstrate a six-fold speed improvement over existing algorithms. In some cases, there exists a degree of freedom which we exploit to quantitatively obtain a placement solution, given the amount of group spreading preferred by the user. We demonstrate the quality and scalability of the algorithm using numerical examples.

1 Introduction

Cloud services have progressed in recent years from provisioning single Virtual Machines (VM) in the physical cloud infrastructure to entire platforms and applications, which have become the new cloud workload. The cloud user specifies a workload consisting of logical entities, such as VMs, data volumes, communication links, and services, and their needs of the underlying physical resources. Moreover, the user specifies requirements on the provisioned topology of such logical entities. Examples of such requirements include physical proximity of the logical entities, availability/reliability concerns, preferred hosting requirements, licensing and cost issues, and migration requirements. On the other hand, the cloud service provider attempts to maximize the use of the physical resources in a way that provides best performance to users, e.g. load balanced resources. When a user request arrives to the cloud management system, the placement engine decides on a mapping of the logical entities in the request to the physical entities in the cloud system, given its current state, in a way to optimize a given objective function which combines user and provider objectives [2]. This placement optimization problem is challenging due to its potentially large size.

There is a growing research interest in solving the cloud placement problem [5]. Posed as a combinatorial optimization problem, we find a variety of solution techniques, ranging from heuristic-based, simulation, to evolutionary algorithms. A technique for attempting to decrease the size of the problem has

© Springer International Publishing Switzerland 2015
J. Campos and B.R. Haverkort (Eds.): QEST 2015, LNCS 9259, pp. 195–210, 2015.
DOI: 10.1007/978-3-319-22264-6_13

been proposed in [4]. A biased sampling algorithm is introduced in [8]. Others include: a network-aware placement and migration algorithm [7], a topology-aware mapping algorithm [9], and a subgraph matching algorithm [10]. Special hierarchical structures of clouds were exploited to devise heuristic placement algorithms [1,3]. Further, hierarchical placement algorithms have been proposed [6].

Cloud computing environments have inherently a hierarchical structure topology. Such a topology results not only from the networking architecture, but also due to availability, security, management, and physical considerations. Cloud providers make such a hierarchical topology, or at least some abstract logical representation of it, available to cloud users in order to take advantage of specifying needs of applications and services. Such needs and requirements may relate to the placement of logical components when deploying and/or migrating a given cloud workload, so as to satisfy some constraints related to non-computational considerations such as availability, legal requirements, and security. In this paper, we consider a hierarchical cloud topology and address the problem of optimally placing a group of logical entities according to some policy constraining the allocation of the members of the group at the various levels of the hierarchy. We introduce a simple group hierarchical placement policy, parametrized by lower and upper bounds, that is generic enough to include several existing policies such as collocation and anti-collocation, among others, as special cases. We present an efficient placement algorithm for this group hierarchical placement policy. In the case where the lower and upper bounds are not equal, there exists a degree of freedom which we exploit to quantitatively obtain a placement solution, given the amount of group spreading preferred by the user.

The paper is organized as follows. The general placement problem and a solution approach based on biased sampling are described in Sect. 2. The inclusion of hierarchical constraints for placing a group is considered in Sect. 3. We present our placement algorithm, first in Sect. 4, where we consider the case of upper bounds on the paced logical entities at a given level in the hierarchy. Then, we address both upper and lower bounds in Sect. 5. The flexibility in spreading the placement is considered in Sect. 6. In Sect. 7, we present numerical results demonstrating the efficiency of our algorithm. Section 8 concludes the paper.

2   The General Placement Problem

2.1   Problem Description

Consider a cloud infrastructure which consists of a set $\mathcal{N}$ of N physical entities (PE) that is subjected to a stream of requests, where each request represents an application (also referred to as pattern, workload, and/or service) to be deployed in the cloud. A PE may represent a Physical Machine (PM) in a virtualized environment, a data storage device, or a bare metal machine. The request specifies the set $\mathcal{M}$ of M logical entities (LE) of the application and includes constraints related to the deployment of the applications. An LE may represent a Virtual

Machine (VM), a data volume, or a Container in an OS-level virtualized environment. Basically, an LE is placed on a PE, given some requirements related to the types of LE and PE, the resources, matching LE demands and PE availability, as well as other constraints related to networking and location among LEs. In particular, some constraints relate to a group of homogeneous LEs, such as collocation (or anti-collocation) of member LEs of the group at some level in the physical topology of the cloud. (Also known as affinity and anti-affinity constraints, respectively.) The result of each request is a placement, i.e. a mapping of LEs in the request to PEs in the cloud. Typically, there are objectives for such a placement dictated by policies addressing both the cloud provider and the cloud user. Hence, with each request, we are faced with an assignment optimization problem. Let X be a variable mapping logical to physical entities and $\mathcal{S}$ the set of possible solutions given the requirement constraints specified in the user request. Let the objective function, $F(X)$, be a scalar function with range $\mathbb{R}$, the set of real numbers. The unconstrained state space for variable X is the cartesian power $\mathcal{N}^M$. Then, the constrained, nonempty state space, is $\mathcal{S} \subseteq \mathcal{N}^M$. The optimization problem is stated as,

$$\min_{X} F(X), \quad X \in \mathcal{S}. \tag{1}$$

The objective function is a weighted sum of system (provider) objective and user objective, as a result of placement X, $F(X) = (w_{sys} \, F_{sys}(X) + w_{usr} \, F_{usr}(X))/ (w_{sys} + w_{usr})$. For example, the system objective may be the standard deviation of the utilization of the prime resource across the cloud system, i.e. the objective is to balance the load across the cloud. And, the pattern objective captures the deviation from the desired location constraints specified in the pattern request.

2.2   Solution Approach: Biased Sampling

Basically, there are two broad classes of algorithms for solving problem 1: deterministic and stochastic. Solution approaches which proceed either numerically or using heuristics are deterministic in their search for a solution in the solution space, e.g. [4,9,10]. Alternatively, other approaches search the space in a probabilistic fashion. Examples of the latter are: stochastic approximation, annealing, evolutionary computation, and statistical sampling. A particular statistical search algorithm which employs biasing and importance sampling is BSA (Biased Sampling Algorithm) [2,8]. There, an $M \times N$ (row) stochastic matrix P, i.e. element $p_{m,n} \in [0,1]$, $\|P_m\| = 1$, $m \in \mathcal{M}$, and $n \in \mathcal{N}$, is used as a generator of assignment solutions to the optimization problem. The elements of P are calculated using biasing functions which are based on conditional probabilities. More specifically, given an initial P, the biased generator, denoted by P', is calculated as follows. Let $\mathbb{B} = \{B(r); \ r \in \mathcal{R}\}$ be a family of R stochastic matrices, each of size $M \times N$, where $\mathcal{R} = \{1, 2, \cdots, R\}$, $R \geq 1$. For R biasing criteria, $B(r)$ represents the basing matrix for criterion r, $r \in \mathcal{R}$. Each criterion represents a type of requirement constraint in the user request, e.g. communication, location, target preference, license usage, and cost constraints. Define a weight vector Ω

of length R, where element $\omega_r \geq 0$ is a weight associated with $\boldsymbol{B}(r)$, $r \in \mathcal{R}$. Let $\boldsymbol{B}$ denote the weighted product of $\boldsymbol{B}(r)$, given by $\boldsymbol{B} = \underset{r \in \mathcal{R}}{\circ} \boldsymbol{B}(r)^{\omega_r}$, where the symbol $\circ$ represents the Hadamart element wise product of matrices, and the exponent ω_r applies to all elements of matrix $\boldsymbol{B}(r)$. Then, we write

$$\boldsymbol{P}' = \mathbf{diag}(\boldsymbol{C}) \, (\boldsymbol{P} \circ \boldsymbol{B}), \tag{2}$$

where $\boldsymbol{C}$ is a normalization constant vector of length M to make $\boldsymbol{P}'$ stochastic.

In this paper, we are mainly concerned with one, $R = 1$, biasing criterion, namely biasing the solution to achieve hierarchical constraints. We present an algorithm for computing such biasing values.

3   Constrained Hierarchical Group Placement

3.1   Physical Tree

Consider a cloud infrastructure consisting of a collection of N PEs. We assume that there exists a hierarchical topology overlaying the PEs, as illustrated in Fig. 1. We refer to such a topology as the physical tree topology. The N PEs are the leaves of the tree, indexed from 0 to $N - 1$. The height of the tree is $L \geq 1$, hence we have $L + 1$ levels, where level 0 represents the leaves and level L is the root of the tree. Examples of levels 0 through L are PMs, Blade Centers (BC), Racks, Zones, and so on, up to the root of the tree, representing the entire cloud. One may also imagine a multi-cloud environment where level $L - 1$ consists of the various clouds, each with its own subtree.

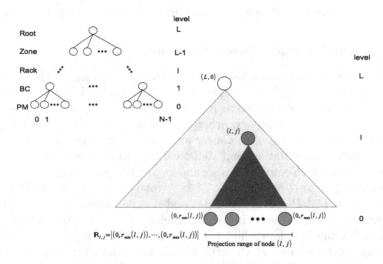

Fig. 1. Hierarchical topology and physical tree

We introduce some notation regarding nodes in the physical tree topology, as depicted in Fig. 1. We define a coordinate system for nodes by labeling the jth

node, $j \geq 0$, ordered left to right, at level l, where $l = 0, \cdots, L$, as node (l, j). Hence, the root node is $(L, 0)$ and the leaves are nodes $(0, 0)$ through $(0, N - 1)$. For convenience, we define the projection range of node (l, j) as the set $\mathcal{R}_{l,j}$ of leaf nodes in the subtree rooted at node (l, j). We denote the indexes of the first (leftmost) and last (rightmost) nodes in the set $\mathcal{R}_{l,j}$ by $r_{min}(l, j)$ and $r_{max}(l, j)$, respectively.

3.2   Logical Tree

Consider a group of M homogeneous LEs, where $M > 1$. This group of homogeneous LEs may represent a scaling group in an application. Here we are concerned with the location constraints among the M elements of the group with respect to their placement onto the physical tree topology.

Let $k_{l,j}$ be the number of LEs placed on PEs in $\mathcal{R}_{l,j}$. We define the logical tree, induced by a particular placement (mapping) of LEs onto PEs, as the subtree of the physical tree with nodes having $k_{l,j} > 0$, as depicted in Fig. 2. We use the same coordinate system of the physical tree to label nodes in the logical tree, where the first coordinate representing the level is the same, and the second coordinate representing the index of the node within a level is given by a mapping function, $\phi(.)$, relating node (l, i) in the logical tree to node (l, j) in the physical tree, as $\phi(l, i) = (l, j)$, where $l = 0, \cdots, L$. In other words, the logical tree is a subset of the physical tree with (1) the same height and (2) having leaves that map to leaf nodes (PEs) in the physical tree where the LEs are placed.

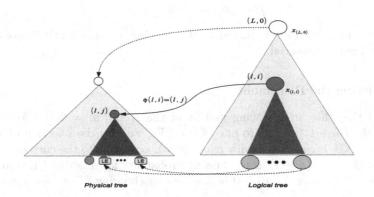

Fig. 2. Mapping of logical to physical tree

Define the state of node (l, i) in the logical tree by $x_{l,i} = k_{l,i}$, $l = 0, \cdots, L$. Hence, $x_{0,i}$ is the number of LEs from the group that are placed on the PE with index j, where $\phi(l, i) = (l, j)$. And, the total number of LEs from the group that are placed in the ith entity at level l is given by $x_{l,i}$. Let $x_{l,i}$ be allowed to be in the range $[\underline{K}_l, \overline{K}_l]$, $1 \leq \underline{K}_l \leq \overline{K}_l \leq M$, i.e. the number of LEs placed under *lnode* (l, i) is constrained between a lower and upper bounds. For brevity, we refer to the physical (logical) tree by *ptree* (*ltree*), respectively. Further, we refer to a node in the physical (logical) tree by *pnode* (*lnode*), respectively.

Let's consider the steps of placing the M LEs in the group. In particular, let one LE be placed at a time. Consider the state of the *ltree* after m LEs are placed on *pnodes* in the set $\mathcal{N}_m$, where $m = 0, \cdots, M$. Then, $\forall$ *pnode* $(0, n) \in \mathcal{N}_m$, $\exists$ *lnode* $(l, i) \in$ *ltree*, s.t. $\phi(l, i) = (l, j)$ and *pnode* (l, j) is an l-ancestor of $(0, n)$, $l = 0, \cdots, L$.

We define the weight of *lnode* (l, i) after m LEs are placed, denoted by $w_{l,i}(m)$, as a potential for placing additional LEs in the subtree rooted at that *lnode*. For convenience, the weight is defined in the range $w_{l,i}(m) \in [0, 1]$. We use a simple linear function given by

$$w_{l,i}(m) = \begin{cases} 1, & x_{l,i}(m) < \underline{K}_l, \\ \frac{\overline{K}_l - x_{l,i}(m)}{\overline{K}_l - \underline{K}_l + 1}, & \underline{K}_l \le x_{l,i}(m) < \overline{K}_l, \\ 0, & x_{l,i}(m) \ge \overline{K}_l, \end{cases} \tag{3}$$

where $m = 0, \cdots, M$, and $l = 0, \cdots, L$. Thus, the higher (lower) the weight of an *lnode*, the more (less) room it has for the placement of the remaining $(M - m)$ LEs. In particular, $w_{l,i}(m) = 0$ means that no more LEs could be placed in the subtree rooted at *lnode* (l, i). Note that if $\overline{K}_l = 1$, we have the intuitive weights, $w_{l,i}(m) = 1 - x_{l,i}(m)/\overline{K}_l$, $x_{l,i}(m) < \overline{K}_l$.

Define a non-negative, non-decreasing function $f()$ of the weight of an *lnode*, which we employ in our placement algorithm. For instance, consider the power function

$$f(w) = \eta \, w^\gamma, \quad \eta, \gamma > 0 \ . \tag{4}$$

The value of η is used for scaling and γ for shaping, i.e. making the impact of w more (or less) pronounced.

3.3   Biasing the Placement

Consider the state after placing m LEs of the group, $m = 0, \cdots, M - 1$. In placing the $(m + 1)^{st}$ LE onto one of the PEs, we bias the selection of PE n, $n = 0, \cdots, N - 1$, associated with *pnode* $(0, n)$, according to the current state of the *ltree*. Define $p_n(m + 1)$ as the biasing probability for selecting PE n to place the $(m + 1)^{st}$ LE. Let $\beta_n(m) \ge 0$ be the bias for choosing PE n. Hence, we have

$$p_n(m + 1) = \frac{\beta_n(m)}{\sum_{j=0}^{N-1} \beta_j(m)}. \tag{5}$$

The value of $\beta_n(m)$, should be calculated in a way to increase the chance of achieving the level constraints. We calculate $\beta_n(m)$ as follows. Let *lnode* $(0, s_0)$ be the *lnode* mapped to *pnode* $(0, n)$, i.e. $\phi(0, s_0) = (0, n)$. If *lnode* $(0, s_0) \notin$ *ltree*, then $\beta_n(m) = 1$, as a *neutral*, unused node. Otherwise, we calculate the bias as a product form as follows. Let *lnode* $(0, s_l)$ be the l-ancestor of $(0, s_0)$ in the *ltree*,

$l = 0, \cdots, L$. If $\exists l$ s.t. $x_{l,s_l}(m) = \overline{K}_l$, then $\beta_n(m) = 0$, since the node is already at its maximum allowed, otherwise we use

$$\beta_n(m) = \prod_{l=0}^{L} f(w_{l,s_l}(m)). \tag{6}$$

For simplicity, we first consider, in Sect. 4, the case of only upper bounds, i.e. $\underline{K}_l = 1$, $l = 0, \cdots, L$. Then, we consider the more general case in Sect. 5.

4   Constrained Placement : Upper Bounds

For brevity, we denote the upper bound $\overline{K}_l$ by K_l, $l = 0, \cdots, L$. Let $k_{l,i} \geq 1$ be the cumulative number of LEs placed on PEs in the projection range of node $\phi(l, i)$. Thus, we have the constraints

$$k_{l,i} \leq K_l, \quad \text{s.t.} \quad 1 \leq K_0 \leq K_1 \cdots \leq K_L = M, \tag{7}$$

where $l = 0, \cdots, L$. In other words, among the M LEs in the group, the number of LEs that are placed at a given level l in the cloud hierarchical topology is upper-bounded by K_l. The above constraints are satisfied iff $x_{l,i}(M) \leq K_l$, $\forall lnodes$ $(l, i) \in ltree$.

Our placement algorithm for this case is outlined in Algorithm 1. The main procedure is PLACEGROUP(), which takes as input the physical tree, *ptree*, the size of the group, M, the array of $L + 1$ upper bounds, $\mathbf{K}$, and the array of N initial probabilities, $\mathbf{p}_{init}$, for selecting a particular PE for placement. The latter may be biased by resource availability or other considerations [8]. Array $\mathbf{p}$ acts as a probability distribution for the selection of a PE as a host for an LE. Basically, we have a loop over the M LEs, where one LE is placed at a time. For a given $m \geq 0$ already placed LEs, the body of the loop determines the placement of the $(m+1)^{st}$ LE and calculates the placement probabilities for the next LE, given the current placement. Once $\mathbf{p}$ is sampled and a particular PE is chosen as a host, the logical tree is updated to reflect the placement and the bias values are computed as a consequence. Updating the logical tree is the function of procedure UPDATELOGICALTREE(), where *lnodes* may be added, the mapping to the *tree* is performed, and the states of the *lnodes* are maintained, as discussed in Sect. 3.2. Bias values are calculated in procedure CALCULATEBIAS(). By default, all *pnodes* are initially given a neutral bias of one. Then, such values are adjusted in procedure ADJUSTBIAS(), visiting *lnodes* top down from the root of the *tree*, according to Eqs. 3, 4, and 6. Condition ISNODEBLOCKED(*lnode*) is *true iff* $(x_{lnode} \geq K_l)$, where l is the level of *lnode*.

The complexity of procedure UPDATELOGICALTREE() is $O(L)$ since the nodes in the *ltree* are visited from a leaf node to the root. ($O(L)$ is typically $O(ln(N))$.) The complexity of procedure CALCULATEBIAS() is $O(N)$ since, in the worst case, the bias values of all N leaf nodes in the *ptree* are set, through the recursive calling of procedure ADJUSTBIAS(). Thus, analyzing the main procedure PLACE-GROUP(), we find that the loop is executed M times, where the complexity of the body of the loop is $O(N)$, hence has an overall complexity $O(MN)$.

Algorithm 1. Placement with upper bounds

1: **procedure** PLACEGROUP($ptree, M, \boldsymbol{K}, \boldsymbol{p}_{init}$)
2:      $ltree \leftarrow$ CREATEEMPTYLOGICALTREE($\boldsymbol{K}$)
3:      $\boldsymbol{p} \leftarrow \boldsymbol{p}_{init}$
4:      **for** $m = 0, 1, \cdots, M - 1$ **do**
5:          $n \leftarrow$ GETSAMPLEFROMDISTRIBUTION($\boldsymbol{p}$)
6:          Place LE m on PE n
7:          UPDATELOGICALTREE($ltree, ptree, (0, n)$)
8:          $\boldsymbol{\beta} \leftarrow$ CALCULATEBIAS($ltree,\ root$)
9:          $\boldsymbol{p} \leftarrow$ NORMALIZE($\boldsymbol{p}_{init} \circ \boldsymbol{\beta}$)
10:
11: **procedure** UPDATELOGICALTREE($ltree, ptree, pnodePlaced$)
12:      $pnode \leftarrow pnodePlaced$
13:      **while** pnode **do**
14:          $lnode \leftarrow \phi^{-1}(pnode)$
15:          **if** $lnode = null$ **then**
16:              $l \leftarrow$ LEVEL($pnode$)
17:              $lnode \leftarrow$ CREATENEWNODE($ltree, l$)
18:              $\phi(lnode) \leftarrow pnode$
19:          $x_{lnode} + +$
20:          $pnode \leftarrow$ PARENT($pnode$)
21:
22: **procedure** CALCULATEBIAS($ltree$)
23:      $\boldsymbol{\beta} \leftarrow \mathbf{1}_N$
24:      $lnode \leftarrow$ GETROOT(ltree)
25:      ADJUSTBIAS($ltree, lnode, \boldsymbol{\beta}$)
26:      **return** $\boldsymbol{\beta}$
27:
28: **procedure** ADJUSTBIAS($ltree, lnode, \boldsymbol{\beta}$)
29:      $(r_{min}, r_{max}) \leftarrow$ GETPROJECTIONRANGE($ltree, lnode$)
30:      **if** ISNODEBLOCKED($lnode$) **then**
31:          $\boldsymbol{\beta}_{[r_{min}:r_{max}]} \leftarrow 0$
32:          **return**
33:      $\boldsymbol{\beta}_{[r_{min}:r_{max}]}* = f(w_{lnode})$
34:      **for all** $cnode \in$ CHILDREN($lnode$) **do**
35:          ADJUSTBIAS($ltree, cnode, \boldsymbol{\beta}$)

4.1 Example: Placement with Upper Bounds

Consider a simple configuration of $N = 9$ homogeneous PEs, $pm0$ through $pm8$, arranged in a balanced tree topology of height $L = 2$. (Larger examples are the subject of Sect. 7.) At level 1, we have $zone0$ through $zone2$, with $zone0 = \{pm0, pm1, pm2\}$ and so on. Let there be a single resource per PE with capacity 8. Consider the problem of placing a group of $M = 12$ LEs, each with a resource demand of 1. Let the upper bounds of the hierarchical policy be $\boldsymbol{K} = [2, 4, 12]$. We use Algorithm 1 with parameters: $\eta = 1$, $\gamma = 2$, and $\boldsymbol{p}_{init}$ proportional to resource availability. The objective function in the placement optimization (minimization) problem is the standard deviation of the resource utilization, hence the goal is to balance the load. First, given an empty system, we obtain the placement result illustrated in Example 1. The circles depict the zones. Note that the upper bounds are uniformly satisfied. Further, note that the LEs are spread out for two reasons. Firstly, the algorithm gives a higher bias value (one) to neutral, unused nodes, compared to a smaller value to an already used node. Secondly, the overall objective is to balance the load.

Example 1. Placement with upper bounds, empty system

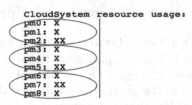

The case of an initially, partially loaded system is illustrated in Example 2, where the placed 12 LEs in the group are shown in red. The upper bounds are satisfied and the load is balanced within the constraints.

Example 2. Placement with upper bounds, busy system

4.2   Special Case: Group Anti-collocation

Conveniently, the group anti-collocation policy is a special case of the upper-bound constrained placement problem. Consider the case where the M LEs in the group are to be anti-collocated at level l, where $l = 0, \cdots, L - 1$. In other words, if an LE is placed in some entity at level l, then no other LE is to be placed in the same entity. This turns out to be a special case of the group placement problem with upper bounds. In particular, the upper bounds are given by

$$K_v = \begin{cases} 1, & v = 0, \cdots, l, \\ M, & v = l+1, \cdots, L. \end{cases} \tag{8}$$

In our example of Sect. 4.1 with $M = 6$ and $\boldsymbol{K} = [1, 6, 6]$, i.e. group anti-collocation at level 0 (PM), we obtain the placement result illustrated in Example 3. Note that the LEs are equally distributed among the zones as well.

Example 3. Placement with group anti-collocation constraint

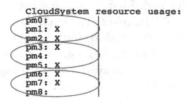

5   Constrained Placement: Lower and Upper Bounds

Now, we consider the general case. Let the constraint for $k_{l,i}$, the cumulative number of LEs placed on PEs in the projection range of node $\phi(l,i)$ be given by $\underline{K}_l \leq k_{l,i} \leq \overline{K}_l$, where $1 \leq \underline{K}_l \leq \overline{K}_l \leq M$ and $l = 0, \cdots, L$. In other words, among the M LEs in the group, the number of LEs that are placed at a given level l in the cloud hierarchical topology is within the bounds $[\underline{K}_l, \overline{K}_l]$. The above constraints are satisfied *iff* $x_{l,i}(M) \in [\underline{K}_l, \overline{K}_l]$, $\forall lnodes\ (l,i) \in ltree$. The values of lower and upper bounds are such that

$$1 \leq \underline{K}_0 \leq \underline{K}_1 \leq \cdots \leq \underline{K}_L = M,$$
$$1 \leq \overline{K}_0 \leq \overline{K}_1 \leq \cdots \leq \overline{K}_L = M, \tag{9}$$
$$\underline{K}_l \leq \overline{K}_l, \qquad l = 0, \cdots, L-1.$$

Our placement algorithm for this bounded case is similar to that of the upper-only bounded case as outlined in Algorithm 1, except that it places LEs until the lower bounds are satisfied first. The algorithm is outlined in Algorithm 2, where procedure PLACEGROUP() is skipped since it is the same as given by Algorithm 1. Also, we assume that variable K includes lower and upper bounds for all levels. We add a pointer *lnodeBelowBound*, pointing at the lowest level *lnode* in the *ltree* whose state is below the lower bound for that level. Thus, we attempt to satisfy the lower bounds in a bottom-up fashion. Note that, after the first placement of an LE from the group, *lnodeBelowBound* will always point to an *lnode*, including the root of the *ltree*, until all LEs in the group are placed. *lnodeBelowBound* is updated in procedure UPDATELOGICALTREE(), consulting condition ISNODESHORT(*lnode*) which is *true iff* $(x_{lnode} < \underline{K}_l)$, where l is the level of *lnode*. Procedure CALCULATEBIAS() is modified so as any node outside the subtree rooted at *lnodeBelowBound* is set to zero, hence making sure to satisfy the lower bounds first. In procedure ADJUSTBIAS(), a condition is handled in lines 29 through 35. Basically, if node availability is not enough to create a new child satisfying its minimum limit, set the bias of all leaves in the projection range that are not already open (with some allocation) to zero. (An efficient implementation may skip lines 32 to 35 and start the product in line 36 with one, instead.) The complexity of Algorithm 2 remains $O(MN)$.

5.1   Special Case: Group Collocation

Conveniently, the group collocation policy is a special case of this constrained placement problem. Consider the case where the M LEs in the group are to

Algorithm 2. Placement with lower and upper bounds
──
1: **procedure** UPDATELOGICALTREE($ltree, ptree, pnodePlaced$)
2:     $pnode \leftarrow pnodePlaced$
3:     $lnodeBelowBound \leftarrow null$
4:     **while** pnode **do**
5:         $lnode \leftarrow \phi^{-1}(pnode)$
6:         **if** $lnode = null$ **then**
7:             $l \leftarrow$ LEVEL($pnode$)
8:             $lnode \leftarrow$ CREATENEWNODE($ltree, l$)
9:             $\phi(lnode) \leftarrow pnode$
10:         $x_{lnode} + +$
11:         **if** $lnodeBelowBound = null$ & ISNODESHORT ($lnode$) **then**
12:             $lnodeBelowBound = lnode$
13:         $pnode \leftarrow$ PARENT($pnode$)

14:
15: **procedure** CALCULATEBIAS($ltree$)
16:     $\beta \leftarrow \mathbf{0}_N$
17:     $lnode \leftarrow lnodeBelowBound$
18:     $(r_{min}, r_{max}) \leftarrow$ GETPROJECTIONRANGE($ltree, lnode$)
19:     $\beta_{[r_{min}:r_{max}]} \leftarrow 1$
20:     ADJUSTBIAS($ltree, lnode, \beta$)
21:     **return** β

22:
23: **procedure** ADJUSTBIAS($ltree, lnode, \beta$)
24:     $(r_{min}, r_{max}) \leftarrow$ GETPROJECTIONRANGE($ltree, lnode$)
25:     **if** ISNODEBLOCKED($lnode$) **then**
26:         $\beta_{[r_{min}:r_{max}]} \leftarrow 0$
27:         **return**
28:     $l \leftarrow$ LEVEL($lnode$)
29:     **if** AVAILABLEONNODE($lnode$) $< \underline{K}_{l-1}$ **then**
30:         $\mathcal{A} \leftarrow$ GETPROJECTIONRANGE($ltree, lnode$)
31:         $\mathcal{B} \leftarrow \emptyset$
32:         **for all** $cnode \in$ CHILDREN($lnode$) **do**
33:             $\mathcal{B} \leftarrow \mathcal{B} \cup$ GETPROJECTIONRANGE(ltree, cnode)
34:         **for all** $pnode \in (\mathcal{A} - \mathcal{B})$ **do**
35:             $\beta_{pnode} \leftarrow 0$
36:     $\beta_{[r_{min}:r_{max}]} * = f(w_{lnode})$
37:     **for all** $cnode \in$ CHILDREN($lnode$) **do**
38:         ADJUSTBIAS($ltree, cnode, \beta$)
──

be collocated at level l, where $l = 0, \cdots, L$. This is a special case of the group placement problem with lower bounds. In particular, the lower and upper bounds are given by

$$\underline{K}_v = \begin{cases} 1, & v = 0, \cdots, l-1, \\ M, & v = l, \cdots, L, \end{cases}$$
$$\overline{K}_v = M, \qquad v = 0, \cdots, L. \tag{10}$$

For illustration, we use the example of Sect. 4.2, except with a group collocation at level 1 (zone). Thus, we have $\boldsymbol{\underline{K}} = [1, 6, 6]$ and $\boldsymbol{\overline{K}} = [6, 6, 6]$. We obtain the placement result illustrated in Example 4 where we notice that all LEs re placed in the same zone, and equally among the PMs in that zone.

Example 4. Placement with group collocation constraint

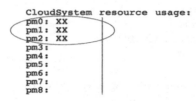

```
CloudSystem resource usage:
pm0: XX
pm1: XX
pm2: XX
pm3:
pm4:
pm5:
pm6:
pm7:
pm8:
```

6 Constrained Placement: Quantitative Placement

Consider the case where the state of *lnode* (l, i), after the placement of the M LEs in the group, is such that $\underline{K}_l \leq x_{l,i} \leq \overline{K}_l$, and $\underline{K}_l < \overline{K}_l$, where $0 \leq l \leq L$. In other words, the range of feasible values $(\overline{K}_l - \underline{K}_l) > 0$ allows for more than one value of $x_{l,i}$. Hence, the placement algorithm has a degree of freedom for choosing a value from the set of feasible values. In this section we address this issue and describe a mechanism for (1) specifying a user preference towards spreading or packing the placement of the LEs in the group, and (2) achieving such a preference through biasing the solution of the constrained placement optimization problem.

Intuitively, the closer the value of $x_{l,i}$ to $\underline{K}_l$, i.e. the lower bound for node occupancy, results in a larger number of nodes, hence more spreading. Alternatively, the closer the value of $x_{l,i}$ to $\overline{K}_l$, i.e. the upper bound for node occupancy, results in a small number of nodes, hence more packing. Thus, depending on the user preference *spread* or *pack*, the target for $x_{l,i}$ is $\underline{K}_l$ or $\overline{K}_l$, respectively. Define the deviation from target by $\delta_{l,i}$, as a normalized value in $[0,1]$, given by

$$\delta_{l,i} = \begin{cases} \frac{x_{l,i} - \underline{K}_l}{\overline{K}_l - \underline{K}_l}, & spread, \\ \frac{\overline{K}_l - x_{l,i}}{\overline{K}_l - \underline{K}_l}, & pack, \end{cases} \tag{11}$$

where $l = 0, \cdots, L$. Further, define a non-negative, non-decreasing function $g()$ of deviation. For instance, consider the power function,

$$g(\delta) = \delta^{\alpha}, \quad \alpha > 0. \tag{12}$$

Let $G(ltree)$ be the cumulative (could also be the average) value of $g()$ for all nodes (l, i) in the *ltree* such that $(\overline{K}_l - \underline{K}_l) > 0$. Thus, an objective of the constrained placement optimization problem is to minimize $G(ltree)$. In order to bias the solution towards that objective, as defined by Eq. 6, we need to modify the function $f()$ so that the bias is larger when the state is closer to the required target. In particular, the weight defined in Eq. 3 yields higher weight when $x_{l,i}$ is closer to $\underline{K_l}$, i.e. realizing the *spread* preference. The opposite, namely the *pack* preference may be realized by using a modified function, denoted by $\hat{f}(w)$, as

$$\hat{f}(w) = \hat{\eta} \left(1 - w^{\gamma}\right), \quad \hat{\eta}, \gamma > 0. \tag{13}$$

The scaling factors should be such that $\hat{\eta} \gg \eta$, so that when compared to a neutral node with a bias of an absolute one, a used node would have a much higher bias in the *pack* case, and a much smaller bias in the *spread* case. Practically, as the bias values are normalized in Eq. 5, we use $\hat{\eta} = N$ and $\eta = 1/N$.

Placement algorithms described in Algorithms 1 and 2 remain intact. However, the function $G()$ described above is added to the objective function of the minimization problem. (As an extension, another term in the objective may be the total number of nodes in the logical tree, compared to the minimum (maximum) number of nodes, as obtained using the upper (lower) bounds, corresponding to the *pack* (*spread*) preference, respectively.)

Consider Example 1 where the preference was *spread*, by default. With the *pack* preference we get the placement result given Example 5. Note that the PEs are packed to the maximum limits.

Example 5. Placement with pack preference

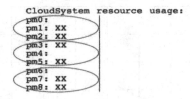

7   Experimental Results

We present results for two sets of experiments. In the first set, we compare the performance of our placement algorithm with hierarchical constraints to the state-of-the-art algorithm when the constraints are specified pairwise within the group. And, in the second set, we explore the range of problem size as far as the system size, N, and group size, M, are concerned and investigate the performance of our placement algorithm with hierarchical constraints

The simulated cloud consists of N homogeneous PEs varying in the range, $N = \{256, 512, 1024, 2048\}$. The cloud hierarchy is a balanced tree topology with height $L = 3$. Level 0 correspond to PMs, level 1 to Racks, and level 2 to Zones. There are 32 PMs in a Rack and 8 Racks in a Zone. The number of Zones varies in the range $\{1, 2, 4, 8\}$, yielding the different cloud sizes. Let there be a single resource per PM with capacity 32. The cloud is assumed to be loaded randomly at an average of 50 %. Consider the problem of placing a group of M homogeneous LEs, varying in the range, $M = \{16, 32, 64, 128\}$, and each with a resource demand of 1. We use the following parameters: $\eta = 1/N$, $\hat{\eta} = N$, $\gamma = 4$, $\alpha = 2$, and $\boldsymbol{p}_{init}$ proportional to resource availability. The objective function in the placement optimization (minimization) problem is the standard deviation of the resource utilization, hence the goal is to balance the load.

The algorithm is coded in C and runs on a MacBook Pro with 2.4 GHz Intel Core 2 Duo and 4 GB RAM, running Mac OS X 10.10.2 with optimized code[1]. The weights in the objective function are set to $w_{sys} = w_{usr} = 1$.

7.1 Performance Improvement

We compare the performance of our algorithm, which we refer to as *hierarchical* to [8] where the constraints are specified pairwise within the group and refer to it as *pairwise*. We consider the setup described above with a group size, $M = 128$, with anti-collocation (also known as anti-affinity) constraints among all members of the group at level 0. We measure the algorithm time as we vary the cloud size N. The comparison results are depicted in Fig. 3. As shown, the algorithm is linear in N with our *hierarchical* algorithm about 6 times faster than the *pairwise* algorithm.

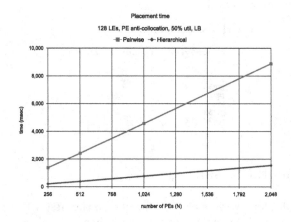

Fig. 3. Performance improvement

7.2 Performance Scalability

Now, we vary both problem size parameters: N and M. The bounds are given in Table 1. In all cases, the group ofis to be placed all in one Zone (level 2). The limits for levels 0 and 1 vary by case. For example, for $M = 64$, the number of members placed on one PM (level 0) is to be between a minimum of 8 and a maximum of 16. And, the number of members placed on one Rack (level 1) is

---

[1] The number of samples generated in each iteration is 20 and a fraction, 0.1, of those is used as important samples. The stopping criterion is a relative improvement in the objective function of less than 0.001, or a maximum number of iterations of 10.

to be between a minimum of 16 and a maximum of 32. In all cases, we ran both options: *spread* and *pack*. The results were such that the placement achieved the upper and lower bounds, respectively.

Table 1. Lower and upper bounds

M	$\underline{K}$	$\overline{K}$
16	[2, 4, 16, 16]	[4, 8, 16, 16]
32	[4, 8, 32, 32]	[8, 16, 32, 32]
64	[8, 16, 64, 64]	[16, 32, 64, 64]
128	[16, 32, 128, 128]	[32, 64, 128, 128]

The performance of the algorithm, measured in execution time, as we vary the cloud size N and the group size M, are illustrated in Fig. 4. The *hierarchical* algorithm presented in the paper as well as the encompassing BSA (Biased Sampling Algorithm)[8] are linear in N. Whereas, as a function of M, the former is linear and the latter is quadratic, hence the shape of the curves in Fig. 4. Uniformly, the *pack* option is slightly longer than the *spread* option. Given that the system is already at 50 % load, it takes more work to find hosts that have enough availability for packing and satisfy the hierarchical constraints. For the *spread* (*pack*) option, the placement was tight to the lower (upper) bounds There was one exception in the case of $N = 1,024$ and $M = 128$ and *pack* option, where yielding the upper bounds as a placement result would place the 128 VMs equally among 4 idle PMs (32 VMs each), such that each pair of PMs on a separate Rack, but all in the same Zone. The probabilistic search could not find such an arrangement in the allotted time, hence the placement was on 6 PMs with an allocation of $\{18, 23, 23\}$ VMs in one Rack and $\{18, 21, 25\}$ on another.

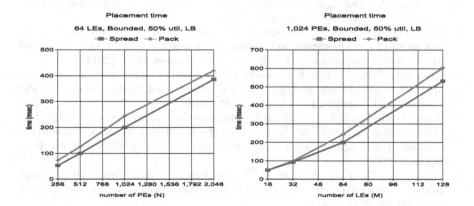

Fig. 4. Scaling system and group sizes

8 Conclusion

We presented a stochastic approach for solving the cloud placement problem with hierarchical group constraints. We introduced a generic bounding policy which covers regular collocation and anti-collocation constraints, as well as more advanced location distribution constraints. We presented an efficient algorithm and illustrated its performance, quality, and scalability behavior. Compared to state-of-the-art pairwise-based constrained algorithm, our hierarchy-based algorithm demonstrated an improvement factor of about six in execution time.

Additional work may consider the case where the hierarchical group constraints are soft, i.e. to be satisfied as best as possible, instead of rejecting requests that do not satisfy the constraints. Also, the complexity of our algorithm could be improved by extending the *ptree* implementation to hold the bias values and probabilities in the internal nodes, instead of the arrays of length N, thus achieving a complexity of $O(M \, ln(N))$.

References

1. Aldhalaan, A., Menasce, D.A.: Autonomic allocation of communicating virtual machines in hierarchical cloud data centers. In: Proceedings of the 2014 IEEE International Conference on Cloud and Autonomic Computing, CAC 2014, IEEE. IEEE Computer Society, London, 8–12 September 2014
2. Arnold, W., Arroyo, D., Segmuller, W., Spreitzer, M., Steinder, M., Tantawi, A.: Workload orchestration and optimization for software defined environments. IBM J. Res. Dev. **58**(2), 1–12 (2014)
3. Espling, D., Larsson, L., Li, W., Tordsson, J., Elmroth, E.: Modeling and placement of cloud services with internal structure. IEEE Trans. Cloud Comput. **PP**(99), 1–1 (2014)
4. Giurgiu, I., Castillo, C., Tantawi, A., Steinder, M.: Enabling efficient placement of virtual infrastructures in the cloud. In: Narasimhan, P., Triantafillou, P. (eds.) Middleware 2012. LNCS, vol. 7662, pp. 332–353. Springer, Heidelberg (2012)
5. Jennings, B., Stadler, R.: Resource management in clouds: survey and research challenges. J. Netw. Syst. Manag. 1–53 (2014)
6. Moens, H., Hanssens, B., Dhoedt, B., De Turck, F.: Hierarchical network-aware placement of service oriented applications in clouds. In: Network Operations and Management Symposium (NOMS), 2014 IEEE, pp. 1–8. IEEE (2014)
7. Piao, J.T., Yan, J.: A network-aware virtual machine placement and migration approach in cloud computing. In: 2010 9th International Conference on Grid and Cooperative Computing (GCC), pp. 87–92 (2010)
8. Tantawi, A.: A scalable algorithm for placement of virtual clusters in large data centers. In: 2012 IEEE 20th International Symposium on Modeling, Analysis & Simulation of Computer and Telecommunication Systems (MASCOTS), pp. 3–10. IEEE (2012)
9. Wei, X., Li, H., Yang, K., Zou, L.: Topology-aware partial virtual cluster mapping algorithm on shared distributed infrastructures. IEEE Trans. Parallel. Distrib. Syst. **25**(10), 2721–2730 (2014)
10. Zong, B., Raghavendra, R., Srivatsa, M., Yan, X., Singh, A.K., Lee, K.W.: Cloud service placement via subgraph matching. In: 2014 IEEE 30th International Conference on Data Engineering (ICDE), pp. 832–843. IEEE (2014)

Characterizing Data Dependence Constraints for Dynamic Reliability Using N-Queens Attack Domains

Ulya Bayram, Kristin Yvonne Rozier, and Eric W.D. Rozier[⊠]

University of Cincinnati, Cincinnati, OH, USA
bayramua@mail.uc.edu, {Kristin.Y.Rozier,Eric.Rozier}@uc.edu

Abstract. As data centers attempt to cope with the exponential growth of data, new techniques for intelligent, software-defined data centers (SDDC) are being developed to confront the scale and pace of changing resources and requirements. For cost-constrained environments, like those increasingly present in scientific research labs, SDDCs also present the possibility to provide better reliability and performability with no additional hardware through the use of dynamic syndrome allocation. To do so the middleware layers of SDDCs must be able to calculate and account for complex dependence relationships to determine an optimal data layout. This challenge is exacerbated by the growth of constraints on the dependence problem when available resources are both large (due to a higher number of syndromes that can be stored) and small (due to the lack of available space for syndrome allocation). We present a quantitative method for characterizing these challenges using an analysis of attack domains for high-dimension variants of the n-queens problem that enables performable solutions via the SMT solver Z3. We demonstrate correctness of our technique, and provide experimental evidence of its efficacy; our implementation is publicly available.

Keywords: Big data · Reliability · Storage · n-queens · Intelligent systems

1 Introduction

One of the largest challenges facing the storage industry is the continued exponential growth of Big Data. The growth of data in the modern world is exceeding the ability of designers and researchers to build appropriate platforms [11,26] but presents a special challenge to scientific labs and non-profit organizations whose budgets have not grown (and often have been cut) as their data needs steeply rise. The NASA Center for Climate Simulation revealed that while their computing needs had increased 300 fold in the last ten years, storage had increased 2,000 fold, and called storage infrastructure one of the largest challenges facing climate scientists [8]. This trend has been driving reliance on commercial off the shelf (COTS) solutions to drive down the cost of data ownership. Despite

© Springer International Publishing Switzerland 2015
J. Campos and B.R. Haverkort (Eds.): QEST 2015, LNCS 9259, pp. 211–227, 2015.
DOI: 10.1007/978-3-319-22264-6_14

its importance, the goal of affordable data curation comes at a cost in terms of reliability creating a difficult to solve system design constraints problem.

To cope with the increase in cost, deduplication techniques are commonly used in many storage systems. Deduplication is a storage efficiency improvement technique that removes the duplicate substrings in a storage system and replaces them with references to the single location storing the duplicate data. While this achieves a higher storage efficiency in terms of reducing the cost of ownership of a system, it can negatively impact the reliability of the underlying storage system since loss of a block with high number of references means a critical number of files being lost unrecoverably [22].

Data reliability was previously improved using enterprise-class storage devices that typically suffer faults as much as two orders of magnitude less often than COTS storage devices. In the face of the exponential growth of the digital universe [28] the cost of this solution has become prohibitively expensive, inspiring a switch to near-line components, thus lessening storage reliability guarantees. While reliability could be improved through the addition of new hardware, today the scale of growth of inexpensive storage is being exceeded by the growth of Big Data.

In most storage systems reliability improvements are achieved through the allocation of additional disks in Redundant Arrays of Independent Disks (RAID) [20]. RAID arrays achieve reliability through the allocation of coding syndromes [21] which create dependence relationships in the storage system to allow recovery of files after failures. While RAID systems are incredibly effective at the task of improving reliability, they add to the cost of the storage systems in which they are deployed.

Methods used to increase reliability also increases the cost of maintaining the storage system, and same is true for the methods that reduce the cost; they also reduce reliability. In order to meet these cost and reliability constraints, and find a way to break the proportional relationship in between, previously we have conducted a study where we have documented that systems are often over-provisioned, and this over-provisioning level is highly predictable using intelligent systems algorithms [23]. Using these models, we have proposed that dynamically allocated reliability syndromes could be created and stored in this excess capacity to improve reliability without the addition of new hardware [3]. Based on this result, it is now possible to modify traditional RAID schemes to dynamically allocate new syndromes for reliability in over-provisioned space through the risk-averse prediction of available storage over the next epoch of operation of a storage system. Furthermore this can be done while maintaining quality of service (QoS) and availability of the storage system while simultaneously providing maximum additional reliability. The only assumption is that the additional syndromes can be placed in a way that respects data dependence constraints. The ability to predict the expected level of over-provisioning allows us to create software defined data centers which can allocate virtual disks made up of free space compiled from across the data center to hold additional reliability syndromes. An unsolved challenge which stands in the way of this technique, however, is the development of algorithms that account for complex data dependencies such as existing reliability syndromes and deduplication, providing a strategy for syndrome storage and new RAID relationships in a performable way that maximizes the additional number of reliability syndromes that can be allocated without violating the dependence constraints on those syndromes.

In order to solve these dependence constraints we cast our problem into a unique variant of the n-Queens problem. We map a RAID array into a mathematical representation of a chess board with a set number of *ranks* (defining the y-axis) and *files* (defining the x-axis). We propose a quantitative solution for virtual disk allocation in software defined data centers, respecting all dependence constraints within the data center, or, when no such configuration exists, identifying the unsatisfiability of the problem. This method allows us to take advantage of the over-provisioned space without constraining our problem to traditional RAID geometries. We propose solving this problem quantitatively by mapping it to an innovative variation of the n-queens problem that utilizes a 3D Latin board configuration [12,16], nontraditional queen types and attack domains, and population limits on the number of queens of a given type placed within certain bounds.

The challenge of defining dynamic syndromes is inherently characterized by a well-defined set of constraints: total number of disks, current disk utilization, distribution of unutilized space, existing dependence relationships due to RAID reliability syndromes, and deduplication relationships. By creating a mapping to n-queens under these constraints, we can intuitively represent the problem in a way that facilitates validation and harness the power of the Satisfiability Modulo Theories (SMT) solver Z3 to return a constraint-satisfying solution or determine that a solution cannot exist. Z3 [6] is a very efficient and freely available solver for SMT, which is a decision problem for logical first order formulas with respect to combinations of background theories including the uninterpreted functions integral to our solution. The n-queens problem is a classic way to represent such a constraint satisfaction problem [17,25] and a common benchmark for such a solver [13]. Classification as a constraint satisfaction problem that can be solved by Z3 has proven to be successful in other design domains, such as automating design of encryption and signature schemes [1].

Our contributions include a new quantitative solution for the problem of dynamic allocation of new reliability syndromes while respecting dependence constraints to improve the reliability of software defined data centers without the addition of new hardware. We define and prove a correct mapping of this problem to a variation of the classic n-queens problem, thus enabling efficient analysis via powerful SMT solvers like Z3. We provide an implementation in Z3 for python and include a case study demonstrating the effectiveness of our technique. This new solution will serve as the core for a dynamic allocation system to be deployed in software-defined data centers which will be deployed at the laboratories of partner organizations.

This paper is organized as follows: Sect. 2 provides background on dependence relationships in storage systems, and related work in novel RAID geometries. Section 3 introduces an encoding for this problem in a variant of n-Queens mapping the problem of data layout strategies which respect all data dependence

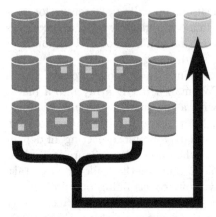 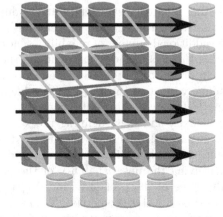

(a) Example of a virtual disk being constructed out of overprovisioned space.

(b) Example of independent syndrome calculation as the XOR parity of diagonals.

Fig. 1. Example allocations of virtual disks from over-provisioned space.

contraints to maximize additional syndrome coverage for any given dataset to
the problem of placing novel queen-types on a Latin chess board; we discuss the
constraints of our problem in Sect. 4, and its implementation in Z3. We provide
experimental results demonstrating the efficacy and efficiency of our approach
in Sect. 5. Finally, Sect. 6 concludes and points to future work.

2   Characterizing File System Dependence

As we have shown in our previous work [3,23], it is possible to predict the future
storage resource needs of the users in a system. In recent work [3], we have
modeled user behaviors using the training data we obtained from a real system
to create and train Markov models, and predicted the future disk usage needs
of the users in an on-line fashion, and compared the results with the test data
we also obtained from the same system to measure the prediction performance.
We have observed that with a good clustering method and fine parameter tun-
ing, it is possible to predict user behaviors and resource requirements. We have
proposed this method for predicting over-provisioning, and further that it could
allow for dynamic improvement of reliability through the allocation of additional

Table 1. Annual rates of block loss (ABL) per system type with varying numbers of
additional syndromes (n_{synd}) allocated.

RAID5 conf.	ABL (no syndromes)	ABL ($n_{synd} = 2$)	ABL ($n_{synd} = 3$)
5+1	1.79×10^5	1.31×10^{-7}	1.92×10^{-15}
8+1	4.60×10^5	1.02×10^{-6}	5.06×10^{-14}
10+1	8.06×10^5	2.76×10^{-6}	1.79×10^{-13}

syndromes by creating new *virtual disks* using any over-provisioned storage that are found to be independent of the current RAID grouping as shown in Fig. 1a. Our experiments on real storage system data have shown that even when being incredibly risk adverse, we can allocate between three and four additional syndromes more than 50 % of the time, and on average allocate two additional syndromes for all of the data and three additional syndromes for more than 90 % of the data, dramatically improving the reliability of the system [3]. We analyzed these improvements on systems with one petabyte of primary storage with initial RAID5 configurations of 5+1, 8+1 and 10+1 over which we introduce two and three additional syndromes after predictions. Changes in reliability are measured using the rate of annual block loss (ABL), when taking into account whole disk failures and latent sector errors. Table 1 illustrates the calculated ABLs for three RAID5 configured primary storage systems, each provisioned for a maximum capacity of one petabyte. The steep increase in the reliability represented by decrease in ABL rates as the number of allocated syndromes increases shows the promise of such predictive analysis and dynamic allocation.

Allocation of new syndromes in order to increase the reliability is possible through the deployment of RAID5 XOR parity syndromes [20] or RAID6 Galois-field based syndromes [2,5]. Additional syndromes can be allocated using techniques such as erasure coding, though they generally have a severe impact on performance, and as a result, lower the QoS of the system [15]. As such, we focus on alternative RAID geometries to make use of additional XOR parity and Galois-field based syndromes.

In this paper we propose an efficient method for allocation of additional syndromes. Additional coverage can be provided using non-traditional RAID geometries as shown in Fig. 1b. While the idea of using non-traditional RAID geometries itself is not new, and have been explored in previous studies [18,19,27], prior work in this field has always maintained the assumption that the layout of the RAID arrays are pre-defined. Instead, we propose the creation of dynamic per-stripe geometries using over-provisioned space in an existing data center.

When creating non-traditional RAID geometries, care must be taken to respect data dependence relationships [24] to ensure that the new RAID strategy improves reliability. We consider two types of data dependence relationships; one resulting from pre-existing RAID groups, and the other from data deduplication [22].

2.1 Dependence Due to RAID

In order to improve reliability, a syndrome allocated as part of a RAID group must be independent of all other syndromes computed within the data center. There are two primary methods for establishing independence with respect to a syndrome, and a set of data on disks. First and most simply, we consider a data center D to be a collection of disks $D = \{d_0, d_1, d_2, d_3, \ldots\}$ protected by a set of RAID strategies $R = \{R_0, R_1, R_2, \ldots\}$ each of which is represented as a set of M disks $R_k = \{d_{k,0}, d_{k,1}, d_{k,2}, \ldots d_{k,(M-1)}\}$ forming a *stripe*. Each stripe consists of a number of data and syndrome disks. For any two disks $d_i \in D$ and $d_j \in D$ there is a mutual dependence relationship $\leftrightarrow$, $d_i \leftrightarrow d_j$ if $\exists R_k \in R$ such that $d_i \in R_k$ and $d_j \in R_k$. If no such R_k exists then d_i and d_j are independent with respect to RAID.

2.2 Dependence Due to Deduplication

Data deduplication is a wide spread technique which improves the storage efficiency of a data center by eliminating redundant data. As shown in Fig. 2 deduplication identifies substrings of redundant data, typically between block boundaries, stored in a data center and replaces the instances of all but one of those substrings with a reference to a single stored instance on another disk. This creates a critical dependence relationship as the loss of that disk, which may be on another RAID group, may render the data which references it useless, resulting in correlated failures. As such data deduplication introduces one way dependence relationships which we represent as $d_i \rightarrow d_j$ when d_i depends on d_j. Given a disk d_i if $\exists d_l \in R_k$ such that $d_i \in R_k$ and d_l contains a reference to deduplicated data on disk d_j then the dependence relationship $d_i \rightarrow d_j$ exists, otherwise d_i is independent of d_j with respect to deduplication.

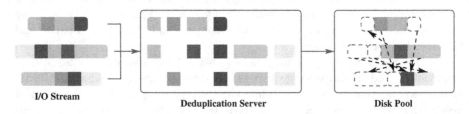

I/O Stream   **Deduplication Server**   **Disk Pool**

Fig. 2. Data deduplication identifies similar blocks within the data center, and eliminates redundant information.

3   N-Queens with Dynamic Domains of Attack

In order to solve our problem and find a data layout that allows us to build virtual disks which are independent of the data they are protecting, we provide a mapping of our problem into a variant on the classical n-Queens [10] constraint satisfaction problem with few alterations.

First, we adopt a Latin board, allowing us to examine our problem in a three dimensional space [12,16]. We define this space according to three axes, the level, rank, and file, as shown in Fig. 3. We further define a column on this Latin board as the set of squares defined by a fixed rank and file across all levels of the board. Each column in our Latin board corresponds to a disk within our data center, with each rank consisting of a traditional RAID group. Levels represent independent sub-problems solving for data independence for each disk in turn. Thus, in practice, given a problem with N ranks and M files, we construct our board with $L = N \cdot M$ levels.

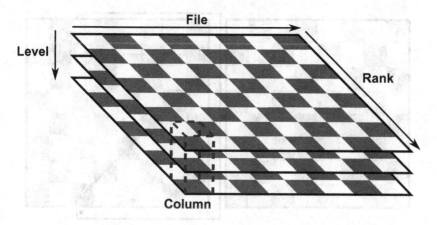

Fig. 3. Example Space with dimensions $3 \times 8 \times 8$.

We represent the state of dependence relationships in a file system by placing Queens on our boards, using their attack domains to represent file dependence relationships. For any level l in our board, this level is used to solve a sub-problem for the lth disk in our data center (numbered in rank major order, such that if the disk is in rank r and file f the level which solves its independence constraints is $l = r*M + f$) the full attack domain of all queens on level l represents those disks on which the lth disk depends. We call this lth disk for level l the principle disk for that level. To represent these dependence relationships, however, we must modify the attack domain definitions for each queen to match the dependence relationships we must represent. We introduce three new queen types each with a unique attack domain.

- **Degenerate Queens** - a degenerate queen is so named because it attacks only a single square, that which it is occupying. Degenerate queens are used to represent the disk being protected, and disks containing deduplicated blocks upon which files on that disk depend. Degenerate queens are used to exclude a square on a level from the solution space of new dynamic RAID groupings. The attack domain of a degenerate queen is illustrated in Fig. 4a.
- **Linear Queens** - a linear queen's attack domain is defined to include both its own square and $M - 2$ squares on the board extending in a line from the queen, potentially wrapping around the board as if it were a toroidal board as discussed originally in the class of Modular n-Queens problems [9]. Linear queens can be used to represent existing RAID groups, or new dynamic RAID groups with more traditional geometries. Two example attack domains for linear queens are illustrated in Fig. 4b[1].

---

[1] While we allow linear queens to attack in any direction as a matter of completeness of our variant n-Queens definition, we note that our method only makes use of linear queens which attack along ranks towards squares in higher numbered files, wrapping toroidally.

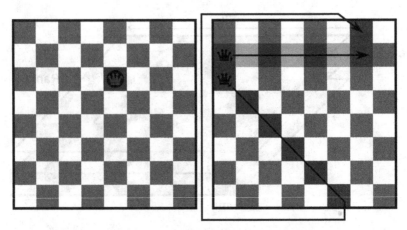

(a) Example attack domain of a single de- (b) Example attack domain of two linear
generate queen.   queens.

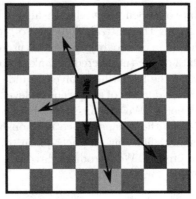

(c) Example attack domain of a single indirect queen.

Fig. 4. Example attack domains of all three queen types.

– **Indirect Queens** - the final type of queen we introduce is an indirect queen, whose attack domain consists of its own square, and $M - 2$ other squares on the board, each within a rank unique to the queen's attack domain. The attack domain of an indirect queen is illustrated in Fig. 4c.

In order to solve the problem of independent syndrome placement, and the creation of new dynamic RAID groupings, we begin with a pre-defined board based on the state of the data center, which contains a number of degenerate and linear queens representing this system state, such as the example shown in Fig. 5a. We then proceed to place Q new indirect queens on the board with each indirect queen representing the storage location of a new pair of XOR and galois field parity syndromes, and the attack domain of that queen representing the independent disks to use to form a new dynamic RAID group associated with those syndromes.

Block to be protected

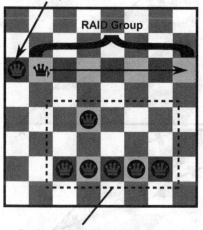

Deduplicated Blocks

(a) Example of initial constraints when protecting a block on disk 0 of RAID group 2 which has references to deduplicated blocks on six other disks. Degenerate queens are used to include the disks containing the initial and deduplicated blocks in the attack domain, and a linear queen is used to include the RAID group in the attack domain.

(b) An example solution with two additional syndromes. Indirect Queens occupy the spaces corresponding to the disks in which the new syndromes will be stored, their attack domains include all disks protected by the new syndrome.

Fig. 5. Representation of a single level of an $8 \times 8\text{x}64$ board.

The initial placement of degenerate and linear queens is accomplished according to the dependence relationships defined in Sect. 2. We begin by placing a single degenerate queen on the square corresponding to our level and the disk it solves for, d_i, i.e. for level l we place a degenerate queen at $(\frac{l}{M}, l\text{mod}M)$. If we placed that degenerate queen in the last file of it's rank we place a linear queen at $(\frac{l}{M}, 0)$, otherwise we place it at $(\frac{l}{M}, (l\text{mod}M) + 1)$. This linear queen accounts for the mutual RAID dependence relationships. We then place a single degenerate queen on each square corresponding to some d_j for which $d_i \rightarrow d_j$ to account for the data deduplication derived dependence relationships.

Once the initial board is fixed for each level, we attempt to solve for placing indirect queens representing new RAID groups such that neither the new queens, nor their attack domains overlap with the existing attack domains on the given level. An example solution with two additional syndrome storage locations (allowing up to four additional syndromes) is shown in Fig. 5b.

Our final modification to the traditional n-Queens problem is the addition of a population constraint board. This board enforces a limit to the number of indirect queens placed on any level in a given column, constraining the possible satisfying assignments which respect dependence relationships, and possibly rules some columns out entirely when it comes to queen placement.

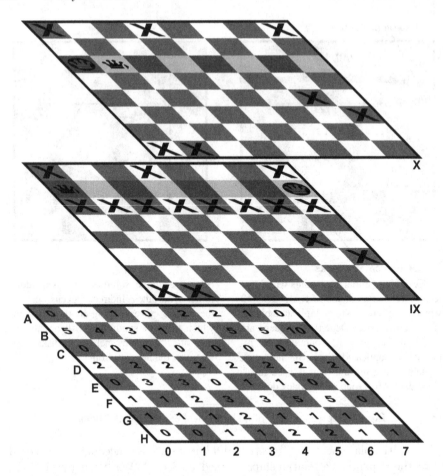

Fig. 6. Example of a Problem with a Population Constraint Board. This board is unsatisfiable for any indirect queen placement due to the population constraints present in rank C. While a placement may exist for level X, no placement can exist for level IX as we need to place an indirect queen and allocate six attack domains, all on different ranks. Note only two levels are shown here, when in a full problem a total of 64 levels would exist.

Figure 6 illustrates an example of a population constraint board being used with a $2x4x4$ Latin board. These population constraints represent the overprovisioned space on each disk that is available for additional reliability syndrome creation as estimated by our predictive models.

While the less restrictive attack domains of our three new queen types would seem to make the problem less difficult than traditional n-Queens, and more equivalent to the trivial n-Rooks problem [4,29], the population constraints board serves to complicate the problem of queen placement, especially as the number of levels we must solve for grows polynomially. The population constraints board has the effect of creating attack domains in the z-axis when enough queens are placed in a column. Figure 7 shows the relative difficulty of

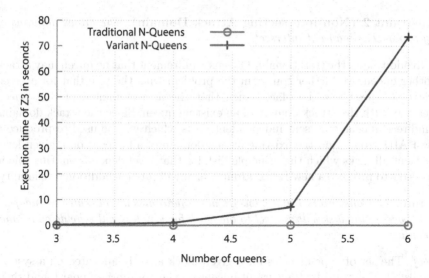

Fig. 7. Comparison of solution times for Z3 given the placement of Y queens on a $Y \times Y$ traditional n-Queens board and a $Y^2 \times Y \times Y$ variant n-Queens board from our problem.

solving this new variant n-Queens problem vs. traditional n-Queens, and high-lights the additional complexity despite the more easily satisfied attack domains of our variant queens. While this graph suggests that scalability is an issue, we will address scalability concerns in Sect. 5 through a proposed compositional approach.

4   Solving Virtual Disk Allocation with N-Queens

Given the encoding of our problem into n-Queens, to solve for the placement of indirect queens representing new virtual disks holding unique syndromes we define a set of constraints which maintain the necessary independence relation-ships to ensure our new RAID groups provide additional reliability. In total we can encapsulate these requirements in five constraints.

Constraint 1 (Standard n-Queens). *No queen may be placed within the attack domain of an existing queen.*

As in standard n-Queens problems no indirect queen may be placed within the attack domain of an existing queen. Since we define the attack domain of an existing queen to also encompass the square in which the queen is placed, no two queens may occupy the same square on a board. This ensures that we do not attempt to allocate a new syndrome on a disk which the principle disk for the level is dependent.

Constraint 2 (Non-intersecting Attack Domain). *The attack domains of any two queens may not intersect.*

In addition to the traditional n-Queens requirement that no queen may attack another queen, we further constrain the problem with the rule that for a given level l no queen may be placed such that the attack domain of that queen intersects with the attack domain of an existing queen. Since the attack domains of indirect queens represent independent disks which will be used to produce a new RAID group, and the existing attack domain of all queens on a given level represent all disks which the principle disk for that level depends on, this rule is necessary to produce a new RAID group which will provide additional reliability.

Lemma 1. *Constraints 1 and 2 ensure new syndromes are allocated in independent space on the disk with respect to the block for which it will provide additional reliability.*

Proof. The set of dependencies for a given block already allocated on a system are described in total by the current queens on the appropriate board, and their attack domains. The block to be protected, and any deduplicated blocks will be represented by degenerate queens, thus constrained by 1, the default RAID grouping for the block in question will be represented by the attack domain of a linear queen, thus constrained by 1, and any additional dynamic syndromes will be represented by the attack domain of an indirect queen, also constrained by 1.

Constraint 3 (Indirect Queen Attack Domain). *Each element of a given indirect queen's attack domain must be on a unique rank.*

We assume that our system contains a number of initial RAID groupings corresponding to each rank in our system. As such given two disks d_i and d_j, if both are of the same rank it must be true that $d_i \leftrightarrow d_j$. As such to form a new RAID group of strictly independent disks, a single indirect queen must not contain two square within the same rank in its attack domain.

Lemma 2. *Constraint 3 guarantees that a new indirect queen will be an independent grouping allowing XOR or Galois parity protection.*

Proof. In order for independent XOR or Galois parity blocks to be computed every element of a RAID grouping must be independent. If two attack domains for an indirect queen are on the same rank, they will not be independent, as they will already be involved in another parity group protecting each other.

Constraint 4 (Column Indirect Queen Population Limit). *Each column may be assigned a population limit $p_z \in \mathbb{N}$. The total number of all indirect queens within that column must not exceed this limit.*

As defined in Sect. 3 we include a population constraint board to represent limits in available over-provisioned space in our data center. These population limits restrict the placement of queens within a given column, creating dependence between all levels of our Latin chess board.

Constraint 5 (Level Protection Requirement). *For a given level l, the sum of the number of all indirect queens on that level must be greater than or equal to the protection requirement P.*

As we want to ensure uniform protection of all queens, we constrain the solutions for our system to those which have equal queens on all boards and find the maximum P for which our problem is satisfiable.

4.1 Translation into Z3

In order to determine if a given data center state and desired protection level is satisfiable, we utilized Z3 and encoded our problem in the form of variables and uninterpreted functions forming an SMT problem. We first defined $N^2 M^2$ integer variables, one for each square in our Latin n-queens board with domains of $\{-1, 0, 1, 2\}$ representing:

-1 : The attack domain of an indirect queen

  0 : A degenerate queen, a linear queen, or the attack domain of a linear queen.

  1 : An indirect queen.

  2 : An empty square.

For the initial board setup we included an uninterpreted function fixing the assignment of the corresponding variables to 0. Population limits were enforced by the inclusion of uninterpreted functions which constrained the count of variables assigned a value of 1 in a given column to the limit for that column (fed as input to the solver) using the If() function of Z3. A protection level of P was enforced for each level by ensuring the count of variables assigned a value of 1 on a level was equal to P using the If() function of Z3. We ensured that each indirect queen could be assigned an appropriate attack domain by ensuring the count of variables assigned a value of -1 on a level was equal to $(M - 2)P$ using the If() function of Z3, and that the count of variables assigned a value of -1 in a given rank for a given level was less than or equal to P. These last two uninterpreted functions also served to make solution more efficient by accounting for symmetric and lump-able attack domains in a single solution [7].

5 Experimental Results and Validation

In order to validate our results we conducted experiments with random initial system states for both population constraints boards, and data deduplication constraints. All experiments were run using a single EC2 c4.large instance with 2 virtual CPUs and 3.75 GiB of RAM. We implemented our solver to print out the resulting boards in a human readable format and hand checked the results, also collecting performance statistics for the Z3 solutions.

Figure 8 along with Fig. 7 provide a summary of the results of our experiments. We found a sharp satisfiability cliff accompanying the population constraint's board which corresponded to the probability of a rank having no available space. This suggests an important observation to account for when moving forward with

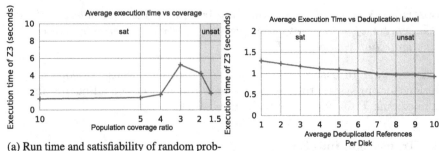

(a) Run time and satisfiability of random problems as the population constraints board is made more restrictive.

(b) Run time and satisfiability of random problems as the deduplication ratio is increased.

Fig. 8. Partial summary of experimental results.

a full implementation of software defined data centers, namely that balancing of over-provisioned space can be critical when such space becomes rare and the data center approaches capacity if the excess space is to be used to improve reliability. This limit is approached even swifter for large systems in which many levels are competing for the same population constraints within a rank.

We found the problem to be less sensitive to deduplication. While we eventually found a region of unsatisfiable problems at higher deduplication ratios, the more random placement of deduplicated references ameliorated their constraints on the solution space. It should also be noted that such constraints only became an issue at very high levels of deduplication, suggesting that deduplication based dependences are not as difficult to account for as might be expected.

The exponential growth in runtimes is somewhat concerning, as it seems to limit this solution technique to smaller storage systems, which presents a problem when confronted with the exponential growth of Big Data. Large-scale systems could potentially take infeasible amounts of time to solve if solved directly. As a consequence of this result we propose that larger systems be solved compositionally. For instance, while a 160 TB system takes 74 s to solve, if the system is blocked into two 80 TB systems by decomposing individual ranks a satisfying solution for each system can be found within 2.5 s each, and can be solved in parallel. The exponential improvements found through compositional solution, coupled with the embarrassingly parallel nature of the SMT sub-problems created by partitioning the system by rank provides a very scalable alternative to attacking the entire problem at once. This method has the advantage of respecting dependence relationships, as when decomposed into separate sub-problems all relationships can be accounted for between sub-models in a trivial fashion since their proposed solutions will include only those ranks within a given sub-problem.

Since the population constraint board is known as part of the system state, we can choose to sort each rank into one of S subproblems based on the rank of the population constraint board associated with the rank of the Latin board. The satisfiability of the subproblems, depending primarily on these population constraints, can be maximized by sorting the ranks on the basis of the population constraints associated with their columns. Using such a solution we are able to scale linearly with the size of our data center.

6   Conclusions

In this paper we have presented a novel formulation of the n-Queens problem using a moudlar Latin board, non-traditional queen variants, and column-based population constraints. This formulation serves as a translation of data dependence constraints and the problem of virtual syndrome creation for software-defined data structures into SMT allowing for efficient solution which allows for improved reliability with no additional hardware in over-provisioned systems. While our problem grows exponentially more difficult for larger storage systems, we provide a scalable way to achieve similar levels of protection through rank-wise decomposition of the problem space using population-constraint sorting into embarrassingly parallel subproblems.

This new method will form the basis for a performable dynamic RAID allocation system for use in large-scale storage systems serving cost-constrained organizations, providing an intelligent software stack which will help to combat the exponential growth of Big Data.

6.1   Future Work

Now that we have an efficient, scalable method for determining whether there exists a dynamic reliability syndrome that satisfies its data dependence constraints, we can move onto looking at other interesting optimizations. Currently we either generate a single strategy for additional syndrome allocation, or prove that no such allocation exists. However, the option is now open for us to harness more of the power of Z3 to query the solution space to optimize for secondary considerations, such as geometries that we find more attractive. For example, we may search for solutions with such features using the solution enumeration capabilities of Z3 [14].

We plan to implement our solution technique in a hardware-based middleware controller that monitors back-end data systems, and reshapes incoming file traffic to build the proposed dynamic allocations of RAID groups in response to predictions for overprovisioning. We can also envision an extension enabling data storage system designers to query Z3 regarding hypothetical disk configurations and data dependence constraints as they design a new storage system, thus enabling them to optimize their designs with respect to the robustness/cost tradeoff before purchasing any hardware.

Availability. We have made our implementation, all associated source code, and data available under the terms of the University of Illinois/NCSA Open Source License[2] at our laboratory website http://trust.dataengineering.org/research/nqueens/.

---

[2] http://opensource.org/licenses/NCSA

References

1. Akinyele, J.A., Green, M., Hohenberger, S.: Using smt solvers to automate design tasks for encryption and signature schemes. In: Proceedings of the 2013 ACM SIGSAC conference on Computer & communications security, pp. 399–410. ACM (2013)
2. Anvin, H.P.: The mathematics of raid-6 (2007)
3. Bayram, U., Rozier, E.W., Zhou, P., Divine, D.: Improving reliability with dynamic syndrome allocation in intelligent software defined data centers. In: IEEE/IFIP International Conference on Dependable Systems & Networks, DSN 2015. IEEE (2015)
4. Bell, J., Stevens, B.: A survey of known results and research areas for n-queens. Discrete Math. **309**(1), 1–31 (2009)
5. Corbett, P., English, B., Goel, A., Grcanac, T., Kleiman, S., Leong, J., Sankar, S.: Row-diagonal parity for double disk failure correction. In: Proceedings of the 3rd USENIX Conference on File and Storage Technologies, pp. 1–14 (2004)
6. de Moura, L., Bjørner, N.S.: Z3: an efficient SMT solver. In: Ramakrishnan, C.R., Rehof, J. (eds.) TACAS 2008. LNCS, vol. 4963, pp. 337–340. Springer, Heidelberg (2008)
7. Derisavi, S., Kemper, P., Sanders, W.H.: Symbolic state-space exploration and numerical analysis of state-sharing composed models. Linear Algebra Appl. **386**, 137–166 (2004)
8. Duffy, D., Schnase, J.: Meeting the big data challenges of climate science through cloud-enabled climate analytics-as-a-service. In: Proceedings of the 30th International Conference on Massive Storage Systems and Technology. IEEE Computer Society (2014)
9. Eickenscheidt, B.: Das n-damen-problem auf dem zylinderbrett. Feenschach **50**, 382–385 (1980)
10. Gu, J., et al.: Efficient local search with conflict minimization: a case study of the n-queens problem. IEEE Trans. Knowl. Data Eng. **6**(5), 661–668 (1994)
11. Hashem, I.A.T., Yaqoob, I., Anuar, N.B., Mokhtar, S., Gani, A.B., Khan, S.U.: The rise of big data on cloud computing: review and open research issues. Inf. Syst. **47**, 98–115 (2015)
12. Klarner, D.A.: Queen squares. J. Recreational Math **12**(3), 177–178 (1979)
13. Klebanov, V., et al.: The 1st verified software competition: experience report. In: Butler, M., Schulte, W. (eds.) FM 2011. LNCS, vol. 6664, pp. 154–168. Springer, Heidelberg (2011)
14. Köksal, A.S., Kuncak, V., Suter, P.: Scala to the power of Z3: Integrating SMT and programming. In: Bjørner, N., Sofronie-Stokkermans, V. (eds.) CADE 2011. LNCS, vol. 6803, pp. 400–406. Springer, Heidelberg (2011)
15. Leventhal, A.: Triple-parity raid and beyond. Queue **7**(11), 30 (2009)
16. McCarty, C.P.: Queen squares. Am. Math. Monthly **85**, 578–580 (1978)
17. Nadel, B.A.: Representation selection for constraint satisfaction: a case study using n-queens. IEEE Intell. Syst. **5**(3), 16–23 (1990)
18. Pâris, J.F., Amer, A., Schwarz, T.J.: Low-redundancy two-dimensional raid arrays. In: 2012 International Conference on Computing, Networking and Communications (ICNC), pp. 507–511. IEEE (2012)
19. Pâris, J.F., Long, D.D., Litwin, W.: Three-dimensional redundancy codes for archival storage. In: 2013 IEEE 21st International Symposium on Modeling, Analysis & Simulation of Computer and Telecommunication Systems (MASCOTS), pp. 328–332. IEEE (2013)

20. Patterson, D.A., Gibson, G., Katz, R.H.: A case for redundant arrays of inexpensive disks (RAID), vol. 17. ACM (1988)
21. Pless, V.: Introduction to the Theory of Error-Correcting Codes, vol. 48. Wiley, New York (2011)
22. Rozier, E.W., Sanders, W.H., Zhou, P., Mandagere, N., Uttamchandani, S.M., Yakushev, M.L.: Modeling the fault tolerance consequences of deduplication. In: 2011 30th IEEE Symposium on Reliable Distributed Systems (SRDS), pp. 75–84. IEEE (2011)
23. Rozier, E.W., Zhou, P., Divine, D.: Building intelligence for software defined data centers: modeling usage patterns. In: Proceedings of the 6th International Systems and Storage Conference, p. 20. ACM (2013)
24. Rozier, E.W.D., Sanders, W.H.: A framework for efficient evaluation of the fault tolerance of deduplicated storage systems. In: 2012 42nd Annual IEEE/IFIP International Conference on Dependable Systems and Networks (DSN), pp. 1–12. IEEE (2012)
25. Salido, M.A., Barber, F.: How to classify hard and soft constraints in non-binary constraint satisfaction problems. In: Coenen, F., Preece, A., Macintosh, A. (eds.) AResearch and Development in Intelligent Systems XX, pp. 213–226. Springer, New York (2004)
26. Schnase, J.L., Duffy, D.Q., Tamkin, G.S., Nadeau, D., Thompson, J.H., Grieg, C.M., McInerney, M.A., Webster, W.P.: Merra analytic services: Meeting the big data challenges of climate science through cloud-enabled climate analytics-as-a-service. Computers, Environment and Urban Systems (2014)
27. Schwarz, S., Long, D.D., Paris, J.F.: Reliability of disk arrays with double parity. In: 2013 IEEE 19th Pacific Rim International Symposium on Dependable Computing (PRDC), pp. 108–117. IEEE (2013)
28. Turner, V., Gantz, J.F., Reinsel, D., Minton, S.: The digital universe of opportunities: Rich data and the increasing value of the internet of things. International Data Corporation, White Paper, IDC_1672 (2014)
29. Weisstein, E.W.: Rooks problem (2002)

Quantitative Analysis of Consistency in NoSQL Key-Value Stores

Si Liu[✉], Son Nguyen, Jatin Ganhotra, Muntasir Raihan Rahman,
Indranil Gupta, and José Meseguer

Department of Computer Science, University of Illinois
at Urbana-Champaign, Champaign, USA
siliu3@illinois.edu

Abstract. The promise of high scalability and availability has prompted many companies to replace traditional relational database management systems (RDBMS) with NoSQL key-value stores. This comes at the cost of relaxed consistency guarantees: key-value stores only guarantee eventual consistency in principle. In practice, however, many key-value stores seem to offer stronger consistency. Quantifying how well consistency properties are met is a non-trivial problem. We address this problem by formally modeling key-value stores as probabilistic systems and quantitatively analyzing their consistency properties by statistical model checking. We present for the first time a formal probabilistic model of Apache Cassandra, a popular NoSQL key-value store, and quantify how much Cassandra achieves various consistency guarantees under various conditions. To validate our model, we evaluate multiple consistency properties using two methods and compare them against each other. The two methods are: (1) an implementation-based evaluation of the source code; and (2) a statistical model checking analysis of our probabilistic model.

1 Introduction

The promise of high scalability and availability has prompted many companies and organizations to replace traditional relational database management systems (RDBMS) with NoSQL key-value stores in order to store large data sets and tackle an increasing number of users. According to DB-Engines Ranking [2] by June 2015, three NoSQL datastores, namely MongoDB [3], Cassandra [1] and Redis [4], have advanced into the top 10 most popular database engines among 277 systems, highlighting the increasing popularity of NoSQL key-value stores. For example, Cassandra is currently being used at Facebook, Netflix, eBay, GitHub, Instagram, Comcast, and over 1500 more companies.

NoSQL key-value stores invariably replicate application data on multiple servers for greater availability in the presence of failures. Brewer's CAP Theorem [12] implies that, under network partitions, a key-value store must

---

Partially supported by NSF CNS 1319527, NSF 1409416, NSF CCF 0964471, and AFOSR/AFRL FA8750-11-2-0084.

J. Campos and B.R. Haverkort (Eds.): QEST 2015, LNCS 9259, pp. 228–243, 2015.
DOI: 10.1007/978-3-319-22264-6_15

choose between consistency (keeping all replicas in sync) and availability (latency). Many key-value stores prefer availability, and thus they provide a relaxed form of consistency guarantees (e.g., eventual consistency [27]). This means key-value store applications can be exposed to stale values. This can negatively impact key-value store user experience. Not surprisingly, in practice, many key-value stores seem to offer stronger consistency than they promise. Therefore there is considerable interest in accurately predicting and quantifying what consistency properties a key-value store actually delivers, and in comparing in an objective, and quantifiable way how well properties of interest are met by different designs.

However, the task of accurately predicting such consistency properties is non-trivial. To begin with, building a large scale distributed key-value store is a challenging task. A key-value store usually embodies a large number of components (e.g., membership management, consistent hashing, and so on), and each component can be thought of as source code which embodies many complex design decisions. Today, if a developer wishes to improve the performance of a system (e.g., to improve consistency guarantees, or reduce operation latency) by implementing an alternative design choice for a component, then the only option currently available is to make changes to huge source code bases (e.g., Apache Cassandra [1] has about 345,000 lines of code). Not only does this require many man months, it also comes with a high risk of introducing new bugs, needs understanding in a huge code base before making changes, and is unfortunately not repeatable. Developers can only afford to explore very few design alternatives, which may in the end fail to lead to a better design.

In this paper we address these challenges by proposing a formally model-based methodology for designing and quantitatively analyzing key-value stores. We formally model key-value stores as probabilistic systems specified by *probabilistic rewrite rules* [5], and quantitatively analyze their properties by *statistical model checking* [24, 30]. We demonstrate the practical usefulness of our methodology by developing, to the best of our knowledge for the first time, a formal probabilistic model of Cassandra, as well as of an alternative Cassandra-like design, in Maude [13]. Our formal probabilistic model extends and improves a nondeterministic one we used in [19] to answer *qualitative* yes/no consistency questions about Cassandra. It models the main components of Cassandra and its environment such as strategies for ordering multiple versions of data and message delay. We have also specified two consistency guarantees that are largely used in industry, *strong consistency* (the strongest consistency guarantee) and *read your writes* (a popular intermediate consistency guarantee) [26], in the QUATEX probabilistic temporal logic [5]. Using the PVESTA [6] statistical model checking tool we have then quantified the satisfaction of such consistency properties in Cassandra under various conditions such as consistency level combination and operation issue time latency. To illustrate the versatility and ease with which different design alternatives can be modeled and analyzed in our methodology, we have also modeled and analyzed the same properties for an alternative Cassandra-like design.

An important question is how much trust can be placed on such models and analysis. That is, how reliable is the predictive power of our proposed methodology? We have been able to answer this question for our case study as follows: (i) we have experimentally evaluated the same consistency properties for both Cassandra and the alternative Cassandra-like design[1]; and (ii) we have compared the results obtained from the formal probabilistic models and the statistical model checking with the experimentally-obtained results. Our analysis indicates that the model-based consistency predictions conform well to consistency evaluations derived experimentally from the real Cassandra deployment, with both showing that Cassandra in fact achieves much higher consistency (sometimes up to strong consistency) than the promised eventual consistency. They also show that the alternative design is not competitive in terms of the consistency guarantees considered. Our entire Maude specification, including the alternative design, has less than 1000 lines of code, which further underlines the versatility and ease of use of our methodology at the software engineering level.

Our main contributions include:

- We present a formal methodology for the quantitative analysis of key-value store designs and develop, to the best of our knowledge for the first time, a *formal executable probabilistic model* for the Cassandra key-value store and for an alternative Cassandra-like design.
- We present, to the best of our knowledge for the first time, a statistical model checking analysis for quantifying consistency guarantees, namely, *strong consistency* and *read your writes*, in Cassandra and the alternative design.
- We demonstrate the good predictive power of our methodology by comparing the model-based consistency predictions with experimental evaluations from a real Cassandra deployment on a real cluster. Our results indicate similar consistency trends for the model and deployment.

2   Preliminaries

2.1   Cassandra Overview

Cassandra [1] is a distributed, scalable, and highly available NoSQL database design. It is distributed over collaborative servers that appear as a single instance to the end client. Data are dynamically assigned to several servers in the cluster (called the *ring*), and each server (called a *replica*) is responsible for different ranges of the data stored as key-value pairs. Each key-value pair is stored at multiple replicas for fault-tolerance.

In Cassandra a client can perform *read* or *write* operations to query or update data. When a client requests a read/write operation to a cluster, the server it is connected to will act as a *coordinator* to forward the request to all replicas that hold copies of the requested key-value pair. According to the specified *consistency level* in the operation, the coordinator will reply to the client with a value/ack after collecting *sufficient* responses from replicas. Cassandra supports tunable

---

[1] We implemented the alternative Cassandra-like design by modifying the source code of Apache Cassandra version 1.2.0.

consistency levels with ONE, QUORUM and ALL the three major ones, e.g., a QUORUM read means that, when a *majority* of replicas respond, the coordinator returns to the client the most recent value. This strategy is thus called *Timestamp-based Strategy* (TB) in the case of processing reads. Note that replicas may return different timestamped values to the coordinator. To ensure that all replicas agree on the most current value, Cassandra uses in the background the *read repair* mechanism to update those replicas holding outdated values.

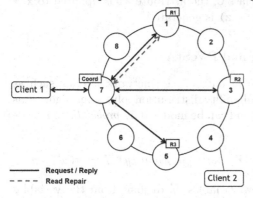

Figure 1 shows an example Cassandra deployed in a single data center cluster of eight nodes with three replicas and consistency level QUORUM. The read/write from client 1 is forwarded to all three replicas 1, 3 and 5. The coordinator 7 then replies to client 1 after receiving the first two responses, e.g., from 1 and 3, to fulfill the request without waiting for the reply from 5. For a read, upon retrieving all three possibly different versions of values, the coordinator 7 then issues a read repair write with the high-

Fig. 1. Cassandra deployed in a single cluster of 8 servers with replication factor 3

est timestamped value to the outdated replica, 1, in this example. Note that various clients may connect to various coordinators in the cluster, but requests from any client on the same key will be forwarded to the same replicas by those coordinators.

2.2 Rewriting Logic and Maude

Rewriting logic [21] is a semantic framework to specify concurrent, object-oriented systems as *rewrite theories* $(\Sigma, E \cup A, R)$, where $(\Sigma, E \cup A)$ is a *member-ship equational logic theory* [13], with Σ an algebraic signature declaring sorts, subsorts, and function symbols, E a set of conditional equations, and A a set of equational axioms. It specifies the system's state space as an algebraic data type; R is a set of *labeled conditional rewrite rules* specifying the system's *local transitions*, each of which has the form $[l] : t \longrightarrow t'$ **if** *cond*, where *cond* is a condition or guard and l is a label. Such a rule specifies a transition from an instance of t to the corresponding instance of t', provided the condition holds.

The Maude system [13] executes rewrite theories specified as Maude modules, and provides a rich collection of analysis tools. In this paper we consider distributed systems such as Cassandra made up of objects communicating with each other via asynchronous message passing. The distributed state of such a system is formalized as a *multiset* of objects and messages, and a state transition is *multiset rewriting*. In an object-oriented module, an object of the form < id : class | a1 : v1, a2 : v2, ..., an : vn > is an instance (with a unique name id) of the class that encapsulates the attributes a1 to an with the current values v1 to vn. Upon receiving a message, an object can change its state and send messages to other objects. For example, the rewrite rule (with label 1)

```
rl [1] :  m(O,z)  < O : C | a1 : x, a2 : O' >
       =>  < O : C | a1 : x + z, a2 : O' >  m'(O',x + z).
```

defines a transition where an incoming message m, with parameters O and z, is consumed by the target object O of class C, the attribute a1 is updated to x + z, and an outgoing message m'(O',x + z) is generated.

2.3 Statistical Model Checking and PVeStA

Distributed systems are probabilistic in nature, e.g., network latency such as message delay may follow a certain probability distribution, plus some algorithms may be probabilistic. Systems of this kind can be modeled by *probabilistic rewrite theories* [5] with rules of the form:

$$[l] : t(\overrightarrow{x}) \rightarrow t'(\overrightarrow{x}, \overrightarrow{y}) \;\; if \;\; cond(\overrightarrow{x}) \;\; with \;\; probability \;\; \overrightarrow{y} := \pi(\overrightarrow{x})$$

where the term t' has additional new variables $\overrightarrow{y}$ disjoint from the variables $\overrightarrow{x}$ in the term t. Since for a given matching instance of the variables $\overrightarrow{x}$ there can be many (often infinite) ways to instantiate the extra variables $\overrightarrow{y}$, such a rule is *non-deterministic*. The probabilistic nature of the rule stems from the probability distribution $\pi(\overrightarrow{x})$, which depends on the matching instance of $\overrightarrow{x}$, and governs the probabilistic choice of the instance of $\overrightarrow{y}$ in the result $t'(\overrightarrow{x}, \overrightarrow{y})$ according to $\pi(\overrightarrow{x})$. In this paper we use the above PMaude [5] notation for probabilistic rewrite rules.

Statistical model checking [24, 30] is an attractive formal approach to analyzing probabilistic systems against temporal logic properties. Instead of offering a yes/no answer, it can verify a property up to a user-specified level of confidence by running Monte-Carlo simulations of the system model. For example, if we consider strong consistency in Cassandra, a statistical model-checking result may be "The Cassandra model satisfies strong consistency 86.87 % of the times with 99 % confidence". Existing statistical verification techniques assume that the system model is purely probabilistic. Using the methodology in [5,15] we can eliminate non-determinism in the choice of firing rules. We then use PVESTA [6], an extension and parallelization of the tool VESTA [25], to statistically model check purely probabilistic systems against properties expressed by QUATEX probabilistic temporal logic [5]. The expected value of a QUATEX expression is iteratively evaluated w.r.t. two parameters α and δ provided as input by sampling until the size of $(1-\alpha)100\%$ confidence interval is bounded by δ. In this paper we will compute the expected probability of satisfying a property based on definitions of the form p() = BExp ; eval E[# p()] ;, where # is the next operator, BExp is a consistency-specific predicate, and p() is a state predicate returning the value either 1.0 or 0.0 after checking whether the current state satisfies BExp or not.

3   Replicated Data Consistency

Distributed key-value stores usually sacrifice consistency for availability (Brewer's CAP theorem [12]), advocating the notion of weak consistency (e.g., Cassandra promises eventual consistency [1]). However, studies on benchmarking eventually consistent systems have shown that those platforms seem in practice to offer more consistency than they promise [11,28]. Thus a natural question derived from those observations is "what consistency does your key-value store actually provide?" We summarize below the prevailing consistency guarantees advocated by Terry [26]. We will focus on two of them (*strong consistency* and *read your writes*) in the rest of this paper.

- Strong Consistency (SC) ensures that each read returns the value of the last write that occurred before that read.
- Read Your Writes (RYW) guarantees that the effects of all writes performed by a client are visible to her subsequent reads.
- Monotonic Reads (MR) ensures a client to observe a key-value store increasingly up to date over time.
- Consistent Prefix (CP) guarantees a client to observe an ordered sequence of writes starting with the first write to the system.
- (Time-) Bounded Staleness (BS) restricts the staleness of values returned by reads within a time period.
- Eventual Consistency (EC) claims that if no new updates are made, eventually all reads will return the last updated value.

Note that SC and EC lie at the two ends of the consistency spectrum, while the other intermediate guarantees are not comparable in general [26].

In [19,20] we investigated SC, RYW and EC from a qualitative perspective using standard model checking, where they were specified using *linear temporal logic* (LTL). The questions we answered there are simply yes/no questions such as "Does Cassandra satisfy strong consistency?" and "In what scenarios does Cassandra violate read your writes?". We indeed showed by counterexamples that Cassandra violates SC and RYW under certain circumstances, e.g., successive write and read with the combinations of lower consistency levels. Regarding EC, the model checking results of our experiments with bounded number of clients, servers and messages conforms to the promise. We refer the reader to [19,20] for details.

In this paper we look into the consistency issue for Cassandra in terms of SC and RYW from a quantitative, statistical model checking perspective. To aid the specification of the two properties (Sect. 5.1), we now restate them more formally. As all operations from different clients can be totally ordered by their issuing times, we can first view, from a client's perspective, a key-value store S as a history $H = o_1, o_2, ..., o_n$ of n read/write operations, where any operation o_i can be expressed as $o_i = (k, v, c, t)$, where t denotes the *global* time when o_i was issued by client c, and v is the value read from or written to on key k. We can then define both consistency properties based on H:

– We say S satisfies SC if for any read $o_i = (k, v_i, c_i, t_i)$, provided there exists a write $o_j = (k, v_j, c_j, t_j)$ with $t_j < t_i$, and without any other write $o_h = (k, v_h, c_h, t_h)$ such that $t_j < t_h < t_i$, we have $v_i = v_j$. Note that c_h, c_i and c_j are not necessarily different;

– We say S satisfies RYW if either (1) S satisfies SC, or (2) for any read $o_i = (k, v_i, c_i, t_i)$, provided there exists a write $o_j = (k, v_j, c_j, t_j)$ with $c_i = c_j$ and $t_j < t_i$, and with any other write $o_h = (k, v_h, c_h, t_h)$ such that $c_i \neq c_h$ and $t_j < t_h < t_i$, we have $v_i = v_j$.

4   Probabilistic Modeling of Cassandra Designs

This section describes a formal probabilistic model of Cassandra including the underlying communication model (Sect. 4.1) as well as an alternative Cassandra-like design (Sect. 4.2). The entire executable Maude specifications are available at https://sites.google.com/site/siliunobi/qest15-cassandra.

4.1   Formalizing Probabilistic Communication in Cassandra

In [19] we built a formal executable model of Cassandra summarized in Sect. 2.1. Specifically, we modeled the ring structure, clients and servers, messages, and Cassandra's dynamics. Moreover, we also introduced a *scheduler* object to schedule messages by maintaining a global clock GlobalTime[2] and a queue of inactive/scheduled messages MsgQueue. By activating those messages, it provides a deterministic total ordering of messages and allows synchronization of all clients and servers, aiding formal analysis of consistency properties (Sect. 5.1).

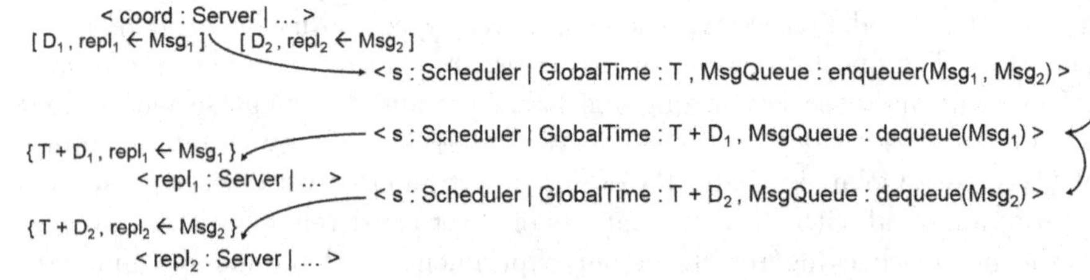

Fig. 2. Visualization of rewrite rules for forwarding requests from a coordinator to the replicas

To illustrate the underlying communication model, Fig. 2 visualizes a segment of the system transitions showing how messages flow between a coordinator and the replicas through the scheduler in terms of rewrite rules. The delayed messages

---

[2] Though in reality synchronization can never be exactly reached due to clock skew [17], cloud system providers use NTP or even expensive GPS devices to keep all clocks synchronized (e.g., Google Spanner). Thus our abstraction of a global clock is reasonable.

(of the form [...]) [D1, repl1 <- Msg1] and [D2, repl2 <- Msg2], targeting replicas repl1 and repl2, are produced by the coordinator at global time T with the respective message delays D1 and D2. The scheduler then enqueues both messages for scheduling. As the global time advances, messages eventually become active (of the form {...}), and are appropriately delivered to the replicas. For example, the scheduler first dequeues Msg1 and then Msg2 at global time T + D1 and T + D2 respectively, assuming D1 < D2. Note that messages can be consumed by the targets only when they are active.

As mentioned in Sect. 2.3, we need to eliminate nondeterminism in our previous Cassandra model prior to statistical model checking. This can be done by transforming nondeterministic rewrite rules to purely probabilistic ones. Below we show an example transformation, where both rules illustrate how the coordinator reacts upon receiving a read reply ReadReplySS from a replica, with KV the returned key-value pair of the form (key,value,timestamp), ID and A the read and client's identifiers, and CL the read's consistency level, respectively. The coordinator S adds KV to its local buffer, and returns to A the highest timestamped value determined by tb via the message ReadReplyCS, provided it has collected the consistency-level number of responses determined by cl?.

In the nondeterministic version [...-nondet], the outgoing message is equipped with a delay D nondeterministically selected from the delay set delays. We keep the set unchanged so that standard model checking will explore all possible choices of delays each time the rule is fired. For example, if delays: (2.0,4.0), two read replies will be generated nondeterministically with the delays 2.0 and 4.0 time units respectively, each of which will lead to an execution path during the state space exploration.

```
crl [on-rec-rrep-coord-nondet] :                    crl [on-rec-rrep-coord-prob] :
  {T, S <- ReadReplySS(ID,KV,CL,A)}                    {T, S <- ReadReplySS(ID,KV,CL,A)}
  < S : Server | buffer: BF, delays: (D,DS), AS >      < S : Server | buffer: BF, AS >
=> < S : Server | buffer: BF', delays: (D,DS), AS > => < S : Server | buffer: BF', AS >
  (if cl?(CL,BF') then                                 (if cl?(CL,BF') then
  [D, A <- ReadReplyCS(ID,tb(BF'))]                    [D, A <- ReadReplyCS(ID,tb(BF'))]
  else none fi)                                        else none fi)
if BF' := add(ID,KV,BF) .                           if BF' := add(ID,KV,BF)
                                                    with probability D := distr(...) .
```

We transform the above rule to the probabilistic version [...-prob], where the delay D is distributed according to the parameterized probability distribution function distr(...). Once the rule fires, only one read reply will be generated with a probabilistic real-valued message delay.

Likewise all nondeterministic rules in our previous model can be transformed to purely probabilistic rewrite rules. Furthermore, as explained in [5, 15], the use of continuous time and the actor-like nature of the specification ensure that *only one probabilistic rule is enabled at each time instant*, thus eliminating any remaining nondeterminism from the firing of rules.

4.2   Alternative Strategy Design

Two major advantages of our model-based approach are: (1) the ease of designing new strategies in an early design stage, and (2) the ability to predict their effects before implementation. Here we illustrates the first part by presenting as an alternative design the *Timestamp-agnostic Strategy* (TA). The key idea is that, instead of using timestamps to decide which value will be returned to the client as TB does (Sect. 2.1), TA uses the values themselves to decide which replica has the latest value. For example, if the replication factor is 3, then for a QUORUM read, the coordinator checks whether the values returned by the first two replicas are identical: if they are, the coordinator returns that value; otherwise it waits for the third replica to return a value. If the third value matches one of the first two values, the coordinator returns the third value. So for a QUORUM read TA guarantees that the coordinator will reply with the value that has been stored at a majority of replicas. For an ALL read, the coordinator compares all three values; if they are all the same, it returns that value. Notice that TA and TB agree on processing ONE reads.

To formalize TA (or other alternative strategies) we only need to specify the corresponding functions of the returned values from the replicas buffered at the coordinator, as we defined tb for TB, without redefining the underlying model. We omit the specification (available online) for simplicity. Note that our component-based model also makes it possible to dynamically choose the optimal strategy in favor of consistency guarantees. More precisely, once we have a pool of strategies and their respective strengths in consistency guarantees (which can be measured by statistical model checking), the coordinator can invoke the corresponding strategy-specific function based on the client's preference. For example, given strategies S1/S2 offering consistency properties C1/C2, if a client issues two consecutive reads with desired consistency C1, C2, respectively, the coordinator will generate, e.g., the C1-consistent value for the preceding read, by calling the strategy function for S1.

5   Quantitative Analysis of Consistency in Cassandra

How well do our Cassandra model and its TA alternative design satisfy SC and RYW? Does TA provide more consistency than TB based on our model? Are those results consistent with reality? We propose to investigate these questions by statistical model checking (Sect. 5.1) and by implementation-based evaluation (Sect. 5.2) of both consistency properties in terms of two strategies.

5.1   Statistical Model Checking Analysis

Scenarios. We define the following setting for our experimental scenarios of statistical model checking:

- We consider a single cluster of 4 servers, and the replication factor of 3.
- All replicas are initialized with default key-value pairs.
- Each read/write can have consistency level ONE, QUORUM or ALL, and all operations concern the same key.
- We consider the lognormal distribution for message delay with the mean $\mu = 0.0$ and standard deviation $\sigma = 1.0$ [10].

– All consistency probabilities are computed with a 99 % confidence level of size at most 0.01 (Sect. 2.3).

Figure 3 shows the two scenarios, with each parallel line denoting one session of one client. Regarding SC, we consider a scenario of three consecutive operations, W1, W2 and R3, issued by three different clients, respectively, where L1 and L2 are the issuing latencies between them. We choose to experiment with consistency level ONE for both W1 and W2 to evaluate different consistency levels for R3. Thus we name each subscenario (TB/TA-O/Q/A) depending on the target strategy and R3's consistency level, e.g., (TB-Q) refers to the case checking SC for TB with R3 of QUORUM.

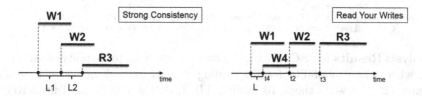

Fig. 3. Experimental scenarios of statistical model checking of SC and RYW

Regarding RYW, we consider a scenario with four operations, where W1, W2 and R3 are issued by one client and *strictly ordered* (a subsequent operation will be blocked until the preceding one on the same key finishes) while W4 is from the other client[3]. The issuing latency L is tunable, which can vary the issuing time of W4. Thus we can derive the corresponding cases in RYW's definition (Sect. 3), and specify and analyze the property accordingly. We choose to experiment with consistency level ONE for both W1 and W4 to evaluate different combinations of consistency levels for W2 and R3. The only possible cases violating RYW are, if we forget W4 for the moment, (R3,W2) = (O,O)/(O,Q)/(Q,O) due to the fact that a read is guaranteed to see its preceding write from the same client, if $R + W > RF$ with R and W the respective consistency levels and RF the replication factor. Thus we name each subscenario (TB/TA-OO/OQ/QO/...) depending on the target strategy and the combination of consistency levels. For simplicity, we let W2 and R3 happen immediately upon their preceding operations finish.

Formalizing Consistency Properties. Based on the consistency definitions (Sect. 3) and the above scenario, SC is satisfied if R3 reads the value of W2. Thus we define a parameterized predicate sc?(A,A',O,O',C) that holds if we can match the value returned by the subsequent read O (R3 in this case) from client A with that in the preceding write O' (W2) from client A'. Note that the attribute store records the associated information of each operation issued by the client: operation O was issued at global time T on key K with returned/written value V for a read/write.

---

[3] Section 3 describes two disjoint cases for RYW, which we mimic with tuneable L: if t4 < t2, only W2 is RYW-consistent; otherwise both W2 and W4 are RYW-consistent.

```
op sc? : Address Address Nat Nat Config -> Bool .
eq sc?(A,A',0,0',< A  : Client | store : ((0, K,V,T), ...), ... >
                 < A' : Client | store : ((0',K,V,T'), ...), ... > REST) = true .
```

Likewise we define for RYW a parameterized predicate ryw?(A,A',O1,O2, O3,C) that holds if we can match the value returned by the subsequent read O2 (R3 in this case) with that in the preceding write O1 by itself (W2 in this case), or in a more recent write O3 (W4 in this case if issued after W2) determined by T3 >= T1).

```
op ryw? : Address Address Nat Nat Nat Config -> Bool .
 eq ryw?(A,A',O1,O2,O3,< A  : Client | store : ((O1,K,V,T1), (O2,K,V, T2), ...),
       ... > REST) = true .
ceq ryw?(A,A',O1,O2,O3,< A  : Client | store : ((O1,K,V,T1), (O2,K,V',T2), ...), ... >
        < A' : Client | store : ((O3,K,V',T3), ...), ... > REST) = true if T3 >= T1 .
```

Analysis Results for SC. Figure 4 shows the resulting probability of satisfying SC, where the probability (of R3 reading W2) is plotted against the issuing latency (L2) between them. Regarding TB, from the results (and intuitively), given the same issuing latency, increasing the consistency level provides higher consistency; given the same consistency level, higher issuing latency results in higher consistency (as the replicas converge, a sufficiently later read (R3) will return the consistent value up to 100 %). Surprisingly, QUORUM and ALL reads start to achieve SC within a very short latency around 0.5 and 1.5 time units respectively (with 5 time units for even ONE reads).

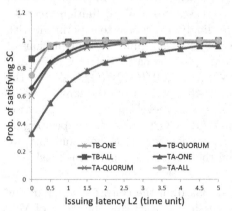

On the other hand, all observations for TB apply to TA in general. In fact, for QUORUM and ALL reads, the two strategies perform almost the same, except that: (1) for ALL reads, TB provides noticeably more consistency than TA within an extremely short latency of 0.5 time units; and (2) for QUORUM reads, TB offers slightly more consistency than TA within 2.5 time units.

Fig. 4. Probability of satisfying SC by statistical model checking

Based on the results it seems fair to say that both TB and TA provide high SC, especially with QUORUM and ALL reads. The consistency difference between the two strategies results from the overlap of R3 and W2. More precisely, since the subsequent read has higher chance to read multiple versions of the key-value pair with lower issuing latency, TA, only relying on the version itself, will return the matched value that is probably stale.

Analysis Results for RYW. Figure 5-(a) shows the resulting probability of satisfying RYW, where the probability (of R3 reading W2 or a more recent value) is plotted against the issuing latency (L) between W1 and W4. From the results it is straightforward to see that scenarios (TB-OA/QQ/AO/AA) guarantee RYW due to the fact "R3 + W2 > RF". Since we have already seen that the Cassandra model satisfied SC quite well, it is also reasonable that all combinations of consistency levels provide high RYW consistency, even with the lowest combination (0,0) that already achieves a probability around 90 %. Surprisingly, it appears that a QUORUM read offers RYW consistency nearly 100 %, even after a preceding write with the low consistency level down to ONE (scenario (TB-OQ)). Another observation is that, in spite of the concurrent write from the other client, the probability of satisfying RYW stays fairly stable.

Figure 5-(b) shows the comparison of TA and TB regarding RYW, where for simplicity we only list three combinations of consistency levels from R3's perspective with W2's consistency level fixed to ONE (in fact, with W2's consistency level increases, the corresponding scenarios will provide even higher consistency). In general, all observations for TB apply to TA, and it seems fair to say that both TA and TB offer high RYW consistency. Certainly TA and TB agree on the combination (0,0). However, TA cannot offer higher consistency than TB in any other scenario, with TA providing slightly lower consistency for some points, even though TA's overall performance is close to TB's over issuing latency. One reason is that TA does not respect the fact "R + W > RF" in general (e.g., two strictly ordered Quorum write and read cannot guarantee RYW).

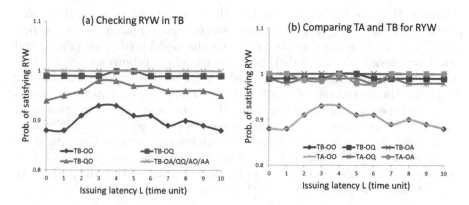

Fig. 5. Probability of satisfying RYW by statistical model checking

Remark. In summary, our Cassandra model actually achieves much higher consistency (up to SC) than the promised EC, with QUORUM reads sufficient to provide up to 100 % consistency in almost all scenarios. Comparing TA and TB, it seems fair to say that TA is not a competitive alternative to TB in terms of SC or RYW, even though TA is close to TB in most cases.

Our model, including the alternative design, is less than 1000 lines of code and the time to compute the probabilities for the consistency guarantees is 15 min (worst-case). The upper bound for model runtime depends on the confidence level of our statistical model checker (99 % confidence level for all our experiments).

5.2 Implementation-Based Evaluation of Consistency

Experimental Setup. We deploy Cassandra on a single Emulab [29] server, which means that the coordinator and replicas are separate processes on the same server. We use YCSB [14] to inject read/write workloads. For RYW tests, we use two separate YCSB clients. Our test workloads are read-heavy (that are representative of many real-world workloads such as Facebook's photo storage [9]) with 90 % reads, and we vary consistency levels between ONE, QUORUM, and ALL. We run Cassandra and YCSB clients for fixed time intervals and log the results. Based on the log we calculate the percentage of reads that satisfy SC/RYW. Note that other configurations follow our setup for statistical model checking.

Analysis Results for SC. We show the resulting, experimentally computed probability of strongly consistent reads against L2 (Fig. 3) for deployment runs regarding the two strategies (only for QUORUM and ALL reads). Overall, the results indicate similar trends for the model predictions (Fig. 4) and real deployment runs (Fig. 6-(a)): for both model predictions and deployment runs, the probability is higher for ALL reads than for QUORUM reads regarding both strategies, especially when L2 is low; consistency does not vary much with different strategies.

Analysis Results for RYW. We show the resulting probability of RYW consistent reads against L (Fig. 3) for deployment runs regarding two strategies. Again, the results indicate similar trends for the model predictions (Fig. 5) and real deployment runs (Fig. 6-(b)). Both the model predictions and deployment runs show very high probability of satisfying RYW. This is expected since for each client the operations are mostly ordered, and for any read operation from a client, we expect any previous write from the same client to be committed to all replicas. For the deployment runs, we observe that we get 100 % RYW consistency, except for scenario (TB-OO), which matches expectations, since ONE is the lowest consistency level and does not guarantee anything more than EC. This also matches our model predictions in Fig. 5, where we see that the probability of satisfying RYW for scenario (TB-OO) is lower compared to other cases.

Remark. Both the model predictions and implementation-based evaluations reach the same conclusion: Cassandra provides much higher consistency than the promised EC, and TA does not improve consistency compared to TB. Note that the actual probability values from both sides might differ due to factors like hard-to-match experimental configurations, the inherent difference between statistical model checking and implementation-based evaluation[4], and processing

---

[4] Implementation-based evaluation is based on a single trace of tens of thousands of operations, while statistical model checking is based on sampling tens of thousands of Monte-Carlo simulations of several operations (that can be considered as a segment of the trace) up to a certain statistical confidence.

delay at client/server side that our model does not include. However, the important observation is that the resulting trends from both sides are similar, leading to the same conclusion w.r.t. consistency measurement and strategy comparison.

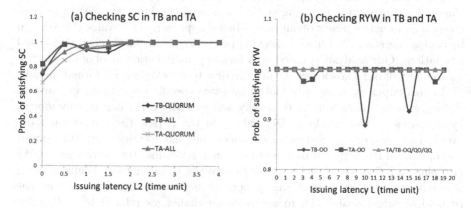

Fig. 6. Probability of satisfying SC/RYW by deployment run

6   Related Work and Concluding Remarks

Model-Based Performance Analysis of NoSQL Stores. Osman and Piazzolla [22] presents a queueing Petri net model of Cassandra parameterized by benchmarking only one server. The model is scaled to represent the characteristics of read workloads for different replication strategies and cluster sizes. Regarding performance, only response times and throughput are considered. Gandini, et al. [16] benchmarks three NoSQL databases, namely Cassandra, MongoDB and HBase, by throughput and operation latency. Two simple high-level queuing network models are presented that are able to capture those performance characteristics. Compared to both, our probabilistic model embodies the major components and features of Cassandra, and serves as the basis of statistical analysis of consistency with multiple clients and servers. Our model is also shown to be able to measure and predict new strategy designs by both statistical model checking and the conformance to the code-based evaluation. Other recent work on model-based performance analysis includes [8], which applies multi-formalism modeling approach to the Apache Hive query language for NoSQL databases.

Experimental Consistency Benchmarking in NoSQL Stores. In [11,23,28] propose active and passive consistency benchmarking approaches, where operation logs are analyzed to find consistency violations. Bailis, et al. [7] proposes probabilistic notions of consistency to predict the data staleness, and uses Monte-Carlo simulations to explore the trade-off between latency and consistency in Dynamo-style partial quorum systems. Their focus is more on developing the theory of consistency models. However, we focus on building a probabilistic model for a key-value store like Cassandra itself, and our objective is to compare the consistency benchmarking results with the model-based predictions from our statistical model checking.

Our main focus in this paper has been twofold: (i) to predict what consistency Cassandra can provide by using statistical model checking; and (ii) to demonstrate the predictive power of our model-based approach in key-value store design by comparing statistical model checking predictions with implementation-based evaluations. Our analysis is based on a formal probabilistic model of Cassandra. To the best of our knowledge, we are the first to develop such a formal model.

In this paper we have only looked into two specific consistency guarantees. A natural next step would be to specify and quantify other consistency models by statistical model checking. Depending on the perspective (key-value store providers, users, or application developers), different metrics (e.g., throughput and operation latency) can be used to measure key-value store performance. We also plan to refine our model in order to quantify those metrics. While showing scalability is not the goal of this paper, we believe our results will continue to hold at larger scales. There are resource challenges related to scaling the model checking to larger scales (e.g., parallelizing it in the proper way), and we plan to solve this in our future work. More broadly, our long-term goal is to develop a library of formally specified executable components embodying the key functionalities of NoSQL key-value stores (not just Cassandra), as well as of distributed transaction systems [18]. We plan to use such components and the formal analysis of their performance to facilitate efficient exploration of the design space for such systems and their compositions with minimal manual effort.

References

1. Cassandra. http://cassandra.apache.org
2. DB-Engines. http://db-engines.com/en/ranking
3. MongoDB. http://www.mongodb.org
4. Redis. http://redis.io
5. Agha, G.A., Meseguer, J., Sen, K.: PMaude: rewrite-based specification language for probabilistic object systems. Electr. Notes Theor. Comput. Sci. **153**(2), 213–239 (2006)
6. AlTurki, M., Meseguer, J.: Pvesta: a parallel statistical model checking and quantitative analysis tool. In: CALCO, pp. 386–392 (2011)
7. Bailis, P., Venkataraman, S., Franklin, M.J., Hellerstein, J.M., Stoica, I.: Probabilistically bounded staleness for practical partial quorums. Proceedings of the VLDB Endowment **5**(8), 776–787 (2012)
8. Barbierato, E., Gribaudo, M., Iacono, M.: Performance evaluation of NoSQL bigdata applications using multi-formalism models. Future Gener. Comp. Syst. **37**, 345–353 (2014)
9. Beaver, D., Kumar, S., Li, H.C., Sobel, J., Vajgel, P.: Finding a needle in haystack: facebook's photo storage. In: OSDI 2010, pp. 47–60 (2010)
10. Benson, T., Akella, A., Maltz, D.A.: Network traffic characteristics of data centers in the wild. In: IMC, pp. 267–280 (2010)

11. Bermbach, D., Tai, S.: Eventual consistency: how soon is eventual? an evaluation of amazon s3's consistency behavior. In: Middleware, p. 1 (2011)
12. Brewer, E.A.: Towards robust distributed systems (abstract). In: PODC, p. 7 (2000)
13. Clavel, M., Durán, F., Eker, S., Lincoln, P., Martí-Oliet, N., Meseguer, J., Talcott, C.: All About Maude - A High-Performance Logical Framework. LNCS, vol. 4350. Springer, Heidelberg (2007)
14. Cooper, B.F., Silberstein, A., Tam, E., Ramakrishnan, R., Sears, R.: Benchmarking cloud serving systems with YCSB. In: SOCC, pp. 143–154 (2010)
15. Eckhardt, J., Mühlbauer, T., Meseguer, J., Wirsing, M.: Statistical model checking for composite actor systems. In: WADT, pp. 143–160 (2012)
16. Gandini, A., Gribaudo, M., Knottenbelt, W.J., Osman, R., Piazzolla, P.: Performance evaluation of NoSQL databases. In: EPEW, pp. 16–29 (2014)
17. Lamport, L.: Time, clocks, and the ordering of events in a distributed system. Commun. ACM 21(7), 558–565 (1978)
18. Liu, S., Rahman, M.R., Ganhotra, J., Ölveczky, P.C., Gupta, I., Meseguer, J.: Formal modeling and analysis of RAMP transaction systems (2015). https://sites. google.com/site/siliunobi/ramp
19. Liu, S., Rahman, M.R., Skeirik, S., Gupta, I., Meseguer, J.: Formal modeling and analysis of cassandra in maude. In: ICFEM, pp. 332–347 (2014)
20. Liu, S., Rahman, M.R., Skeirik, S., Gupta, I., Meseguer, J.: Formal modeling and analysis of cassandra in maude (2014). https://sites.google.com/site/siliunobi/ icfem-cassandra
21. Meseguer, J.: Conditional rewriting logic as a unified model of concurrency. Theor. Comput. Sci. 96, 73–155 (1992)
22. Osman, R., Piazzolla, P.: Modelling replication in nosql datastores. In: QEST, pp. 194–209 (2014)
23. Rahman, M.R., Golab, W., AuYoung, A., Keeton, K., Wylie, J.J.: Toward a principled framework for benchmarking consistency. In: HotDep (2012)
24. Sen, K., Viswanathan, M., Agha, G.: On statistical model checking of stochastic systems. In: CAV, pp. 266–280 (2005)
25. Sen, K., Viswanathan, M., Agha, G.A.: VESTA: a statistical model-checker and analyzer for probabilistic systems. In: QEST, pp. 251–252 (2005)
26. Terry, D.: Replicated data consistency explained through baseball. Commun. ACM 56(12), 82–89 (2013)
27. Vogels, W.: Eventually consistent. Commun. ACM 52(1), 40–44 (2009)
28. Wada, H., Fekete, A., Zhao, L., Lee, K., Liu, A.: Data consistency properties and the trade-offs in commercial cloud storage: the consumers' perspective. In: CIDR, pp. 134–143 (2011)
29. White, B., Lepreau, J., Stoller, L., Ricci, R., Guruprasad, S., Newbold, M., Hibler, M., Barb, C., Joglekar, A.: An integrated experimental environment for distributed systems and networks. In: OSDI, pp. 255–270 (2002)
30. Younes, H.L.S., Simmons, R.G.: Statistical probabilistic model checking with a focus on time-bounded properties. Inf. Comput. 204(9), 1368–1409 (2006)

Impact of Policy Design on Workflow Resiliency Computation Time

John C. Mace$^{(\boxtimes)}$, Charles Morisset, and Aad van Moorsel

School of Computing Science, Newcastle University,
Newcastle upon Tyne NE1 7RU, UK
{john.mace,charles.morisset,aad.vanmoorsel}@ncl.ac.uk

Abstract. Workflows are complex operational processes that include security constraints restricting which users can perform which tasks. An improper user-task assignment may prevent the completion of the workflow, and deciding such an assignment at runtime is known to be complex, especially when considering user unavailability (known as the resiliency problem). Therefore, design tools are required that allow fast evaluation of workflow resiliency. In this paper, we propose a methodology for workflow designers to assess the impact of the security policy on computing the resiliency of a workflow. Our approach relies on encoding a workflow into the probabilistic model-checker PRISM, allowing its resiliency to be evaluated by solving a Markov Decision Process. We observe and illustrate that adding or removing some constraints has a clear impact on the resiliency computation time, and we compute the set of security constraints that can be artificially added to a security policy in order to reduce the computation time while maintaining the resiliency.

Keywords: Workflow satisfiability problem · Probabilistic model checker · User availability

1 Introduction

Workflows are used in multiple domains, for instance business environments, to represent complex operational processes [7,14], or healthcare environments, to represent the different protocols that must be respected [26]. There is also an increasing interest in scientific environments, where platforms like eScience Central [17] allow domain experts to define scientific processes, which are then automatically deployed and executed. Although the exact definition can change from one context to another, a workflow typically consists of a partially ordered set of tasks, where each task must be executed by a user [1]. Workflow designers may have to impose complex security policies, restricting which users can perform which tasks. This includes static user-task permissions but also dynamic constraints, such as separation or binding of duty constraints, which indicate tasks that cannot be performed by the same user [19], or tasks that must be performed by the same user [9], respectively.

© Springer International Publishing Switzerland 2015
J. Campos and B.R. Haverkort (Eds.): QEST 2015, LNCS 9259, pp. 244–259, 2015.
DOI: 10.1007/978-3-319-22264-6_16

In general, purely granting an assignment request for a task based on its user permissions and constraints with previously executed tasks may not be enough. Assigning a specific user to a task can prevent the completion of the workflow at a later stage, meaning in general, all possible options have to be considered. Checking that a particular user-task assignment is both valid and allows the workflow to finish is known as the workflow satisfiability problem (WSP) and has been shown to be NP-hard [12,28], indicating the runtime assignment process may be computationally demanding.

Workflow resiliency extends the WSP by considering users may become unavailable at runtime, a concept first introduced by Wang and Li [28]. This problem was later refined by Mace et al. [22], who considered a more quantitative approach, where each user is associated with a probability to become unavailable, and showed that calculating the resiliency of a workflow was equivalent to finding the optimal policy of a Markov Decision Process (MDP). The value returned by the value function of the MDP provides a measure of likely workflow completion. Therefore, evaluating resiliency at runtime can ensure assignments are granted only if the rest of the workflow can be satisfied with a probability above a given resiliency threshold.

Indeed, contrary to the WSP, which can be solved at design time, maximising the resiliency of workflow requires to re-compute at each step the expected resiliency, in order to adjust the user-task assignment to the current availability of the users. Hence, evaluating resiliency for assignments at runtime has itself an impact on workflow execution time. Recent optimising approaches for the WSP, such as [12], and algorithms and tools, such as model-checking [2], have been proposed, however, they are not directly concerned with user availability.

In this paper, we investigate how to improve the computation time for the resiliency of a workflow at runtime. In particular, we observe that adding or removing security components to the security policy has a clear impact on the resiliency computation time, that can be either increased or decreased. We therefore propose a methodology to help a workflow designer assess the impact of such policy changes. We apply this methodology to show how to compute the set of security constraints that can be added to a workflow, without impacting the actual resiliency while significantly decreasing the resiliency computation time.

After discussing the related work (Sect. 2) and formally defining the notion of workflow resiliency (Sect. 3), we present the contributions of this paper, which are: the automated analysis of workflow resiliency, using an encoding in the probabilistic model checker PRISM [20] of the theoretical approach presented in [22] (Sect. 4); the empirical assessment of policy changes on the resiliency computation time (Sect. 5); the methodology to calculate a set of *artificial* security policy constraints, in order to reduce the resiliency computation time while maintaining the actual resiliency value, and its illustration on an example (Sect. 6). We believe that building efficient tools for the analysis of workflows will be helpful to workflow designers, by helping them understanding the complexity of the workflow they are building, and estimating the potential runtime impact of their security policy designs.

2   Related Work

A number of previous studies on workflow resiliency appear in the literature. Wang and Li took a first step in [28] to quantify resiliency by addressing the problem of whether a workflow can still complete in the absence of users and defined a workflow as k resilient to all failures of up to k users across an entire workflow. Lowalekar et al. in [21] use security attributes to choose the most favourable between multiple assignments exhibiting the same level of k resiliency.

Basin et al. consider the impact of security on resiliency by allowing user-task permission changes to overcome user failure induced workflow blocks, at a quantifiable cost [5,6]. Wainer et al. also consider in [27] the explicit overriding of security constraints in workflows, by defining a notion of privilege. Similarly, Bakkali [4] suggests overcoming user unavailability through selected delegation and the placement of criticality values over workflows.

A mechanism for the specification and enforcement of workflow authorisation constraints is given by Bertino et al. in [8] whilst Ayed et al. discuss security policy definition and deployment for workflow management systems in [3]. Model checking has been used by Armando et al. [2] to formally model and automatically analyse security constrained business processes to ensure they meet given security properties. He et al. in [15] also use modelling techniques to analyse security constraint impact in terms of computational time and resources on workflow execution.

Herbert et al. in [16] model workflows expressed in BPMN as MDPs. The probabilistic model checker PRISM is utilised to check various probabilistic properties such as reaching particular states of interest, or the occurrence and ordering of certain events. Calinescu et al. use PRISM to evaluate the Quality of Service (QoS) delivered by dynamically composed service-based systems [11]. PRISM has also been used for identifying and recovering from runtime requirement violations in dynamically adaptable application software [10]. Quantitative access control using partially-observable MDPs is presented by Martinelli et al. in [25] which under uncertainty, aims to optimise the decision process for a sequence of access requests.

However, to the best of our knowledge, there is no current literature neither on automatic analysis of workflow resiliency, nor on the analysis of how changes to a workflow's security policy impact resiliency computation, which is the focus of this paper.

3   Workflow

In this section we provide our working definition of a workflow and describe the process of assigning users to tasks whilst respecting the security policy, known as the workflow satisfiability problem (WSP). We then describe the notion of workflow resiliency which looks to solve the WSP under the assumption users may become unavailable for future task assignments.

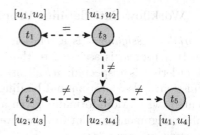

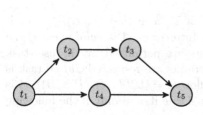

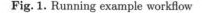

Fig. 1. Running example workflow   **Fig. 2.** Running example security policy

3.1 Workflow Definition

We define here a workflow in similar fashion to Wang and Li [28] and Crampton et al. [12]. A workflow firstly consists of a partially ordered set of tasks $(T, <)$, such that for any two tasks $t, t' \in T$, if $t < t'$, then t must be performed before t' in any given instance of the workflow.

Each task needs to be assigned to a user in a given set U, and such an assignment must respect a *security policy*. In general, a policy is a triple $p = (P, S, B)$ where:

- $P \subseteq U \times T$ is a set of *user-task permissions*, such that $(u, t) \in P$ if, and only if u is allowed to perform t.
- $S \subseteq T \times T$ is a set of *separations of duty*, such that $(t, t') \in S$ if, and only if the users assigned to t and t' are distinct.
- $B \subseteq T \times T$ is a set of *bindings of duty*, such that $(t, t') \in B$ if, and only if the same user is assigned to t and t'.

Definition 1 (Workflow). *A workflow is a tuple* $w = (U, (T, <), p)$, *where* U *is a set of users,* T *is a partially ordered set of tasks, and* p *is a security policy.*

Running Example. *As a running example to illustrate the different concepts presented here, we consider the workflow* $w_1 = (U_1, (T_1, <), p_1)$, *where* $U_1 = \{u_1, u_2, u_3, u_4\}$, $T_1 = \{t_1, t_2, t_3, t_4, t_5\}$ *such that* $t_1 < t_2 < t_3 < t_5$ *and* $t_1 < t_4 < t_5$, *and the* p_1 *is defined as the triple* (P_1, S_1, B_1) *where:*

- $P_1 = \{(u_1, t_1), (u_2, t_1), (u_2, t_2), (u_3, t_2), (u_1, t_3),$
  $(u_2, t_3), (u_2, t_4), (u_4, t_4), (u_1, t_5), (u_4, t_5)\}$
- $S_1 = \{(t_2, t_4), (t_3, t_4), (t_4, t_5)\}$
- $B_1 = \{(t_1, t_3)\}$

Figure 1 illustrates the task ordering over T_1 *and Fig. 2 illustrates this security policy, where a dotted arrow signifies a constraint between the tasks* t *and* t' *labelled* $\neq$ *to indicate a separation of duty, and* $=$ *to indicate a binding of duty. A label* $[u_i, \ldots, u_j]$ *states the users that are authorised by* P_1 *to execute* t.

3.2   Workflow Satisfiability Problem

A *workflow assignment* is a relation $A \subseteq U \times T$, such that $(u_i, t_i) \in A$ indicates that user u_i is assigned to the task t_i. Intuitively, A is *valid* when *i)* the task ordering is respected; *ii)* all assignments are permitted by the user-task permission; *iii)* separation and binding constraints are respected; *iv)* no task is executed twice. More formally, given a workflow $w = (U, (T, <), (P, S, B))$, A is a valid assignment, and in this case we write $A \vdash w$ if, and only if the following five conditions are met:

$$\forall (u, t) \in A \ \forall t' \in T \ t' < t \Rightarrow \exists u' \in U \ (u', t') \in A \tag{1}$$

$$A \subseteq P \tag{2}$$

$$\forall (t, t') \in S \ \exists (u, t) \in A \ \exists (u', t') \in A \Rightarrow u' \neq u \tag{3}$$

$$\forall (t, t') \in B \ \exists (u, t) \in A \ \exists (u', t') \in A \Rightarrow u' = u \tag{4}$$

$$\forall t \in T \ \forall u, u' \in U \ (u, t) \in A \wedge (u', t) \in A \Rightarrow u = u' \tag{5}$$

A workflow assignment A is said to be a *partial* if it does not include an assignment for every task in the workflow. For instance, in our running example, $\{(u_1, t_1), (u_3, t_2), (u_2, t_4)\}$ is a valid partial assignment whereas $\{(u_1, t_1), (u_2, t_2), (u_2, t_4)\}$ is not as it violates the separation of duty constraint between tasks t_2 and t_4. For a workflow to complete successfully, *every* task needs to be assigned a user for execution. A workflow assignment A is therefore said to be *complete*, if, and only if:

$$\forall t \in T \ \exists u \in U \ (u, t) \in A \tag{6}$$

The workflow satisfiability problem (WSP) consists of finding a complete and valid assignment, and in some cases can be relatively simple. For instance, consider a policy where $S = B = \emptyset$, i.e., where there are no separations or bindings of duty. In this case, it is enough to assign each task t with a user u such that $(u, t) \in P$. If there is no such user, the workflow is unsatisfiable. However, in general, the WSP has been shown to be NP hard [28], i.e., roughly speaking, finding a complete and valid assignment might require to check all possible assignments. With our running example, imagine we want to find a complete assignment for w_1 and begin assigning users to tasks t_1, t_2 and t_4 to form the partial assignment $A = \{(u_2, t_1), (u_3, t_2), (u_2, t_4)\}$. Although this assignment is valid, there is no user u such that $A \cup \{(u, t_3)\}$ is also valid, meaning that the workflow cannot finish. However with the partial assignment $\{(u_2, t_1), (u_3, t_2), (u_4, t_4)\}$, we can add (u_2, t_3) and (u_1, t_5) to form a valid and complete assignment.

3.3   Workflow Resiliency

Solving the WSP assumes users will always be available for future tasks, however in practice, sickness, vacation, heavy workloads, etc., can cause users to be unavailable for a given user-task assignment. It is important to take this into account when finding A for a given workflow. This is called the resiliency problem, whether a workflow can be satisfied even when some users become absent.

Wang and Li defined an approach to calculate a valid assignment if one exists, that is resilient to up to k users failing, in other words declaring a workflow to be either k resilient or not [28]. This approach is rather *binary* as in many cases, finding an assignment for a workflow that is resilient to every combination of k user failures may be impossible. Yet finding a valid assignment that is resilient in 9 out of 10 cases is arguably better than choosing a valid assignment that is resilient in only 1 out of 10 cases.

The problem of resiliency adds another level of complexity to the WSP. For instance, consider $\{(u_1, t_1), (u_2, t_2), (u_1, t_3), (u_4, t_4), (u_1, t_5)\} \vdash w_1$ in our running example, where u_4 has a very high probability of failing at or before t_4. If u_4 does fail, t_4 cannot be reassigned to any other user meaning the workflow cannot finish. If we chose a different assignment $\{(u_1, t_1), (u_3, t_2), (u_1, t_3), (u_4, t_4), (u_1, t_5)\} \vdash w_1$, intuitively the workflow is more resilient as t_4 can be reassigned to u_2 and still finish if u_4 did indeed fail. In [22], Mace et al. introduce probabilistic user failures and show that computing the optimal policy of an MDP is equivalent to finding $A \vdash w$ that maximises the value function. The value function returns $0 < v \leq 1$ if there exists $A \vdash w$ where v indicates the probability of the workflow to finish, or 0 otherwise.

Moreover, Mace et al. define in [24] several user availability models and discuss the effects model choice can have on workflow resiliency analysis. In this paper we consider a dynamic user availability model meaning any user who becomes unavailable for a task may become available again at any step later in the workflow.

4   Computing Workflow Resiliency at Runtime

Although user availability is modelled in a probabilistic way, at runtime, a user is either available or not. In other words, the resiliency of a workflow denotes a prediction of completion, and not a level a completion: a workflow only terminates if all tasks have been assigned to a user available for that task. When the availability of users does not change at runtime, any valid assignment computed before execution remains valid throughout execution. However, when user availability is dynamic, the validity of an assignment might change during the execution, and therefore a new assignment might need to be found.

According to Crampton and Khambhammettu [13], there are two main workflow execution models: *workflow-driven execution model* (WDEM), where users are automatically assigned tasks to execute, and *user-driven execution model* (UDEM), where users initiate requests to be assigned tasks at runtime. The impact of dynamic user availability is slightly different between the two models: with WDEM, intuitively, we want to continuously compute the most resilient assignment, adapting to changes in user availability; Whereas with UDEM, we want to ensure that a user asking to execute a specific task either belongs to the most resilient assignment, or satisfies a threshold of resiliency.

With either model, resiliency might then need to be recomputed at runtime, which can be done by solving a Markov Decision Process (MDP) [22]. There are

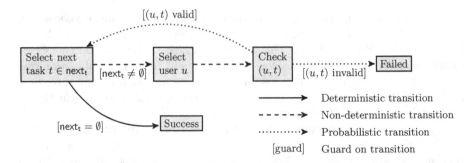

Fig. 3. Process computing the resiliency of a workflow, where next$_t$ denotes the set of tasks remaining to be executed.

many ways to solve an MDP including dynamic programming (e.g. value iteration) [18]. This technique is provided by the probabilistic model checking tool PRISM, which enables the specification, construction and analysis of probabilistic models such as MDPs [20]. PRISM is an intuitive choice as it can model both probabilistic and non-deterministic choice, and gives an efficient way to solve an MDP whilst providing analysis data regarding computation overheads.

The workflow assignment process is shown as a state diagram in Fig. 3. Each node represents a process state while each directed arrow between two states s and s' indicates a transition from state s to state s'. The assignment process works as follows: 1) given a starting state the next unassigned task t in the workflow is selected, where the set of possible tasks is represented by next$_t$. Task selection is in general non-deterministic since several tasks can be the next one (i.e. in the case of parallel execution). If all tasks have been assigned then next$_t = \emptyset$; 2) When a task t is selected, an arbitrary user u is selected to be assigned to t. The selection of u is non-deterministic as the MDP will essentially try every user for each task assignment. 3) The user-task assignment (u, t) is then checked to see whether it is valid; in other words whether u available and (u, t) satisfies the workflow's security policy p. This check is probabilistic, since user availability is probabilistic. If (u, t) is valid, u is assigned to t and the process starts again with the next task, otherwise the workflow terminates early.

The resiliency of the workflow is therefore computed as the maximal probability of reaching the state **Success**. We provide an overview of the PRISM modelling language and the full PRISM encoding of our running example in a technical report [23]. In our running example the resiliency is computed to be 51.16 % with the probabilistic user availabilities given in Table 4, Appendix A.

5   Empirical Assessment of Policy Changes

In this section we provide an empirical assessment of resiliency computation time to help understanding of how it can be improved at runtime. In doing so we investigate the impact upon computation time of adding security constraints to the security policy.

Table 1. Result averages when applying randomly generated security policies to a workflow with 10 tasks and 5 users

	0	1~5	6~10	11~15	16~20	21~25	26~30	31~35	36~40	41~45
Resiliency (%)	58.23	57.97	55.73	52.78	50.49	46.02	34.85	15.31	0.89	0
0 % resiliency	0	0	0	1	0	11	90	305	488	500
Computation (s)	0.11	0.38	1.56	2.24	1.80	1.08	0.52	0.20	0.07	0.04
Build time (s)	0.56	2.83	16.12	25.91	21.72	13.81	7.52	4.38	2.55	1.78
Total time (s)	0.67	3.21	17.68	28.15	23.52	14.89	8.04	4.58	2.62	1.82
States	3893	58246	346992	600287	522850	332259	171627	89361	47140	29387
Transitions	73249	758351	3352889	4754705	3649065	2171394	1090709	561534	294596	182751

5.1 Assessment Methodology

We first consider one workflow with 10 tasks and 5 users. For simplicity we only consider the addition of separation of duty constraints which is sufficient to show the changes to resiliency computation time. The maximum number of separation of duty constraints for a workflow of 10 tasks is 45 constraints. For each i where $0 \leq i \leq 45$ we generate 100 random security policies such that the permissions policy P contains between 2 and 5 users for each task, the separations of duty policy S contains i constraints, and the bindings of duty policy B has 0 constraints. Each policy is applied to the workflow meaning in all we analyse 4500 workflows using a computing platform incorporating a 2.40Ghz i7-4500U Intel processor and 8GB RAM. To take into account any influence the computing platform may have on analysis time, each analysis is repeated 50 times for each workflow and the average values taken.

In terms of user availability we use a dynamic availability model with probabilities of between 0.8 and 1.0 for each user u to be available for a task t (Table 5, Appendix A) . The resiliency of each workflow is calculated with an unmodified version 4.2.1 of the PRISM model checker using the *explicit* engine. This is suitable for models with a potentially very large state space, only a fraction of which is actually reachable. A test program has been implemented which, given a number of inputs (number of workflows, tasks, users, etc.) creates the required workflows with randomly generated security policies and generates the corresponding PRISM definition files. Each file is passed in turn to the PRISM model checker which logs the output composed of the resiliency value and other computational values including computation time.

5.2 Results

The results shown in Table 1 are given for our workflow with random separation of duty constraints applied, from 0 to 45. To place the workflow resiliency value and its computation time into perspective the following result averages are provided:

- **Resiliency** : workflow resiliency value
- **0 % resiliency** : number of workflows unable to complete
- **Computation** : time to verify the state **Success** is reachable (Sect. 4)

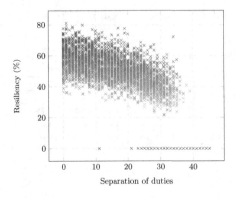
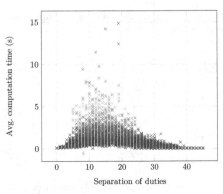

Fig. 4. Resiliency values for a workflow with 10 tasks and 5 users

Fig. 5. Average computation times for a workflow with 10 tasks and 5 users

- **Build time** : time to build PRISM model
- **Total time** : computation + build time
- **States** : number of reachable states in PRISM model
- **Transitions** : number of transitions between states

The following sections provide more detailed analysis of resiliency and computation time.

Resiliency Analysis. Note the resiliency returned by a single execution of a workflow is the same as the average resiliency of 50 executions, in other words the resiliency does not change due to fixed parameters. Each plot in Fig. 4 therefore represents a set of workflows with the same resiliency value and the same number of separation of duty constraints (although each workflow in the set comes with a different set of constraints).

Figure 4 shows how in general, resiliency steadily reduces following an incremental introduction of separation of duty constraints. For example, with no constraints, the workflows generally have between 40 and 80 % resiliency and with 20 constraints between 30 and 70 %. All the workflows with between 0 and 10 constraints are shown to be resilient to some degree and up to the point where 20 constraints are applied, all except 1 workflow have some resiliency.

Each plot where resiliency is zero indicates a set of workflows with the same number of separation of duty constraints whose security policy prevents completion. For example, 2 out of 100 workflows with 21 constraints are unable to complete, whilst 36 out of 100 are unable to complete with 30 constraints. No workflow is resilient once 40 constraints have been applied, however some workflows do exist which have some resiliency even with up to 39 out of a possible 45 constraints. The results indicate that some separation of duty constraints can be added or removed with no effect on resiliency.

Computation Time. The overall time to compute the resiliency of a workflow can be separated into the time it takes to construct the PRISM model from the workflow definition (*build time*), and the time it takes to verify the finishing property holds in model (*computation time*). If a change is made to the definition before verification, PRISM automatically rebuilds and verifies the model so the total time must be considered. However, once a model has been built and no changes are made, it need not be rebuilt. This is useful where cached, pre-built models can be imagined meaning only computation time need be taken into account when making runtime assignments. It is this time we are interested in improving.

Figure 5 shows how in general, the computation time increases and then decreases despite an incremental introduction of separation of duty constraints. The actual times measured are of course somewhat dependent on the efficiency of the model checker used, in this case PRISM. The maximum average computation time is 2.24 s with 11~15 separation of duty constraints. With zero constraints and 41~45 constraints the average computation time is 0.11 and 0.04 s respectively. The latter results can intuitively be attributed to the average 0 % resiliency value when all 45 constraints are applied. However, even with 26~30 constraints and an average 34.85 % resiliency, the average computation time is lower at 0.52 s than the time with 11~15 constraints. This would indicate the workflows are on average at their most complex in terms of longest resiliency computation time when approximately 11~15 separation of duty constraints have been applied.

By observing the size of the model that PRISM must solve, in terms of the number of states and transitions, the computation time can be put into context. The maximum average of 2.24 s is the computation time taken by PRISM to solve a model with an average 600287 states and 4.75 million transitions. These two values are the maximum average values recorded for states and transitions respectively. As one would expect, computation time appears to be closely related to the size of the model meaning in order to reduce computation time we must look to reduce the size of the model without losing resiliency. The results do indicate that in some cases separation of duty constraints can be added or removed to a workflow without any loss of resiliency.

6   Reducing Computation Time

In this section we provide a methodology to calculate a set of *dummy* security policy constraints (e.g., *redundant* separation-of-duty constraints or removing unused user-task permissions), in order to reduce the resiliency computation time while maintaining the actual resiliency value.

It was shown in Sect. 5 that in some cases, separation of duty constraints could be added to or removed from a workflow security policy. We are not in a position to say which constraints can be removed as this may weaken the security policy. Therefore we only consider strengthening the policy, in other words adding separation of duty constraints and removing user-task permissions which in effect can be removed at a later stage if necessary without any loss of security.

Table 2. Average computation times and resiliency values when adding a single separation of duty constraint or removing a single permission from the running example policy p_1

	p_1	$+(t_2, t_3)$	$+(t_2, t_5)$	$+(t_3, t_5)$	$+(t_1, t_4)$	$-(u_4, t_4)$	$-(u_4, t_5)$	$-(u_2, t_4)$	$-(u_1, t_1)$
Resiliency (%)	51.16	47.89	51.16	51.16	51.16	39.47	51.16	51.16	51.16
Computation (s)	0.11	0.109	0.141	0.11	0.063	0.047	0.121	0.11	0.062

6.1 Adding Separations of Duty

In our running example workflow w_1 coming with probabilistic user availabilities (Table 4, Appendix A), the resiliency is computed to be 51.16 % at an average computation time of 0.11 s, based on the average of 50 resiliency calculations. Imagine we now add a new separation of duty constraint (t_2, t_3) to give a new policy $p_2 = (P_2, S_2, B_2)$ where $P_2 = P_1$, $S_2 = S_1 \cup \{(t_2, t_3)\}$, and $B_2 = B_1$. The resiliency of w_1 coming with p_2 is now computed to be 47.89 % at an average computation time of 0.109 s. In other words, the computation time has reduced by 0.001 s but with a loss of 3.27 % resiliency.

Now consider adding in turn some alternative separation of duty constraints (t_2, t_5),(t_3, t_5) and (t_1, t_4) to p_1 to give new policies p_3, p_4 and p_5 respectively. The resiliency values and average computation times are given in Table 2 where $+(t, t')$ denotes the addition of a separation of duty constraint to p_1, whilst $-(u, t)$ denotes the removal of a user-task permission from p_1. The addition of (t_2, t_5) to p_1 (p_3) results in no loss to resiliency but increases the average computation time by 0.031 s. Adding (t_3, t_5) to p_1 (p_4) results in no loss to resiliency nor any reduction of average computation time. However, adding (t_1, t_4) to p_1 (p_5) results in no loss to resiliency yet a reduction to the average computation time of 0.047 s.

6.2 Removing User Permissions

Similarly we now consider removing a user-task permission (u_4, t_4) to give a new policy $p_6 = (P_6, S_6, B_6)$ where $P_6 = P_1 \setminus \{(u_4, t_4)\}$, $S_6 = S_1$, and $B_6 = B_1$. The resiliency of w_1 coming with p_6 is now computed to be 39.47 % at an average computation time of 0.047 s. In other words, the computation time has reduced by 0.063 s but with a loss of 11.69 % resiliency.

We now consider removing in turn some alternative user-task permissions (u_4, t_5),(u_2, t_4) and (u_1, t_1) from p_1 to give new policies p_7, p_8 and p_9 respectively. The resiliency values and average computation times are given in Table 2. The removal of (u_4, t_5) from p_1 (p_7) results in no loss to resiliency but increases the average computation time by 0.011 s to 0.121 s. Removing (u_2, t_4) from p_1 (p_8) results in no loss to resiliency nor any reduction of average computation time. However, removing (u_1, t_1) from p_1 (p_9) results in no loss to resiliency yet reduces the average computation time by 0.048 s to 0.062 s. These results indicate that a selective addition of separation of duty constraints, or removal of user-task permissions can reduce the resiliency computation time without any loss to the actual resiliency value.

Table 3. Average computation times and resiliency values when adding separation of duty constraints or removing permissions from w_B

	w_B	$+s_1$	$+s_2$	$+s_3$	$-p_1$	$-p_2$	$-p_3$
Resiliency (%)	63.96	63.42	62.84	62.21	62.52	60.54	57.92
Computation (s)	6.53	4.28	3.55	3.29	5.09	3.83	1.76

6.3 Calculating Dummy Constraints

With the aid of a larger workflow example, we provide a method of calculating an optimal set of *dummy* security policy constraints that minimises resiliency computation time without any reduction to the resiliency value. For clarity we calculate two optimal sets, one of redundant separation of duty constraints that can be added to the policy, and one of user-task permissions that can be removed. Our method could easily be modified to calculate a single set of optimal dummy constraints composed of separation and binding of duty constraints, and user-task permissions.

We consider a single base workflow w_B with 10 tasks and 5 users, coming with a randomly selected security policy p_B composed of 15 separation of duty constraints, 0 binding of duty constraints, and permissions for each task of up to 4 users (29 permissions in total). We use a dynamic user availability model such that each user has an availability for each task of between 0.8 and 1.0 (Table 5, Appendix A). The resiliency values and computation times of w_B and all forthcoming variations of it are analysed 50 times and the average values taken. We also use the same computing platform and PRISM model checker set-up as described in Sect. 5.1. The resiliency of w_B is calculated as 63.96 % with an average computation time of 6.53 s.

Separations of Duty. A test program has been implemented which, given a base workflow, e.g. w_B, calculates monotonically all possible separation of duty constraint combinations that can be added to the workflow security policy. All combinations include only constraints not already included in the base workflow's security policy. In the case of w_B, the maximum number of constraints is 45 meaning up to 30 can be added. All possible combinations of between 1 and 30 constraints are therefore computed. For each of these a PRISM definition file is automatically generated and analysed by the PRISM model checker. Results are logged for resiliency value, computation time and the set of constraints added to w_B.

The average results of this analysis step are given in Table 3 where the values given for $+s_i$ indicate the average resiliency and computation time for i separation of duty constraints added to w_B. Similarly, values for $-p_i$ indicate the average resiliency and computation time for i user-task permissions removed from w_B. For clarity we only show the impact on computation time of up to 3 additional separation of duty constraints and the removal of up to 3 permissions. In general, adding arbitrary separation of duty constraints in a monotonic fashion is shown to reduce the resiliency computation time but this comes with a reduction in resiliency.

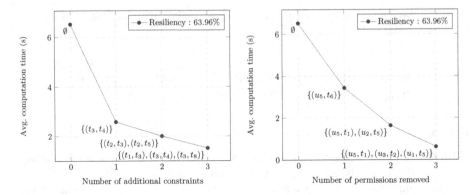

Fig. 6. Impact to resiliency computation time of adding dummy constraints to w_B

Fig. 7. Impact to resiliency computation time of removing permissions from w_B

Finding a set of dummy constraints that reduces computation time without reducing resiliency value is found from performing an automatic double sort on the results, first by resiliency value (largest to smallest) and then by time (smallest to largest). The set of dummy constraints that does not change resiliency yet gives the lowest computation time for each i additional constraints is shown in Fig. 6. For example, adding the constraint (t_3, t_4) has on average the minimum computation time for one change, that of 2.58 s. Notice for three additional constraints, adding $\{(t_1, t_3), (t_3, t_4), (t_3, t_8)\}$ achieves the minimal computation time of 1.52 s, thus reducing the original computation time for w_B by 5.01 s without lowering its resiliency value.

User-Task Permissions. Similarly to adding separation of duty constraints, the results in Table 3 show in general removing arbitrary user-task permissions in a monotonic fashion reduces the resiliency computation time but with a reduction in resiliency. The set of removable permissions shown to give the lowest computation time for each i permissions removed is given in Fig. 7. For example, removing the permission (u_5, t_6) has on average the minimum computation time for one change, that of 3.44 s. Notice for three permissions, removing $\{(u_5, t_1), (u_3, t_2), (u_1, t_5)\}$ achieves the minimal computation time of 0.63 s, thus reducing the original computation time by 5.90 s without lowering the resiliency value.

7   Conclusion

We have shown that the way a workflow security policy is designed has a clear impact on the time required to compute the workflow resiliency, which might need to be done at runtime before the execution of each task, in order to ensure that the user-task assignment is suitable. Our results rely on a systematic encoding of a workflow as a probabilistic model and use the ability of the model checker PRISM to efficiently compute resiliency.

We consider this process to be useful in two settings, firstly the workflow design process allowing domain and security experts to assess how resiliency computation time would be impacted following restrictive and unrestrictive changes to the security policy. Secondly, we have proposed an approach adding dummy or artificial security constraints, in order to reduce the computation time. The gain in time can be significant, for instance in our experiment reducing the computation time from 6.53 s to 0.63 s.

Our experimental results are based on a synthetic workflow, and although it is inspired by real workflows, nothing guarantees that the same efficiency can be gained for all workflows. Hence, we believe our main contribution is the methodology to encode a workflow and to automatically assess its resiliency, based on one of the major probabilistic model-checkers. In the end, the trade-off between resiliency and efficiency can only be resolved by the workflow designer/executer, and we believe that our approach can be helpful in that direction.

In terms of future work we aim to introduce more complex security constraints including cardinality, restricting the number of times a user can be assigned a specific task, to assess their impact on resiliency computation time. We also plan to consider more complex workflows, for instance including loops and choice tasks. These features tend to introduce another level of non-determinism in the execution of the workflow itself, and present as such some challenging aspects in their analysis.

A Probabilities for User Availability

Table 4. User probabilistic availabilities used to compute resiliency for running example workflow w_1 in Sects. 4, 6.1 and 6.2

	t_1	t_2	t_3	t_4	t_5
u_1	0.9568	0.8338	0.7206	0.7231	0.7099
u_2	0.8565	0.9210	0.8016	0.8091	0.9460
u_3	0.8263	0.8617	0.7705	0.7192	0.7117
u_4	0.7238	0.8999	0.9486	0.8413	0.8063

Table 5. User probabilistic availabilities used to compute resiliency when assessing policy changes (Sect. 5) and for the base workflow w_B (Sect. 6.3)

	t_1	t_2	t_3	t_4	t_5	t_6	t_7	t_8	t_9	t_{10}
u_1	0.9297	0.9996	0.8506	0.8737	0.9057	0.8365	0.9514	0.8555	0.9665	0.9875
u_2	0.8381	0.8883	0.8231	0.8726	0.8099	0.9732	0.9852	0.8506	0.9825	0.8089
u_3	0.9653	0.9246	0.8429	0.9491	0.9597	0.8394	0.8560	0.9585	0.8304	0.8330
u_4	0.9263	0.9691	0.8241	0.9932	0.9868	0.9792	0.9162	0.9339	0.9868	0.8049
u_5	0.9724	0.8817	0.9401	0.8261	0.9339	0.8432	0.9329	0.8682	0.8231	0.8842

References

1. Workflow handbook 1997. chapter The Workflow Reference Model, pp. 243–293. John Wiley and Sons Inc, New York (1997)
2. Armando, A., Ponta, S.E.: Model checking authorization requirements in business processes. Comput. Secur. **40**, 1–22 (2014)
3. Ayed, S., Cuppens-Boulahia, N., Cuppens, F.: Deploying security policy in intra and inter workflow management systems. In: International Conference on Availability, Reliability and Security (ARES 2009), pp. 58–65, March 2009
4. Bakkali, H.E.: Enhancing workflow systems resiliency by using delegation and priority concepts. J. Digital Inf. Manage. **11**(4), 267–276 (2013)
5. Basin, D., Burri, S.J., Karjoth, G.: Obstruction-free authorization enforcement: aligning security with business objectives. In: Proceedings of the 2011 IEEE 24th Computer Security Foundations Symposium (CSF 2011), pp. 99–113. IEEE Computer Society, Washington (2011)
6. Basin, D., Burri, S.J., Karjoth, G.: Optimal workflow-aware authorizations. In: Proceedings of SACMAT 2012, pp. 93–102. ACM, New York (2012)
7. Basu, A., Kumar, A.: Research commentary: workflow management issues in e-business. Info. Sys. Res. **13**(1), 1–14 (2002)
8. Bertino, E., Ferrari, E., Atluri, V.: The specification and enforcement of authorization constraints in workflow management systems. ACM Trans. Inf. Syst. Secur. **2**(1), 65–104 (1999)
9. Botha, R., Eloff, J.H.P.: Separation of duties for access control enforcement in workflow environments. IBM Sys. J. **40**(3), 666–682 (2001)
10. Calinescu, R., Ghezzi, C., Kwiatkowska, M., Mirandola, R.: Self-adaptive software needs quantitative verification at runtime. Commun. ACM **55**(9), 69–77 (2012)
11. Calinescu, R., Grunske, L., Kwiatkowska, M., Mirandola, R., Tamburrelli, G.: Dynamic QoS management and optimisation in service-based systems. IEEE Trans. Softw. Eng. **37**(3), 387–409 (2011)
12. Crampton, J., Gutin, G., Yeo, A.: On the parameterized complexity and kernelization of the workflow satisfiability problem. ACM Trans. Inf. Syst. Secur. **16**(1), 4 (2013)
13. Crampton, J., Khambhammettu, H.: Delegation and satisfiability in workflow systems. In: Proceedings of the 13th ACM symposium on Access control models and technologies, pp. 31–40. ACM (2008)
14. Georgakopoulos, D., Hornick, M., Sheth, A.: An overview of workflow management: From process modeling to workflow automation infrastructure. Distrib. Parallel Databases **3**(2), 119–153 (1995)
15. He, L., Huang, C., Duan, K., Li, K., Chen, H., Sun, J., Jarvis, S.A.: Modeling and analyzing the impact of authorization on workflow executions. Future Gener. Comput. Sys. **28**(8), 1177–1193 (2012)
16. Herbert, L., Sharp, R.: Precise quantitative analysis of probabilistic business process model and notation workflows. J. Comput. Inf. Sci. Eng. **13**(1), 011007 (2013)
17. Hiden, H., Woodman, S., Watson, P., Cala, J.: Developing cloud applications using the e-science central platform. Philos. Trans. R. Soc. A : Math. Phys. Eng. Sci. **371**(1983), 20120085 (2013)
18. Howard, R.A.: Dynamic Programming and Markov Processes. MIT Press, Cambridge (1960)

19. Kohler, M., Liesegang, C., Schaad, A.: Classification model for access control constraints. In: IEEE International on Performance, Computing, and Communications Conference (IPCCC 2007) pp. 410–417, April 2007
20. Kwiatkowska, M., Norman, G., Parker, D.: PRISM 4.0: verification of probabilistic real-time systems. In: Gopalakrishnan, G., Qadeer, S. (eds.) CAV 2011. LNCS, vol. 6806, pp. 585–591. Springer, Heidelberg (2011)
21. Lowalekar, M., Tiwari, R.K., Karlapalem, K.: Security policy satisfiability and failure resilience in workflows. In: Matyáš, V., Fischer-Hübner, S., Cvrček, D., Švenda, P. (eds.) The Future of Identity. IFIP AICT, vol. 298, pp. 197–210. Springer, Heidelberg (2009)
22. Mace, J.C., Morisset, C., van Moorsel, A.: Quantitative workflow resiliency. In: Kutyłowski, M., Vaidya, J. (eds.) ICAIS 2014, Part I. LNCS, vol. 8712, pp. 344–361. Springer, Heidelberg (2014)
23. Mace, J.C., Morisset, C., van Moorsel, A.: Impact of policy design on workflow resiliency computation time. Technical report CS-TR-1469, School of Computing Science, Newcastle University, UK, May 2015
24. Mace, J.C., Morisset, C., van Moorsel, A.: Modelling user availability in workflow resiliency analysis. In: Proceedings of the 2015 Symposium and Bootcamp on the Science of Security (HotSoS 2015), pp. 7:1–7:10. ACM, New York (2015)
25. Martinelli, F., Morisset, C.: Quantitative access control with partially-observable markov decision processes. In: Proceedings of CODASPY 2012, pp. 169–180. ACM, New York (2012)
26. Unertl, K.M., Johnson, K.B., Lorenzi, N.M.: Health information exchange technology on the front lines of healthcare: workflow factors and patterns of use. J. Am. Med. Inform. Assoc. **19**(3), 392–400 (2012)
27. Wainer, J., Barthelmess, P., Kumar, A.: W-rbac - a workflow security model incorporating controlled overriding of constraints. Int. J. Coop. Inf. Sys. **12**, 2003 (2003)
28. Wang, Q., Li, N.: Satisfiability and resiliency in workflow authorization systems. ACM Trans. Inf. Syst. Secur. **13**(4), 40:1–40:35 (2010)

Queueing Systems and Hybrid Systems

Perfect Sampling for Multiclass Closed Queueing Networks

Anne Bouillard, Ana Bušić, and Christelle Rovetta[(✉)]

École Normale Supérieure and Inria Paris-Rocquencourt, Paris, France
christelle.rovetta@inria.fr

Abstract. In this paper we present an exact sampling method for multiclass closed queuing networks. We consider networks for which stationary distribution does not necessarily have a product form. The proposed method uses a compact representation of sets of states, that is used to derive a bounding chain with significantly lower complexity of one-step transition in the coupling from the past scheme. The coupling time of this bounding chain can be larger than the coupling time of the exact chain, but it is finite in expectation. Numerical experiments show that coupling time is close to that of the exact chain. Moreover, the running time of the proposed algorithm outperforms the classical algorithm.

Keywords: Multiclass queueing networks · Simulation · Coupling from the past

1 Introduction

Closed queueing networks are largely used in various application domains due to their modeling simplicity and product-form stationary distribution [11]. In the multiclass case, the product-form structure remains valid only under restrictive conditions on the service policy [3].

When the stationary distribution does not have a product-form, the exact analysis may not be computationally tractable, and we may turn to approximations [2,4,16,18], bounds [8,20], or simulation [1].

This paper concerns simulation, with a focus on stopping criteria. The asymptotic variance appearing in the Central Limit Theorem has been the most common metric to devise stopping rules, while mixing times have become a standard alternative [1,15]. Unfortunately, there are no generic and tractable techniques to compute or bound either the asymptotic variance or the mixing time for non-reversible Markov chains.

Propp and Wilson introduced a method for sampling a random variable according to the stationary distribution of a finite ergodic Markov chain [17]: the coupling from the past (CFTP) algorithm. The CFTP algorithm automatically detects and stops when the sample has the correct distribution. In this way it is possible to generate i.i.d. samples from the chain, and the asymptotic variance of the resulting simulator is the standard variance of the random variable whose mean we wish to estimate.

J. Campos and B.R. Haverkort (Eds.): QEST 2015, LNCS 9259, pp. 263–278, 2015.
DOI: 10.1007/978-3-319-22264-6_17

The main drawback of the original CFTP is that it considers a coupling from all initial conditions. In the case of closed queueing networks the cardinality of the state space is exponential in the number of queues, which is intractable.

Different techniques can be used to efficiently compute one step of the CFTP algorithm: the simplest solution, for monotone Markov chains, is to compute the minimal and maximal trajectories only [17]. For Markov chains with no monotone representations, new techniques have been developed to approximate each step of the computation, at the cost of slightly increasing the number of iterations of the algorithm. Bounding chains have been constructed to detect coalescence for state spaces with lattice structure [10,13], or for models with short range local interactions, such as interacting particle systems [12]. For applications to queueing networks, see for instance [9,19].

The main difficulty with closed queueing networks is that the customer population is constant. This imposes a global constraint on the model, so the approach of [13] cannot be applied directly. Without monotonicity, the complexity of one iteration of the original CFTP algorithm by Propp and Wilson [17] depends on the cardinality of the state space, which is exponential in the number of queues.

For the single class closed queueing networks, Kijima and Matsui [14] proposed a perfect sampling algorithm with overall complexity $O(K^3 \ln(KM))$, where K is the number of queues and M the total number of customers. However, their method strongly relies on the product form representation of the stationary distribution and it cannot be applied to the general case of multiclass networks. In Bouillard and al. [7], a new representation of the sets of states has been proposed. This representation is used to derive a bounding chain for the CFTP algorithm for closed queueing networks, that enables exact sampling from the stationary distribution without considering all initial conditions in the CFTP. This method is far more general, as it does not rely on the product-form property.

In this paper, we propose a generalization of the compact state space representation in [7] for the multiclass closed queueing networks. Each state is represented by a path in a multidimensional diagram. This diagram is a directed graph with nodes in $\{0, \ldots, K\} \times \prod_{z=1}^{Z} \{0, \ldots, M_z\}$, where K denotes the number of queues, Z the number of classes, and M_z the total number of customers of class z (the detailed description of diagrams is given in Sect. 3). The diagram transition function used in our MDCFTP (multiclass diagram CFTP) algorithm needs to read only once this multidimensional diagram, so the complexity of one step of our algorithm is in $O(K \prod_{z=1}^{Z} M_z)$. On the other hand, multiclass diagrams are an over-representation of the states (a set of paths that represents a set of states may represent more states than just the desired subset). Thus the coupling time of the MDCFTP algorithm is in general larger than the coupling time of the classical CFTP. Numerical experiments (Sect. 4) suggest that the coupling times of the two algorithms are very close. Overall, the proposed MDCFTP algorithm significantly outperforms the classical one.

A major difficulty in the multiclass case is the fact that the class to be served depends on the current state of the system, so the coupling construction needs a very careful design of the event representation of the dynamic of the system.

We provide such an event description under the assumption that the class to be served in queue i depends only on the current state of queue i.

The paper is organized as follows. In Sect. 2 we present the model and an event representation of the system that ensures the convergence of the CFTP scheme in finite expected time. In Sect. 3, we describe the multiclass diagram representation of the state space, propose the MDCFTP algorithm and discuss its complexity. Numerical experiments are given in Sect. 4. Final remarks and conclusions are contained in Sect. 5.

2 Model

2.1 Description

We consider a multiclass closed queueing network of K queues and Z classes of customers. Each queue $k \in \{1, \ldots, K\}$ has an infinite buffer capacity, a single exponential server and a service discipline that will be detailed in Sect. 2.3. For simplicity of exposition, we assume a class-independent service rate μ_k. The generalization to class-dependent service times is straightforward, by adding fake transitions that do not modify the state for the classes that have smaller service rates. The set of classes that can be in queue k is denoted by $Z(k) \subseteq \{1, \ldots, Z\}$.

Customers are not allowed to change class, thus the number of customers in each class remains constant. For each class $z \in \{1, \ldots, Z\}$ of customer, the total number of customers in this class is denoted by M_z, the set of queues visited by customers of class z by $K(z) \subseteq \{1, \ldots, K\}$ and $P^z \in \mathbb{R}^{K,K}$ is the routing matrix for class z: $P^z_{i,j}$ is the probability that a customer of class z leaving queue i is routed to queue j, with the convention that if $i \notin K(z)$, then $P^z_{i,i} = 1$ and $P^z_{i,j} = 0$ if $j \neq i$. Matrix P^z is stochastic, so for all $i, j \in \{1, \ldots, K\}$, $P^z_{i,j} \geq 0$ and $\sum_{j=1}^{K} P^z_{ij} = 1$.

We assume that the directed graph $G_z = (K(z), R_z)$ where $R_z = \{(i, j) \in K(z)^2 \text{ such that } P^z_{ij} > 0\}$ is strongly connected. The total number of customers in the system is denoted by $M = \sum_{z=1}^{Z} M_z$.

2.2 State Space

A state of the network is a matrix $\mathbf{x} = (x_{z,k}) \in \mathbb{N}^{Z \times K}$ where $x_{z,k}$ represents the number of customers of class z in queue k. A state $\mathbf{x}$ must satisfy the following constraints:

$$\sum_{k=1}^{K} x_{z,k} = M_z \quad \text{and} \quad \forall k \notin K(z), \ x_{z,k} = 0. \tag{1}$$

Throughout the paper, we will use the following notations.

For $k \in \{1, \ldots, K\}$ and $z \in \{1, \ldots, Z\}$, we denote by:

- $\mathbf{x}_{z,*} = (x_{z,k}, \ldots, x_{z,K}) \in \mathbb{N}^K$ is the queue repartition (row)-vector of class z;
- $\mathbf{x}_{*,k} = (x_{1,k}, \ldots, x_{Z,k})^t \in \mathbb{N}^Z$ is the class repartition (row)-vector in queue k;
- $|\mathbf{x}_{*,k}| = \sum_{z=1}^{Z} x_{z,k}$ is the total number of customers in queue k.

The set of all possible state matrices is denoted by $\mathcal{S} = \{\mathbf{x} \in \mathbb{N}^{Z \times K}$ satisfying (1)$\}$. Its cardinality is

$$|\mathcal{S}| = \prod_{z=1}^{Z} \binom{M_z}{M_z + |K(z)| - 1}.$$

If for each class z, $K(z) \ll M_z$ then $|\mathcal{S}| = O(\prod_{z=1}^{Z} M_z^{|K(z)|}) = O(M_\times^K)$, where $M_\times = \prod_{z=1}^{Z} M_z$.

Example 1. Consider the multiclass queueing network of Fig. 1 having 5 queues, 2 classes, with $M_1 = 2$ and $M_2 = 3$, and routing matrices

$$P^1 = \begin{pmatrix} 0 & 0.5 & 0.5 & 0 & 0 \\ 1 & 0 & 0 & 0 & 0 \\ 1 & 0 & 0 & 0 & 0 \\ 0 & 0 & 0 & 1 & 0 \\ 0 & 0 & 0 & 0 & 1 \end{pmatrix} \quad \text{and} \quad P^2 = \begin{pmatrix} 1 & 0 & 0 & 0 & 0 \\ 0 & 0 & 0.3 & 0 & 0.7 \\ 0 & 0 & 0 & 1 & 0 \\ 0 & 0.8 & 0.2 & 0 & 0 \\ 0 & 0 & 0.5 & 0.5 & 0 \end{pmatrix}.$$

The total number of customers is $M = 5$ and the queues visited by each class are $K(1) = \{1, 2, 3\}$ and $K(2) = \{2, 3, 4, 5\}$. The cardinality of the state space is $|\mathcal{S}| = 120$ and an example of such a state is

$$\mathbf{x} = \begin{pmatrix} 1 & 1 & 0 & 0 & 0 \\ 0 & 2 & 0 & 0 & 1 \end{pmatrix} \in \mathcal{S}.$$

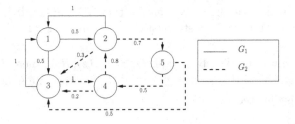

Fig. 1. The multiclass network described in Example 1.

2.3   Service Discipline and Transitions

In order to perform the perfect sampling algorithm for multiclass queueing networks, we need to define transitions on sets of states. These transitions depend on the service discipline of the queues.

The function $\tilde{t}_{i,j,z} : \mathcal{S} \to \mathcal{S}$ describes the routing of a class z customer from queue i to queue j:

$$\tilde{t}_{i,j,z}(\mathbf{x}) = \mathbf{x} - \mathbb{1}_{\{x_{z,i} > 0\}} \mathbf{e}_{zi} + \mathbb{1}_{\{x_{z,i} > 0\}} \mathbf{e}_{zj},$$

where $\mathbf{e}_{zk} \in \mathbb{N}^{Z \times K}$ is the matrix having all its coefficients equal to 0 except $(\mathbf{e}_{zk})_{z,k} = 1$.

When there are several classes of customers in a queue, then the queue discipline determines the class of customers to serve. For each queue $i \in \{1, \ldots, K\}$, $f_i : \mathbb{N}^Z \times [0,1] \to Z(i) \cup \{0\}$ is the function that describes the discipline in queue i. We assume that f_i has the following properties:

Assumption 1. *For a state* $\mathbf{x}$ *and a parameter* $\theta \in [0,1]$:

1. *The service discipline is Markovian and* f_i *only depends on* θ *and* $\mathbf{x}_{*,i}$.
2. $|\mathbf{x}_{*,i}| = 0$ *if and only if* $f_i(\mathbf{x}_{*,i}, \theta) = 0$ *(the service is greedy)*.
3. *If* $|\mathbf{x}_{*,i}| > 0$ *then* $f_i(\mathbf{x}_{*,i}, \theta) \in \{z \text{ such that } x_{z,i} > 0\}$ *(there is a customer of the chosen class in queue* i*)*.

We give two examples of disciplines that satisfy Assumption 1:

– PRIORITY gives the preemptive priority to the class with the smallest index. It can be defined to any total order $\preceq$ on the classes.

$$f_i(\mathbf{x}_{*,i}, \theta) = \min\{z \mid x_{z,i} > 0\} \mathbb{1}_{\{|\mathbf{x}_{*,i}|>0\}}.$$

– LONGEST (serve-the-longest-queue)

$$f_i(\mathbf{x}_{*,i}, \theta) \in \arg\max\{x_{i,z} \mid z \in Z(i)\} \cup \{0\}.$$

PRIORITY does not depend on θ, but this parameter is used to break ties in LONGEST between the classes where the number of customers is maximal.

For $i \in \{1, \ldots, K\}$, $J = (j_1, \ldots, j_Z) \in \{1, \ldots, K\}^Z$ a vector of queues, $\mathbf{x}$ a state, and $\theta \in [0,1]$, we define a transition on state $\mathbf{x}$ by the function:

$$t_{i,J,\theta}(\mathbf{x}) = \tilde{t}_{i,j_z,z}(\mathbf{x}),$$

where $z = f_i(\mathbf{x}_{*,i}, \theta)$ is the class chosen by queue i. This transition describes the routing of a customer from queue i and J gives the destination queue according to the class that is served in queue i. If the transition does not depend on θ (as in PRIORITY), we will write $t_{i,J,\theta}(\mathbf{x}) = t_{i,J}(\mathbf{x})$ to alleviate the notations.

Example 2. Consider state $\mathbf{x} = \begin{pmatrix} 1 & 1 & 0 & 0 & 0 \\ 0 & 2 & 0 & 0 & 1 \end{pmatrix}$ in Example 1. Queue 2 contains 3 customers ($\mathbf{x}_{*,2} = (1,2)$) and function $t_{2,(1,5),\theta}$ describes a possible routing for a customer in queue 2: if a customer of class 1 is served, then it is routed to queue 1; a customer of class 2 would be routed to queue 5. For $\theta = 0$, the class of the served customer is given by $z = f_2(\mathbf{x}_{*,2}, 0)$.

– If queue 2 has the PRIORITY discipline then $z = 1$ and

$$t_{2,(1,5)}(\mathbf{x}) = \tilde{t}_{2,1,1}(\mathbf{x}) = \begin{pmatrix} 2 & 0 & 0 & 0 & 0 \\ 0 & 2 & 0 & 0 & 1 \end{pmatrix}.$$

– If queue 2 has the LONGEST discipline then $z = 2$ and

$$t_{2,(1,5),0}(\mathbf{x}) = \widetilde{t}_{2,5,2}(\mathbf{x}) = \begin{pmatrix} 1\ 1\ 0\ 0\ 0 \\ 0\ 1\ 0\ 0\ 2 \end{pmatrix}.$$

This definition of a transition enables to define it for sets of states: for all $S \subseteq \mathcal{S}$, $\theta \in [0,1]$, $i \in \{1, \dots, K\}$ and $J = (j_1, \dots, j_Z) \in \{1, \dots, K\}^Z$, $t_{i,J,\theta}(S) :=$ $\cup_{\mathbf{x} \in S} t_{i,J,\theta}(\mathbf{x})$ and $t_{i,J}(S) := \cup_{\mathbf{x} \in S} t_{i,J}(\mathbf{x})$ if the transition do not depend on θ.

2.4   Markov Chain and Perfect Sampling

Denote by $(W_n)_{n \in \mathbb{N}}$ an i.i.d. sequence of random variables with distribution

$$\mathbb{P}(W_n = (i, J)) = \frac{\mu_i}{\sum_{k=1}^{K} \mu_k} \prod_{z=1}^{Z} P^z_{i,j_z},$$

where $i \in \{1, \dots, K\}$ and $J = (j_1, \dots, j_Z) \in \{1, \dots, K\}^Z$. It can easily be checked that the probability that the first component of W_n is i equals $\frac{\mu_i}{\sum_{k=1}^{K} \mu_k}$, the probability that the next service occurs at queue i. Moreover, denoting $W_n = (k, J)$, for each class $z \in \{1, \dots, K\}$, $\mathbb{P}(j_z = j \mid k = i) = P^z_{i,j}$, which is the probability of routing a customer of class z from queue i to queue j.

Let $(\Theta_n)_{n \in \mathbb{N}}$ be an i.i.d sequence of random variables, uniformly distributed on $[0,1]$ and independent of $(W_n)_{n \in \mathbb{N}}$. We set $(U_n)_{n \in \mathbb{N}} = (W_n, \Theta_n)_{n \in \mathbb{N}}$.

Let $(X_n)_n$ a random sequence such that $X_0 \in \mathcal{S}$ and

$$X_{n+1} = t_{U_n}(X_n).$$

This equation describes the Markov chain of our model of multiclass network. This Markov chain is ergodic, due to the assumption that the routing graph of each class is strongly connected and Assumption 1. Our objective is to sample the stationary distribution of $(X_n)_{n \in \mathbb{N}}$ with the perfect sampling technique [17][1]. The following theorem concerns the termination and correctness of Algorithm 1.

Theorem 1. *If for each class z the routing graph G_z is strongly connected, and the service policies satisfy Assumption 1, then Algorithm 1 terminates in finite time with probability 1 and the termination time has a finite expectation. The state s_0 is distributed according to the stationary distribution of $\{X_n\}$.*

The proof is a straightforward corollary of the result by Propp and Wilson [17] and Theorem 2, that gives the existence of a coupling sequence, i.e. a sequence that leads to coalescence of the trajectories started in all the initial conditions.

Theorem 2. *If the service discipline of each queue satisfies Assumptions 1 then there exists a finite sequence of transitions $t = t_{U_1} \circ \cdots \circ t_{U_n}$ such that $|t(\mathcal{S})| = 1$.*

The proof of Theorem 2 can be found in the extended version of this paper [5]. Note that the complexity of this algorithm is at least linear in $|\mathcal{S}|$.

---

[1] Algorithm 1 is the variant of the CFTP algorithm from [17] that doubles the value of n at each iteration. In the case when all the states are considered, the basic CFTP algorithm can be used, but this choice has been made to simplify comparison with Algorithm 3 in Sect. 3.

Algorithm 1. CFTP using sets of states

Data: $(U_{-n} = (i_{-n}, J_{-n}, \Theta_{-n}))_{n \in \mathbb{N}}$ an i.i.d sequence of r.v
Result: $\mathbf{x} \in \mathcal{S}$

1 **begin**
2      $n \leftarrow 1$;
3      $t \leftarrow t_{U_{-1}}$;
4   **while** $|t(\mathcal{S})| \neq 1$ **do**
5          $n \leftarrow 2n$;
6          $t \leftarrow t_{U_{-1}} \circ \cdots \circ t_{U_{-n}}$;
7   **return x**, the unique element of $t(\mathcal{S})$

---

3 Diagram Representation

The cardinality of the state space is exponential in K and Z, so it is not possible to perform the perfect sampling algorithm directly, as one must first enumerate all the state space in order to compute a transition. In this section, we present multiclass *diagrams*, a more compact way to describe sets of states. In most of the cases, a diagram representing a given set of states will in fact represent more states, but we show that this representation ensures the termination of the perfect sampling algorithm in finite expected time, and the complexity of the transition becomes linear in K (while still exponential in Z).

3.1 Definition

Let $D = (N, A)$ be a directed graph where $N \subseteq \{0, \ldots, K\} \times \mathbb{N}^Z$ is the set of nodes and A the set of arcs. Let $g : \mathcal{S} \to \mathcal{P}(N^2)$ denote the function which associates a set of arcs to a state $\mathbf{x} \in \mathcal{S}$:

$$g(\mathbf{x}) = \bigcup_{i=1}^{K} \left\{ \left([i-1, (\sum_{k=1}^{i-1} x_{1,k}, \ldots, \sum_{k=1}^{i-1} x_{Z,k})], [i, (\sum_{k=1}^{i} x_{1,k}, \ldots, \sum_{k=1}^{i} x_{Z,k})] \right) \right\}.$$

Graphically, $g(\mathbf{x})$ can be seen in D as a path from node $[0, (0, \ldots, 0)]$ to node $[K, (M_1, \ldots, M_Z)]$ (Fig. 2). Moreover, consider an arc $a = \big([k-1, \mathbf{s}], [k, \mathbf{d}]\big) \in g(\mathbf{x})$ where $\mathbf{s} = (s_1, \ldots, s_Z) \in \mathbb{N}^Z$ and $\mathbf{d} = (d_1, \ldots, d_Z) \in \mathbb{N}^Z$ are two row-vectors. The *slope* of a on its second component can be considered as a class-repartition vector in queue k. Indeed,

$$\mathbf{d} - \mathbf{s} = (\sum_{i=1}^{k} x_{1,i} - \sum_{i=1}^{k-1} x_{1,i}, \ldots, \sum_{i=1}^{k} x_{Z,i} - \sum_{i=1}^{k-1} x_{Z,i}) = (x_{1,k}, \ldots, x_{Z,k}) = \mathbf{x}_{*,k}.$$

Definition 1. *A directed graph $D = (N, A)$ is called a **diagram** if there exists $S \subseteq \mathcal{S}$ such that*

$$A = g(S) := \bigcup_{\mathbf{x} \in S} g(\mathbf{x}).$$

*A diagram is said to be **complete** if $A = g(\mathcal{S})$. It is denoted $\mathcal{D} = (N, A)$.*

For an arc $a = ([k-1,\mathbf{s}],[k,\mathbf{d}]) \in A$, vector $v(a) = \mathbf{d} - \mathbf{s} \in \mathbb{N}^Z$ is called the *value* of a. It represents the class-repartition vector in queue k. Subset $A_k = \{([k-1,\mathbf{s}],[k,\mathbf{d}]) \in A\}$ denotes the set of all arcs in column k.

Example 3. Consider $\mathcal{S}$ the state space of Example 1 and state $\mathbf{x} = \begin{pmatrix} 1\ 1\ 0\ 0\ 0 \\ 0\ 2\ 0\ 0\ 1 \end{pmatrix}$. Diagram $D = (N, g(\{\mathbf{x}\}))$ is given in Fig. 2 and has 5 arcs:

$$([0,(0,0)],[1,(1,0)]),\ ([1,(1,0)],[2,(2,2)]),\ ([2,(2,2)],[3,(2,2)]),$$
$$([3,(2,2)],[4,(2,2)]),\ ([4,(2,2)],[5,(2,3)]).$$

The value of a_2 is $v(a_2) = (1,2) = \mathbf{x}_{*,2}$. The complete diagram $D = (N, g(\mathcal{S}))$ is depicted on Fig. 3 and has $|g(\mathcal{S})| = 71$ arcs.

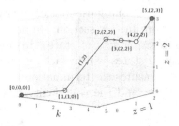

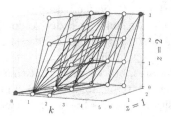

Fig. 2. Diagram $D = (N, g(\{\mathbf{x}\}))$.   **Fig. 3.** Complete diagram $D = (N, g(\mathcal{S}))$.

Lemma 1. *Let $D = (N, A)$ be a diagram. If $K \geq 2$, then*

$$|A| \leq 2 \prod_{z=1}^{Z}(M_z + 1) + (K-2)\prod_{z=1}^{Z}\frac{(M_z+1)(M_z+2)}{2}.$$

Proof (Sketch of proof). The first term bounds the number of arcs in the first and last columns ($\leq M_z + 1$ each), the second term bounds the number of arcs in the other $K-2$ columns ($\leq \prod_{z=1}^{Z}\sum_{m=0}^{M_z} m$).

The equality in Lemma 1 holds for the complete diagram when each class can visit every queue.

A consequence of Lemma 1 is that the space needed for the representation of a diagram is $|A| = O(KM_x^2)$.

Diagrams and Sets of States. In order to perform the transition on the diagrams, we first need to define functions that transform a set of states into a diagram and the reverse. For $S \subseteq \mathcal{S}$, ϕ associates to a set of states $S \in \mathcal{S}$ the diagram $\phi(S) = (N, g(S))$. For $D = (N, A)$, ψ transforms diagram $D = (N, A)$ into the largest set of states $S \subseteq \mathcal{S}$ such that $g(S) = A$:

$$\psi(D) = \bigcup_{S \subseteq \mathcal{S}, \ A = g(S)} S.$$

The followings properties are straightforward from the definitions of ϕ and ψ.

Lemma 2. *For $S \subseteq \mathcal{S}$ a set of states and $D = (N, A)$ a diagram:*

1. *D is complete if and only if $\psi(D) = \mathcal{S}$.*
2. *If D contains only one path (i.e. $|A| = K$) then $|\psi(D)| = 1$.*
3. *If $|S| = 1$ then $\phi(S)$ contains only one path.*
4. *If $S \subseteq \mathcal{S}$ such that $A = g(S)$ then $S \subseteq \psi(D)$.*

3.2 Transition Algorithm

We now extend the transitions to diagrams. Let $(i, J) \in \{1, \ldots, K\}^{Z+1}$ and $\theta \in [0, 1]$. Function $T_{i,J,\theta}$ is defined for each diagram D as

$$T_{i,J,\theta}(D) = \phi \circ t_{i,J,\theta} \circ \psi(D).$$

Lemma 3. *Let $S \subseteq \mathcal{S}$ be a set of states and D be a diagram. For all $(i, J) \in \{1, \ldots, K\}^{Z+1}$ and $\theta \in [0, 1]$,*

1. *if $S \subseteq \psi(D)$ then $t_{i,J,\theta}(S) \subseteq \psi(T_{i,J,\theta}(D))$;*
2. *if $|\psi(D)| = 1$ then $|\psi(T_{i,J,\theta}(D))| = 1$.*

We now present the algorithm to compute $T_{i,J,\theta}(D)$ directly without having to use $t_{i,J,\theta}(\phi(D))$, which would be too costly. The intuition is that the transformation will be similar for sets of paths, and then it can be done simultaneously for them all. The algorithm that computes $T_{i,j,\theta}(D)$ is given as Algorithm 2.

Before describing it, we adapt the definition of the class to be served to diagrams. As the service disciplines we consider satisfy Assumption 1, the way a path is transformed according to (i, J, θ) depends only on the value of the arc in column i. For $a \in A_i$ and $\theta \in [0, 1]$, we can define $F_i(a, \theta) = f_i(v(a), \theta)$, which selects the class to be served for arc a. This class will be the same for every path going through that arc.

For a diagram $D = (N, A)$ and $b \in A$, we denote by $\mathcal{P}\text{aths}(b, A) \subseteq A$ the subset of arcs on paths through arc b:

$$\mathcal{P}\text{aths}(b, A) = \{a \in A \mid \exists \mathbf{x} \in \mathcal{S} \text{ s.t. } a \in g(\mathbf{x}) \text{ AND } b \in g(\mathbf{x})\},$$

and for $B \subseteq A$, $\mathcal{P}\text{aths}(B, A)$ denotes the subset of arcs on paths through an arc $b \in B$:

$$\mathcal{P}\text{aths}(B, A) = \bigcup_{b \in B} \mathcal{P}\text{aths}(b, A).$$

Example 4. Consider $D = (N, g(\mathcal{S}))$ the complete diagram of Example 3 and $a = ([2, (2, 0)], [3, (2, 0)]) \in A$, $b = ([2, (1, 0)], [3, (2, 0)]) \in A$ two arcs in column 3. For $B = \{a, b\} \subseteq A_k$. Subset $\mathcal{P}\text{aths}(b, A)$ is given in Fig. 4 and subset $\mathcal{P}\text{aths}(B, A)$ in Fig. 5.

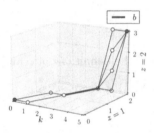

Fig. 4. Subset $\mathcal{P}aths(b, A)$

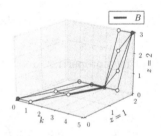

Fig. 5. Subset $\mathcal{P}aths(B, A)$

For each arc $a \in A_i$, function F_i gives the class for which the transition will be performed. As there are $|Z(i)| + 1$ different possible values for $F_i(a, \theta)$, we will compute $|Z(i)| + 1$ different types of transition. For all $z \in Z(i) \cup \{0\}$, set

$$P[z] = \{a \in A_i \mid F_i(a, \theta) = z\} \quad \text{and} \quad \mathcal{S}erve[z] = \mathcal{P}aths(P[z], A).$$

$\mathcal{S}erve[0]$ corresponds to the sub-diagram of the states where queue i is empty. It will remain a sub-diagram of $T_{i,J,\theta}(D)$. For all $z \in Z(i)$, $\mathcal{S}erve[z]$ is the sub-diagram corresponding to states where a customer of class z is served, and routed to queue $j = J[z]$. This sub-diagram is transformed into $\mathcal{S}erve'[z]$. We focus on the case $i < j$ (the case $i > j$ is similar). Each arc $b = ([k-1, \mathbf{s}], [k, \mathbf{d}]) \in \mathcal{S}erve[z]$ is transformed into c the following way:

- If $k < i$ or $k > j$, b is not affected by the transformation, so $c = b$.
- If $k = i$ then b corresponds to a class-repartition vector of queue i, where one customer of class z is served. Then $c = ([k-1, \mathbf{s}], [k, \mathbf{d} - \mathbf{e}_z])$.
- If $i < k < j$ then b corresponds a class-repartition vector that is not affected by the service. However, the origin of the arcs it was connected to has changed from $\mathbf{s}$ to $\mathbf{s} - \mathbf{e}_z$, so its destination must change similarly. So $c = ([k-1, \mathbf{s} - \mathbf{e}_z], [k, \mathbf{d} - \mathbf{e}_z])$.
- If $k = j$, b corresponds to a class-repartition vector of queue j, where one customer of class z arrives. As the origin of the arc it was connected to has changed to $\mathbf{s} - \mathbf{e}_z$, we have $c = ([k-1, \mathbf{s} - \mathbf{e}_z], [k, \mathbf{d}])$.

Diagram $T_{i,J,\theta}(D) = (N, A')$ with $A' = \bigcup_{z \in Z(i) \cup \{0\}} \mathcal{S}erve'[z]$. Indeed, this construction ensures that for all $\mathbf{x} \in \psi(D)$, $t_{iJ,\theta}(\mathbf{x}) \in \psi(T_{i,J,\theta}(D))$.

Suppose that the complexity of computing $F_i(a, \theta)$ is C. Then the complexity of Algorithm 2 is $C|A_i| + Z|A| = O((C + KZ)M_\times^2)$. For PRIORITY and LONGEST service discipline, $C = O(Z)$, so the overall complexity is $O(KZM^{2Z})$.

Example 5. Consider the complete diagram of Example 3 and transition $T_{2,(1,5)}$. Suppose that queue 2 has a PRIORITY discipline (class 1 has the priority). Figure 6 illustrates the partition of arcs in column 2 into $P[0]$, $P[1]$ and $P[2]$.

- $P[0] = \{a \in A_2 \mid v(a) = (0,0)\}$ and $\mathcal{S}erve[0] = \mathcal{P}aths(P[0], A)$

Algorithm 2. Algorithm $T_{i,J,\theta}$

Data: $D = (N, A), i \in \{1, \ldots, K\}, J \in \mathbb{N}^Z, \theta \in [0,1]$
Result: $T_{i,J,\theta}(D)$

```
 1 begin
 2 │   for z = 0 to Z do  P[z] ← {a ∈ A_i | F_i(a, θ) = z};
 3 │   ;
 4 │   Serve'[0] ← Paths(P[0], A);
 5 │   for z = 1 to Z do
 6 │   │   Serve[z] ← Paths(P[z], A);
 7 │   │   Serve'[z] ← ∅;
 8 │   │   j ← J[z] ;
 9 │   │   if i < j then
10 │   │   │   foreach b = ([k − 1, s], [k, d]) ∈ Serve[z] do
11 │   │   │   │   c = ([k − 1, s − 1_{i<k≤j}e_z], [k, d − 1_{i≤k<j}e_z]);
12 │   │   │   │   Serve'[z] ← Serve'[z] ∪ {c};
13 │   │   if i > j then
14 │   │   │   foreach b = ([k − 1, s], [k, d]) ∈ Serve[z] do
15 │   │   │   │   c = ([k − 1, s + 1_{j<k≤i}e_z], [k, d + 1_{j≤k<i}e_z]);
16 │   │   │   │   Serve'[z] ← Serve'[z] ∪ {c};
17 │   A' ← ⋃_{z=0}^{Z} Serve'[z] ;
18 │   return  (N, A')
```

- $P[1] = \{a \in A_2 \mid v(a) = (v1, v2), \; v1 > 0\}$ and $Serve[1] = Paths(P[1], A)$
- $P[2] = \{a \in A_2 \mid v(a) = (0, v2), \; v2 > 0\}$ and $Serve[2] = Paths(P[2], A)$

Notice that sets $Serve[0]$, $Serve[1]$ and $Serve[2]$ are not necessary disjoint. For example, arc $b = ([4, 2, 3], [5, 2, 3]) \in Serve[1] \cap Serve[2]$ (see Figs. 7 and 8). Then for all $z \in \{1, \ldots, Z\}$ we will compute $Serve'[z]$ from $Serve[z]$ according to Algorithm 2. For class 1, the transition is performed for arcs in $Serve[1]$ from queue 2 to queue 1. For class 2, the transition is performed for arcs in $Serve[2]$ from queue 2 to queue 5.

Finaly, $T_{i,J}(D, \theta) = (N, A')$, with $A' = Serve[0] \cup Serve'[1] \cup Serve'[2]$.

3.3   Perfect Sampling with Diagrams

In this section, we show that the diagram representation can be used in the perfect sampling algorithm. First, as in Sect. 2.4, we need to ensure that the complete diagram can be reduced to a diagram containing only one state by a finite sequence of transitions.

Theorem 3. *If each queue discipline satisfies Assumptions 1 then there exists a finite sequence of transitions T such that $|\psi(T(\mathcal{D}))| = 1$.*

The proof can be found in the extended version of this paper [5].

Theorem 4. *Algorithm 3 samples a state according to the stationary distribution on the states and terminates in finite expected time.*

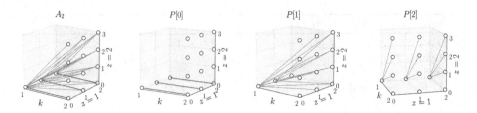

Fig. 6. Arc repartition in column 2 with PRIORITY discipline.

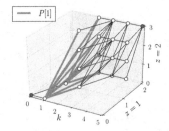

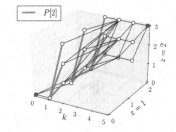

Fig. 7. $\mathcal{S}$erve[1] $= \mathcal{P}$aths$(P[1], A)$.   **Fig. 8.** $\mathcal{S}$erve[2] $= \mathcal{P}$aths$(P[2], A)$.

Proof. Theorem 3 implies that Algorithm 3 ends in finite expected time. Let $N < \infty$ be the value of n when Algorithm 3 ends. Consider $t = t_{U_{-1}} \circ \cdots \circ t_{U_{-N}}$ with the same random sequence $(U_{-n})_{n \in \mathbb{N}}$. Lemma 3 implies that $t(\mathcal{S}) \subseteq \psi(T(\mathcal{D})) \subseteq \mathcal{S}$. But $|\psi(T(\mathcal{D}))| = 1$, which means, by Lemma 2, that $|t(\mathcal{S})| = 1$. Let $\mathbf{x}$ be the unique state of $\psi(T(\mathcal{D})) = t(\mathcal{S})$, the state returned by Algorithm 3. State $\mathbf{x}$ is also the result returned by Algorithm 1. So it samples the stationary distribution.

4   Numerical Experiments

In this section, we compare our MDCFTP (Algorithm 3) with the classical CFTP algorithm (Algorithm 1).

We compare the size of the state representation, the coupling times and the running times. The coupling time is the value n when the algorithm stops. Throughout the experiments, we use the PRIORITY discipline where class 1 is given the priority over class 2. The algorithms have been implemented in Python and performed on a laptop.

4.1   Network of Example 1

We first consider the network of Example 1 with m customers in each class ($M = 2m$), with $m \in [1, 20]$. In Fig. 9 we compare the performance of Algorithms 1 and 3 (using the same sequences (U_{-n})). Each value is the mean of 100 random samples.

Algorithm 3. Multiclass Diagram CFTP

Data: $(U_{-n} = (i_{-n}, J_{-n}), \Theta_{-n})_{n \in \mathbb{N}}$ an i.i.d sequence of r.v
Result: $\mathbf{x} \in \mathcal{S}$
1 **begin**
2     $n \leftarrow 1$;
3     $T \leftarrow T_{U_{-1}}$;
4   **while** $|\psi(T(\mathcal{D}))| \neq 1$ **do**
5        $n \leftarrow 2n$;
6        $T \leftarrow T_{U_{-1}} \circ \cdots \circ T_{U_{-n}}$;
7   **return** $\mathbf{x}$, *the unique state of* $\psi(T(\mathcal{D}))$

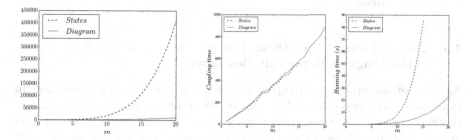

Fig. 9. Comparisons for the network of Example 1. Left: cardinality of the state space $|\mathcal{S}|$ vs. number of arcs in the diagram $|g(\mathcal{S})|$; center: comparison of the coupling times; right: comparison of the running times.

As expected, the cardinality of the state space grows exponentially with m, while $|g(\mathcal{S})|$ only grows polynomially. The coupling times are very close, which indicates that our representation is precise enough to ensure a reasonable coupling time. Algorithm 3 significantly outperforms Algorithm 1 in terms of the running times. For Algorithm 1, the sampling could not be computed in less than six hours for $m > 15$.

4.2 Bidirectional Ring Network

We consider a bidirectional ring network with K queues and 2 classes, and matrices transitions $P^1_{i,i \mod (K)+1} = 0.9$, $P^1_{i \mod (K)+1,i} = 0.1$, $P^2_{i,i \mod (K)+1} = 0.1$ and $P^2_{i \mod (K)+1,i} = 0.9$. The number of customers in each class is 5.

We perform exactly the same experiments as in the previous example, but make the number of queues vary between 2 and 20 (Fig. 10).

Not surprisingly, the cardinality of the state space grows exponentially with K. We were able to perform Algorithm 1 only for very small values ($K \leq 7$). Indeed, the number of states is in $O(m^{2K})$ whereas the state representation with diagrams is in $O(Km^4)$. In those cases, the coupling times of the two algorithms are very close and the running time correlated with the cardinality of the representation.

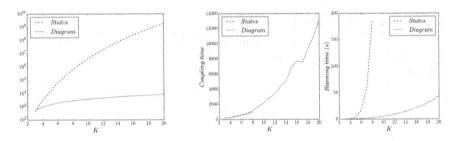

Fig. 10. Bidirectional network: left: cardinality of the state space $|S|$ vs. number of arcs in the diagram $|g(S)|$; center: comparison of the coupling times; right: comparison of the running times.

4.3 Comparisons of the Number of States $|S|$ and the Size of the Diagram Representation $|g(S)|$

As our representation is still exponential in Z, we have restricted ourselves to experiments with two classes of customers. We compare the size of the representation of the state space for Algorithms 1 and 3. The size of the representation only depends on the sets of queues visited by each class, and not on the exact topology of the network. For the sake of simplicity, we assume that each class has m customers and visits every queue. We make m vary for several values of Z and K. The results are depicted in Fig. 11.

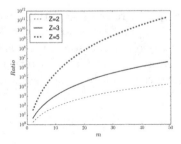 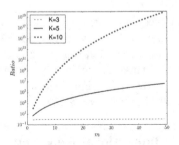

Fig. 11. Ratio $\frac{|S|}{|g(S)|}$. Left: $K = 5$, $Z \in \{2, 3, 5\}$; right: $Z = 3$, $K \in \{3, 5, 10\}$.

5   Conclusion

The main contribution of the paper is the derivation of the CFTP algorithm for multiclass closed queueing networks under various service policies. The only assumption made is that each station i choses the class to serve using only local information - the current state of queue i.

Our CFTP algorithm uses multiclass diagrams, a more compact representation of the state space. As in the monoclass case, using diagrams allows reduction of the complexity of the one-step transition function in the CFTP scheme from exponential to linear, in terms of the number of queues in the system. Unfortunately, diagram CFTP is still exponential in terms of the number of classes, so it is efficient only when the number of classes stays relatively small (less than 5). The main open question is the existence of a compact representation that is also polynomial in the number of classes, while keeping the coupling time of the bounding chain close to that of the original one.

We plan to investigate more closely the implementation of the transition function for the multiclass diagrams and the possible generalizations of the gap free diagrams in [6] that allowed a complexity reduction from $O(KM^2)$ to $O(KM)$ in the monoclass case.

For simplicity of exposition, we focused here on infinite capacity case. Extension to buffers with finite capacity does not represent any major difficulty, but it is not straightforward.

The major theoretical challenge is the study of the coupling times.

Acknowledgments. The work presented in this paper has been carried out at LINCS (www.lincs.fr) and has been founded by the French National Research Agency grant ANR-12-MONU-0019.

References

1. Asmussen, S., Glynn, P.W.: Stochastic simulation: algorithms and analysis. Springer, New York (2007)
2. Balsamo, S.: Queueing networks with blocking: analysis, solution algorithms and properties. In: Kouvatsos, D.D. (ed.) Next Generation Internet: Performance Evaluation and Applications. LNCS, vol. 5233, pp. 233–257. Springer, Heidelberg (2011)
3. Baskett, F., Chandy, K.M., Muntz, R.R., Palacios, F.G.: Open, closed, and mixed networks of queues with different classes of customers. J. Assoc. Comput. Mach. **22**, 248–260 (1975)
4. Baynat, B., Dallery, Y.: A product-form approximation method for general closed queueing networks with several classes of customers. Perform. Eval. **24**(3), 165–188 (1996)
5. Bouillard, A., Busic, A., Rovetta, C.: Perfect sampling for multiclass closed queueing networks. hal-01159962. June 2015
6. Bouillard, A., Bušić, A., Rovetta, C.: Clones: closed queueing networks exact sampling. In: 8th International Conference on Performance Evaluation Methodologies and Tools, VALUETOOLS 2014, ICST (2014)
7. Bouillard, A., Bušić, A., Rovetta, C.: Perfect sampling for closed queueing networks. Perform. Eval. **79**, 146–159 (2014)
8. Bušić, A., Vliegen, I., Scheller-Wolf, A.: Comparing Markov chains: aggregation and precedence relations applied to sets of states, with applications to assemble-to-order systems. Math. Oper. Res. **37**(2), 259–287 (2012)
9. Bušić, A., Gaujal, B., Perronnin, F.: Perfect sampling of networks with finite and infinite capacity queues. In: Al-Begain, K., Fiems, D., Vincent, J.-M. (eds.) ASMTA 2012. LNCS, vol. 7314, pp. 136–149. Springer, Heidelberg (2012)

10. Bušić, A., Gaujal, B., Pin, F.: Perfect sampling of Markov chains with piecewise homogeneous events. .Perform. Eval. **69**(6), 247–266 (2012)
11. Gordon, W., Newel, G.: Closed queueing systems with exponential servers. Oper. Res. **15**(2), 254–265 (1967)
12. Huber, M.: Perfect sampling using bounding chains. Ann. Appl. Probab. **14**(2), 734–753 (2004)
13. Kendall, W., Møller, J.: Perfect simulation using dominating processes on ordered spaces, with application to locally stable point processes. Adv. Appl. Probab. **32**(3), 844–865 (2000)
14. Kijima, S., Matsui, T.: Randomized approximation scheme and perfect sampler for closed Jackson networks with multiple servers. Ann. Oper. Res. **162**(1), 35–55 (2008)
15. Levin, D., Peres, Y., Wilmer, E.: Markov Chains and Mixing Times. American Mathematical Society, New York (2009)
16. Marie, R.: An approximate analytical method for general queueing networks. IEEE Trans. Softw. Eng. **5**(5), 530–538 (1979)
17. Propp, J.G., Wilson, D.B.: Exact sampling with coupled Markov chains and applications to statistical mechanics. Random Struct. Algorithms **9**(1–2), 223–252 (1996)
18. Satyam, K., Krishnamurthy, A., Kamath, M.: Solving general multi-class closed queuing networks using parametric decomposition. Comput. Oper. Res. **40**, 1777–1789 (2013)
19. Sigman, K.: Exact simulation of the stationary distribution of the FIFO M/G/c queue: the general case for $\rho < c$. Queueing Sys. **70**(1), 37–43 (2012)
20. van Dijk, N.M.: Bounds and error bounds for queueing networks. Ann. Oper. Res. **79**, 295–319 (1998)

Power and Effectiveness in Queueing Systems

Gerardo Rubino[✉]

Inria, Campus de Beaulieu, 35042 Rennes Cedex, France
gerardo.rubino@inria.fr

Abstract. In a series of papers, Kleinrock proposed a performance metric called *power* for stable queueing systems in equilibrium, which captures the tradeoff every queue makes between work done and time needed to. In particular, he proved that in the $M/GI/1$ family of models, the maximal power is obtained when the mean number of customers in the system is exactly one (a *keep the pipe empty* property). Since then, this metric has been used in different works, in the analysis of communication systems. In this paper we add some results about power, showing in particular that the concept extends naturally to Jackson product form queueing networks, and that, again, the *keep the pipe empty* property holds. We also show that this property does not hold in general for single-node models of the $GI/GI/1$ type, or when the storage capacity of the system is finite. The power metric takes into account the cost in time to provide service, but not the cost associated with the server itself. For this purpose, we define a different metric, which we call *effectiveness*, whose aim is to also include this aspect of systems. Both metrics coincide on single server models, but differ on multi-server ones, and on networks. We provide arguments in support of this new metric and, in particular, we show that in the case of a Jackson network, the *keep the pipe empty* property again holds.

1 Introduction

Performance evaluation theory is built around a set of metrics quantifying the way a system works, in general assuming it is perfect (that is, ignoring failures and possible repairs). As most systems must share resources (we live in a finite world), the main difficulty performance evaluation faces is to understand the interactions between the users of those resources and their competition for them, and finally, to quantify the whole system behavior. In networking, for instance, typical shared resources are bandwidth, channels, devices such as routers or switches, etc. Some performance metrics are related to the work done by the system, for instance, the mean throughput of a video streaming application, or the utilization rate of a server. Other metrics focus on the price to pay for access to the resources; typical examples are the delays, or the backlog size at a node in the network.

This paper works on an idea of Kleinrock, proposed back in the late 70s, consisting of using for system analysis a metric combining the two aspects previously mentioned, that is, how much work the system does, and at which cost

© Springer International Publishing Switzerland 2015
J. Campos and B.R. Haverkort (Eds.): QEST 2015, LNCS 9259, pp. 279–294, 2015.
DOI: 10.1007/978-3-319-22264-6_18

in time. These two aspects are competitors. For instance, if the traffic offered to the system increases, then the system will work more, a good point, but the delay will also increase, a bad one. Many quantitative aspects of communication systems are highly non linear, and the way the two aspects of its operation evolve are not necessarily similar. This suggests the existence of optimal operating points for the system, taking into account both types of views about its performance. Kleinrock explored the use of the ratio between the throughput of the system and its mean response time, what was called the system's *power* in [2], first in [3]. He and co-workers published some further results about it in [4], [5]. For examples of use, see [6,7,9–14,16,17]; see also [8].

This paper reviews some aspects of this metric that started to be explored in [15]. It adds supplementary results and discusses its application to networks of nodes in the tractable case, that is, when the network possesses the product form property, following the main result in [15]. We propose here a slightly different definition of power in networks, and a simpler proof of the mentioned result. Then, we show that the power metric ignores a different but relevant side of the cost previously discussed, when there are many servers in the system and also when we deal with a queueing network. To take this aspect into account, we propose a new metric which we call *effectiveness*, and we provide a preliminary analysis of its properties. This is the main contribution of the paper.

In the next Section the main definitions and properties of the power are presented. The optimal operating conditions for different basic types of queues are discussed, together with other properties of the metric. Then, Sect. 3 analyzes the case of a product form queueing network and its optimal operation point from the power point of view. Section 4 shows that the power doesn't capture the global processing capacity of the system, and that this aspect can be relevant in specific analysis problems. Then, Sect. 5 proposes the new effectiveness metric. Section 6 proves that the same "keep the pipe empty" that holds for power in the case of a Jackson product-form networks, also holds for effectiveness. In the same section we show how this metric solves the problem that power exhibits when the system is a multi-server queue, or a queueing network. Section 7 concludes the paper.

2 Power

Consider a stable queue in equilibrium, representing for instance the contention for some resource in a communication system, say a $GI/GI/1$ model. In the case, for instance, of the classic $M/M/1$ queue with arrival rate λ and service rate μ, roughly speaking, "stable" means that $\lambda < \mu$ and "in equilibrium" means that we look at the queue at time $t = \infty$. The throughput is denoted by T and the mean service time is denoted by S. Since the model is stable, we have $TS < 1$ (except in the $D/D/1$ case, where we may also have $TS = 1$, see next Section). The product TS is the *load* of the system, which we denote by a. It is also the probability that the server is busy, and also the mean number of units in service (the *server's utilization* or *occupancy*). We denote by R the mean response time of the queue (including the time in service).

In general, the system's manager wants to see a large value of T (indicating that the node deserves a large amount of traffic, that is, that the node works intensively and thus, that a large amount of revenue is generated) and R small, meaning that the mean delay experienced by the customers is small. These two goals are obviously contradictory. A good way to take them both into account is to say that we want the ratio T/R as large as possible. To eliminate the dependency on the dimensions, we can normalize the numerator T by multiplying it by the mean service time S, and the denominator R by dividing it by S. The result is the power P proposed by Kleinrock in [3], following the ideas of [2]: denoting by N the mean number of units in the system (including the one in service, if any), we have

$$P = \frac{TS}{R/S} = \frac{aS}{N/T} = \frac{a^2}{N},$$

where we used the fact that $R = N/T$ (Little's formula).

Observe that $a = TS < 1$ and that $R \geq S$, leading to $P < 1$. See that we can say more, namely, that $P \leq a$ since $N \geq a$. Thus, with this single metric, we can take into account simultaneously the production rate of the system, represented by TS, and the associated cost in terms of delay, given by the *normalized mean response time* $\widetilde{R} = R/S$. This naturally leads to defining the *optimal operating point* for the system, from the power viewpoint, as the situation where power has its maximum value. We will denote this optimal value by P^*, and by m^* the value of any other metric m evaluated at that optimal operating point.

The M/M/1 Case. Let us take as our first example the classical $M/M/1$ model (as Kleinrock does), where the arrival rate is λ, the service rate is μ, and $\lambda < \mu$. We have $T = \lambda$, $S = 1/\mu$, $a = \lambda/\mu$, $N = a/(1-a)$, which gives $P = a(1-a)$. The maximal power value is obtained for $a^* = 1/2$ and its value is $P^* = 1/4$. Kleinrock naturally asks for the value of the mean backlog in the system, N, when $a = a^*$, that is, in optimal conditions from the point of view of the power. The result is $N^* = N|_{a=a^*=1/2} = 1$. Concerning the *normalized mean response time* $\widetilde{R} = R/S$, we have $\widetilde{R}^* = 2$.

The M/GI/1 Case. More generally, consider a $M/GI/1$ queue in equilibrium, where the arrival rate is λ, the mean service time is S (leading to $a = \lambda S < 1$), and the coefficient of variation (or normalized standard deviation) of the service time is C. The analysis of this queue [1] gives $N = a + a^2(1+C^2)/(2(1-a))$. We then obtain

$$P = \frac{2a(1-a)}{2 - a + aC^2}.$$

Observe first that with respect to C, the power is maximal when $C = 0$, that is, in the $M/D/1$ case, leading to the power expression $2a(1-a)/(2-a)$. Its maximum value on $[0, 1]$ is $P^* = 6 - 4\sqrt{2} \approx 0.3431$, obtained for $a^* = 2 - \sqrt{2} \approx 0.5858$. Looking again at the optimal mean backlog in the queue, we get

$$N^* = N\,|_{C=0,\,a=a^*} = a + \frac{a^2}{1-a}\frac{1+C^2}{2}\bigg|_{C=0,\,a=2-\sqrt{2}} = 1.$$

This was obtained by Kleinrock in [3] following a different path, and it's a nice result (it has been called the *keep the pipe empty* property[1]). It says that the best possible system in the $M/GI/1$ family, from the point of view of the power metric, corresponds to the case of exactly one customer in the server and nobody waiting, *on the average*. This property does not extend to any queuing system, as, for instance, the next example will show. Concerning the scaled mean response time at the optimal operating point, we have $\widetilde{R}^* = a^*/P^* = (2-\sqrt{2})/(6-4\sqrt{2}) = 1+\sqrt{2}/2 \approx 1.7071$.

The $E_2/M/1$ Case. Consider now an $E_2/M/1$ model in equilibrium. The distribution of the inter-arrivals is Erlang with rank or order 2 and real parameter ν. The service rate is μ. The mean arrival rate, equal to the system's throughput, is $T = \nu/2$. The system's load is $a = \nu/(2\mu)$ (the stability condition is $\nu < 2\mu$).

To solve for the main performance metrics of this model (see [1]), we must first solve the equation in $x \in (0,1)$, $x = L(\mu(1-x))$, where $L()$ is the Laplace transform of the inter-arrival distribution (here, $L(s) = (\nu/(s+\nu))^2$). We know that this has a single solution, given in this case by $\beta = (1 + 4a - \sqrt{1+8a})/2$. From the general analysis of the $GI/M/1$ model, we have that

$$N = \frac{a}{1-\beta} = \frac{2a}{1-4a-\sqrt{1+8a}}.$$

This gives $P = a(1-\beta)$, that is,

$$P = \frac{a(1-4a+\sqrt{1+8a})}{2}.$$

Power is maximal for $a = a^* = (13 + 5\sqrt{17})/64 \approx 0.5252$. Its maximal value is $P^* = (107 + 51\sqrt{17})/1024 \approx 0.3098$ and $N^* = (3 + \sqrt{17})/8 \approx 0.8904$. This says that at the optimal operating point of this model, there is a bit less than one customer, on the average, in the system. Concerning the scaled mean response time, we obtain $\widetilde{R}^* = (23 + \sqrt{17})/16 \approx 1.69512$. So, no "keep the pipe empty" property here.

A Family of $H_2/M/1$ Models. Consider now an $H_2/M/1$ model, where we are first given the mean inter arrival time $1/\lambda$ and the coefficient of variation of the inter-arrival time, $C \geq 1$. Then, if $f(t) = p_1 e^{-\nu_1 t} + p_2 e^{-\nu_2 t}$ is the density of the inter-arrival distribution, with ν_1, ν_2, p_1, $p_2 > 0$, $p_1 + p_2 = 1$, a frequent choice is to use the symmetric expressions

$$p_1 = \frac{1}{2}\left(1 - \sqrt{\frac{C-1}{C+1}}\right), \quad p_2 = \frac{1}{2}\left(1 + \sqrt{\frac{C-1}{C+1}}\right),$$

plus $\nu_1 = 2p_1\lambda$ and $\nu_2 = 2p_2\lambda$. This gives mean inter-arrival time $1/\lambda$ and coefficient of variation of the inter-arrival time $C \geq 1$. The Laplace transform

---

[1] Of course, this doesn't mean that the mean number of units waiting in the queue is 0: if $a = 2 - \sqrt{2}$, the mean length of the waiting queue is $1 - a = \sqrt{2} - 1$.

of the service time distribution is $L(s) = p_1\nu_1/(\nu_1 + s) + p_2\nu_2/(\nu_2 + s)$. The equation $x = L\big((1 - x)/S\big)$ has in $(0, 1)$ the only solution

$$\beta = \frac{d_1 + d_2 + 1 - \sqrt{(d_1 - d_2)(d_1 - d_2 - 4p_1 + 2) + 1}}{2},$$

where $d_i = \nu_i/\mu$, $i = 1, 2$. Writing now everything in terms of $a = \lambda/\mu$ and C, we get, after some algebra,

$$P = a(1 - \beta) = \frac{a}{2}\left(1 - 2a + \sqrt{Q}\right), \qquad Q = \frac{4a^2 C - 4a^2 + C + 1 - 4aC + 4a}{C + 1}.$$

Since $\partial Q/\partial C = -8a(1 - a)/(C + 1)^2 \leq 0$, we see that for any load a, P is max for $C = 1$ (that is, when the Hyperexponential becomes Exponential):

$$P = \frac{a(1 - 2a)}{2} + \frac{a}{2} = a(1 - a).$$

It follows that in this (important in practice) family of $H_2/M/1$ queues, we have again $P^* = 1/4$, and $a^* = 1/2$. Since $P = a^2/N$, we then have $N^* = 1$. So, the result $N^* = 1$ also holds for this class of $GI/M/1$ models. Obviously, the scaled mean response time at the optimal operating point is again $\widetilde{R}^* = 2$, since we are back at the $M/M/1$ model.

Extreme Values for Power. For the $M/M/1$ model, we saw that the power P satisfies $P \leq 0.25$. In the larger $M/GI/1$ family, we always have $P \leq 0.3431$. In the $E_2/M/1$ case, we found $P \leq 0.3098$ and in the $H_2/M/1$ case, we have again $P \leq 0.25$. We can wonder if, in the general $GI/GI/1$ family, the power is upper-bounded by some number in the interval $(0, 1)$. The answer is negative. Just consider an Unif/Unif/1 model, where the inter-arrival distributions are i.i.d. with the uniform law in the interval $(f - \eta, f + \eta)$ and where the service times are also i.i.d., uniformly distributed on $(g - \delta, g + \delta)$. We assume that $0 < g - \delta$ and that $g + \delta < f - \eta$. We have $a = g/f < 1$. With these assumptions, the system is stable and, if the system starts empty, there is never a customer waiting. If (π_n) is the stationary distribution of the number of units in this model, we have $\pi_1 = a$, $\pi_0 = 1 - a$ and $\pi_n = 0$ if $n \geq 2$. So, $N = a$, leading to $P = a$. We thus see that P can be as close as desired to 1. See that in this family of systems, $N = P$.

Extension to the Finite Capacity $GI/GI/1/H$. Kleinrock proposes a natural extension of the power concept to queues with finite storage capacity. For the $GI/GI/1/H$ model in equilibrium, if we denote by λ the arrival rate (the arrival process not necessarily being Poisson) and by β the loss or blocking probability, the mean throughput in equilibrium is $T = \lambda(1 - \beta)$. The power can be defined as in the $GI/GI/1$ case by the ratio $TS/(R/S) = T^2S^2/N$. See that using Little applied to the server, TS is the mean number of units in service, and thus, $TS < 1$, so, the power is again < 1 as in the open case. Denoting $a = \lambda S$, now more appropriately called the *carried load*, we have $TS = a(1-\beta)$ and the power would be $a^2(1-\beta)^2/N$. Now, Kleinrock suggests to use instead

$$P = \frac{T^2 S^2}{N}(1-\beta) = \frac{a^2(1-\beta)^3}{N},$$

that is, to use a supplementary factor $1-\beta$, to reinforce the fact that we don't want to loose too many customers.

Argument to Support Kleinrock's Definition. A striking argument supporting this definition appears if we just consider the $M/M/1/1$ model. The initial definition leads to a power equal to the ratio $a/(1+a)$, meaning that the power increases with a. When $a \longrightarrow \infty$, the power goes to 1. This is clearly unsatisfactory, since as a gets close to 1, the throughput has little changes, but we loose almost all the offered traffic. With the proposed modification by multiplying by $1-\beta$, we get $P = a/(1+a)^2$, which now goes to 0 as $a \longrightarrow \infty$. This function has a maximum for the offered load $a^* = 1$, and its value is $P^* = 1/4$. At this optimal operating point, the mean number of customers in the system is $N^* = 1/2$.

In the $M/M/1/H$ case, with arrival rate λ and service rate $\mu \neq \lambda$, we obtain, after some algebra,

$$P = a(1-a)\frac{(1-a^H)^3}{(1-a^{H+1})^2[1-(H+1)a^H + Ha^{H+1}]}.$$

If $\lambda = \mu$, we have $\beta = 1/(H+1)$, $N = H/2$, and $P = 2H^2(H+1)^{-3}$. It is easy to see that P is a continuous function of a on the positive reals. It behaves similarly as previously considered power functions when a varies. Let us give just a couple of examples, where we use H as a subscript for indexing the metrics.

$$P_2 = \frac{a(1+a)^3}{(1+a+a^2)^2(1+2a)}, \quad a^* \approx 0.8132, \quad P_2^* \approx 0.3014, \quad N_2^* \approx 0.8631;$$

$$P_3 = \frac{a(1+a+a^2)^3}{(1+a+a^2+a^3)^2(1+2a+3a^2)}, \quad a^* \approx 0.7291, \quad P_3^* \approx 0.2964, \quad N_3^* = 1.116.$$

As we see, the optimal operating point in these examples, from the power's viewpoint, does not behave as in the open $M/M/1$ case (or in the $M/GI/1$ one). In Fig. 1 we see the behavior of the power P_H for $H = 1, 2, 3, 5$ and 10. We also plot the limiting function corresponding in fact to the $M/M/1$ model. Then, in Fig. 2, we plot the key parameters in the analysis of the optimal operating point, P_H^*, a_H^* and N_H^*, as a function of H, for $H = 1, 2, \ldots, 20$.

Extension to multi-server queues. Another extension discussed by Kleinrock is to multi-server queues. Consider a stable $GI/GI/c$ model in equilibrium, with throughput T and mean service time S. Here, the load $a = TS$ satisfies $TS < c$. The utilization of each of the c servers is $\varrho = a/c$. Using Little applied to the set of servers of the system, we have that a is equal to the mean number of busy servers in equilibrium. Kleinrock proposes to define the power as

$$P = \frac{TS}{cR/S} = \frac{a^2}{cN},$$

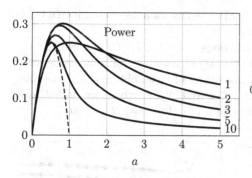

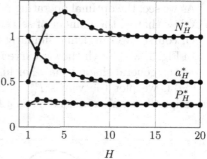

Fig. 1. Power of the $M/M/1/H$ model, for a few values of H, as a function of $a = \lambda/\mu$ where λ is the arrival rate and μ is the service rate; at the right of each curve, the corresponding value of H; see the particular case of P_1; the dashed curve corresponds to the power of the $M/M/1$ (that is, to the case of $H = \infty$)

Fig. 2. Max value of power (P_H^*), argmax value (a_H^*) and corresponding mean backlog of the system (N_H^*) in the $M/M/1/H$ model, as a function of H; the three corresponding horizontal asymptotes are indicated as dashed lines (at values 1, 1/2 and 1/4)

leading again to an index < 1. For instance, in the $M/M/c$ case with arrival rate λ and service rate μ, after some algebra, an expression of P is

$$P = \frac{a(c - a)}{c(c - a + p_W)},$$

where p_W is the probability of waiting. Recall that if the probability of an empty $M/M/c$ in equilibrium is π_0, we have

$$\pi_0 = \left[\frac{a^c}{(c - 1)!(c - a)} + \sum_{j=0}^{c-1} \frac{a^j}{j!} \right]^{-1}, \text{ and } p_W = \frac{a^c}{(c - 1)!(c - a)}\pi_0.$$

The given expression of P allows to see, for instance, that

$$\frac{a(c - a)}{c(c - a + 1)} < P < \frac{a}{c},$$

obvious useful if a is not very close to c. See that if $c = 1$, we obtain the same definition given before for open single server models. Concerning the mean backlog at the optimal operating point, Kleinrock showed in [3] that $N^* \neq c$. Just consider the power in the subclass of $M/M/c$ models. For instance (denoting by P_c the power when there are c servers),

$$P_2 = \frac{a(4 - a^2)}{8}, \quad P_2^* \approx 0.3849, \quad a^* \approx 1.1547, \quad N^* = \sqrt{3} \approx 1.7321;$$

$$P_3 = \frac{a(3 - a)(6 + 4a + a^2)}{3(18 + 6a - a^2)}, \quad P_3^* \approx 0.4649, \quad a^* \approx 1.8865, \quad N^* \approx 2.5515.$$

As we see, the optimal operating point in these models corresponds to a value of N less than c, the number of servers.

Of course, we can define similarly the power of $GI/GI/c/H$ models.

In Fig. 3, we plot the power curve in the $M/M/c$ system for some values of c, to see the general shape it has. For higher values of c, we obtain similar results. In Fig. 4, we plot the values P_c^*, $\varrho_c^* = a_c^*/c$ and N_c^*/c, where we added as a subscript the number of servers, as a function of c, for $c = 1, 2, \ldots, 20$.

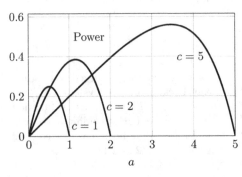

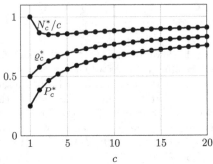

Fig. 3. Power of the $M/M/c$ model, with arrival rate λ and service rate μ, for a few values of c, as a function of the load $a = \lambda/\mu < c$

Fig. 4. For the $M/M/c$ model, we plot the maximal value of the power (P_c^*), the argmax in terms of the parameter ϱ (that is, $\varrho^* = a^*/c$) and the corresponding mean backlog of the system, scaled by c (N_c^*/c), as a function of c

See that as $c \to \infty$, that is, as we get "close" to an $M/M/\infty$ model, $N \to a$, and thus, $P = a^2/(cN) \sim a/c$, coherent with the visual information provided in Figs. 3 and 4. For instance, using numerical tools, we have that $P_{400}^* \approx 0.941613$, reached at $a_{400}^* \approx 383.493$, that is, $\varrho_{400}^* \approx 0.958733$. If $c = 1000$, we have $P_{1000}^* \approx 0.962807$, reached at $a_{1000}^* \approx 973.707$, that is, $\varrho_{1000}^* \approx 0.973707$.

3   The Case of an Open Product Form Queueing Network

Consider now an open queueing network $\mathcal{N}$ with n single server nodes, of the Jackson type. As we know, the equilibrium behavior is determined by the routing probabilities, the arrival rates and the service rates. Denote by μ_i the service rate at node i, and assume you solved the traffic equations leading to the computation of the throughput T_i at i (a function of the routing probabilities and the arrival rates only). Then, under the stability condition $T_i < \mu_i$ for all i, the equilibrium distribution of the network is given by the well-known product-form result: if X_i is the number of customers at node i in equilibrium, and if we denote by a_i the load at i, that is, $a_i = T_i/\mu_i < 1$, then we have

$$\mathbb{P}\big((X_1, X_2, \ldots, X_n) = (x_1, x_2, \ldots, x_n)\big) = \prod_{i=1}^{n} (1 - a_i) a_i^{x_i},$$

for all vector of non-negative integers $(x_1, \ldots, x_n)$. From this theorem due to Jackson, we derive all the performance metrics we are usually interested in.

Now, the *mean total service* received by a generic customer is $S = \sum_{i=1}^{n} v_i S_i$, where $S_i = 1/\mu_i$ and where v_i is the mean number of visits made to node i before leaving the network. We have $v_i = T_i/\lambda$, where λ is the total arrival rate to $\mathcal{N}$. Similarly, the *mean response time* or *mean sojourn time* of a generic customer in $\mathcal{N}$ is $R = \sum_{i=1}^{n} v_i R_i$, where $R_i = (\mu_i - T_i)^{-1}$ is the mean response time at node i.

With these elements, the first natural idea to extend the power to these models is to define it as the ratio $\lambda S/(R/S)$. Observing that $\lambda S < n$ and $R > S$, we would obtain a power metric $< n$. This makes natural to add the factor $1/n$ to get again a normalized metric with values in $[0, 1]$ (as for multi-servers models). We obtain $P = \lambda S/(nR/S) = (\lambda S)^2/(nN)$, that is,

$$P = \frac{1}{n} \frac{\left(\displaystyle\sum_{i=1}^{n} a_i \right)^2}{\displaystyle\sum_{i=1}^{n} \frac{a_i}{1 - a_i}}.$$

We are now going to look at the maximal value of P when $(a_1, \ldots, a_n) \in [0, 1)^n$.

Theorem 1. *In a stable Jackson network, the optimal value of the power is reached when the load at each node is 1/2. In that case, the mean number of customers in equilibrium at any node is 1, and the power of the network is 1/4.*

Proof. Instead of working with these a_is in $[0, 1)$, let us change the variables in the following way: for all i, $a_i/(1 - a_i) = x_i$; then, $a_i = x_i/(1 + x_i)$, and now, we can work with unbounded positive variables. The case of $x_i \to 0$ corresponds to the limiting case $a_i \to 1$. The power function has now the form

$$P_n = \frac{1}{n} \frac{\left(\dfrac{x_1}{1 + x_1} + \cdots + \dfrac{x_n}{1 + x_n} \right)^2}{x_1 + \cdots + x_n}.$$

Let us first look at the case of $n = 1$. We have

$$P_1(x) = \frac{x}{(1 + x)^2}, \qquad P_1'(x) = \frac{1 - x}{(1 + x)^3}, \qquad P_1''(x) = \frac{2x - 4}{(1 + x)^4}.$$

So, argmax $P = x^* = 1/2$, and $P_1^* = 1/4$.

Now, considering $P_n(x_1, \ldots, x_n)$, we immediately see that the only critical point in the interior of its domain is the point $(1, \ldots, 1)$, and that there, the function has value $1/4$. To see it, write $U = x_1(1 + x_1)^{-1} + \cdots + x_n(1 + x_n)^{-1}$ and $V = x_1 + \cdots + x_n$ so that $P = U^2/(nV)$. Then, write $\partial P/\partial x_i = 0$ for all node i, that is,

$$\frac{\partial P}{\partial x_i} = \frac{U}{nV^2} \left[\frac{2V}{(1 + x_i)^2} - U \right] = 0.$$

But this means that $x_i = w$, independent of i. This gives $U = nw(1+w)^{-1}$ and $V = nw$. Using again previous equation, we have

$$\frac{2nw}{(1+w)^2} - \frac{nw}{1+w} = 0,$$

giving the only solution $w = 1$. To see that this is an absolute maximum, we can proceed by induction. We can start from $n = 2$, by looking at

$$P_2(x,y) = \frac{1}{2} \frac{\left(\dfrac{x}{1+x} + \dfrac{y}{1+y}\right)^2}{x+y}.$$

After observing that $P_2(0,0) = P_2(\infty,\infty) = P_2(\infty,y) = P_2(x,\infty) = 0$, we must look at $P_2(t,0) = P_2(0,t) = P_1(t)/2$, for which the absolute maximum is $1/8$. Since at the critical point, $P_2(1,1) = 1/4 > 1/8$, we know that P_2 has an absolute maximum at $(1,1)$ and that this maximum is $1/4$.

Then, assuming that $P_j^* = 1/4$ for $j = 1, 2, \ldots, n-1$, and looking at P_n now, we first have that $P_n(0,\ldots,0) = \lim_{x_i \to \infty} P_n(x_1,\ldots,x_n) = 0$. So, P_n is zero when all the variables are 0, or when at least one of them is ∞. The last point is to see what happens when some of the x_is but not all of them take value 0. Let's assume that $x_1 = x_2 = \cdots = x_k = 0$, for some k verifying $1 \le k < n$. Then,

$$P_n(0,\ldots,0,x_{k+1},\ldots,x_n) = \frac{n-k}{n} P_{n-k}(x_{k+1},\ldots,x_n).$$

But we know that the maximal value of P_{n-k} is, by hypothesis, $1/4$, so, when $x_1 = x_2 = \cdots = x_k = 0$, the maximal value of P_n is

$$\frac{n-k}{n}\frac{1}{4} < P_n(1,\ldots,1) = \frac{1}{4}.$$

We conclude that $P_n^* = 1/4$ and $x_i^* = 1/2$ for all i. ∎

This was already proven in [15], basically by looking at the type of critical point first, then showing that the Hessian of P as a function of the a_is was strictly definite negative, and concluding by showing that the absolute maximum belongs to the interior of D.

4   Problems with the Power Metric

The definition of power when applied to systems having many servers exhibits some features that suggest to look for another metric. Consider the typical "scholar" example where the student must compare the performances of the three following systems, denoted σ_i, $i = 1, 2, 3$, considered in equilibrium:

– σ_1: a Jackson network composed of two identical and parallel $M/M/1$ queues (thus, isolated from each other), each having arrival rate α and service rate β;

- σ_2: an $M/M/2$ model with arrival rate 2α and service rate β at each of the two servers;
- σ_3: an $M/M/1$ model with arrival rate 2α and service rate 2β.

We assume $\alpha < \beta$, to be able to consider the three systems in equilibrium. Denote $b = \alpha/\beta < 1$. Denote by T_i, N_i and R_i the throughput, mean backlog and mean delay of σ_i. Observe that $T_1 = T_2 = T_3 = 2\alpha$. Then,

$$N_1 = \frac{2b}{1-b}, \qquad N_2 = \frac{2b}{1-b^2}, \qquad N_3 = \frac{b}{1-b},$$

$$R_1 = \frac{1}{\beta}\frac{1}{1-b}, \qquad R_2 = \frac{1}{\beta}\frac{1}{1-b^2}, \qquad R_3 = \frac{1}{2\beta}\frac{1}{1-b}.$$

We have $N_1 \geq N_2 \geq N_3$ and, equivalently, $R_1 \geq R_2 \geq R_3$ (since by Little, $N_i = 2\alpha R_i$, $i = 1, 2, 3$. The student must conclude that σ_3 is the best system (it has the smallest mean delay and mean backlog for the same throughput). Then comes σ_2, and finally σ_1 (and she must understand why). What says the power metric for the three systems? After making the computations, we get $P_1 = b(1-b)$ (a Jackson case), $P_2 = b(1-b^2)$ (an $M/M/c$ case) and $P_3 = b(1-b)$ again (the $M/M/1$ case). We have $P_1 = P_3 \geq P_2$, so, from the power viewpoint, σ_1 and σ_3 behave the same, both better than σ_2; bad.

The problem comes from the scaling used in power's definition, and its impact in systems σ_1 and σ_2. See that we are implicitly using the fact that the comparison is "fair" because the three systems deploy the same potential capacity (represented by the number 2β). To further illustrate the point, observe that when two systems have the same throughput, we prefer the one with the smallest response time, not with the smallest *scaled* response time! For instance, here, $R_1 \geq R_2 \geq R_3$ but $\widetilde{R}_1 = \widetilde{R}_3 = 1/(1-b) \geq \widetilde{R}_2 = 1/(1-b^2)$.

To have another illustration of the problems, just consider a tandem network of n identical $./M/1$ queues, with common service rate μ and arrival rate λ to the tandem. Denoting the load of any of the nodes by $b = \lambda/\mu < 1$, we have that $P = a^2/(nN) = (nb)^2/(n^2 b/(1-b)) = b(1-b)$, the same as for the $M/M/1$. This also bothers our intuition, since the largest is n, the largest the mean response time, for the same throughput.

5   Effectiveness

We propose here a new metric that solves the types of problems mentioned in previous section. Let us consider here single or multiple server queues, or open networks of queues. These can be Jackson networks, but we can relax some assumptions (service times with other distributions, perhaps state dependent routing, etc.). For our goal, we don't need to formalize this. The reader can just think of a Jackson network, if useful. Let us start by naming the sum of all the service rates in the system, that is, the sum of the inverses of all the mean service times.

Definition 1. *The* mean total capacity *of the system is the sum of the service rates of all its servers, and will be denoted here by* γ.

Then, observe first that, necessarily, $T \leq \gamma$ if the system is in equilibrium. We can also lower bound R now using γ.

Lemma 1. *The mean system response time R satisfies $R \geq 1/\gamma$.*

Proof. The result is trivial in case of single queues. Consider a network, let v_i be the mean number of visits a customer makes to node i and let μ_j be the service rate of node j. Then,

$$R\gamma = \sum_{i=1}^{n} v_i R_i \sum_{j=1}^{n} \frac{1}{\mu_j} \geq \sum_{i=1}^{n} v_i S_i \sum_{j=1}^{n} \frac{1}{\mu_j} > \sum_{k=1}^{n} v_k S_k \frac{1}{\mu_k} = \sum_{k=1}^{n} v_k = 1,$$

which concludes the proof. ∎

We are now ready to define the *effectiveness* of a queueing system.

Definition 2. *The* effectiveness *of an open and stable queueing system in equilibrium, where T is the throughput, R the response time, N the mean backlog and γ the mean total capacity, is given by*

$$E = \frac{T/\gamma}{R\gamma}, = \frac{T}{\gamma^2 N/T} = \frac{T^2}{\gamma^2 N}.$$

If the system has a finite storage capacity, leading to a loss or blocking probability β, we can extend previous definition as for the power, by adding a supplementary factor $1 - \beta$. This will not be explored further in this paper.

Let us look at some immediate consequences of the previous definition. First, due to Lemma 1, effectiveness is always ≤ 1. Second, if the model is of the $GI/GI/1$ type, then power and effectiveness coincide. If the model is a $GI/GI/c$ queue, then $E = P/c$. For instance, consider the $M/M/c$ model discussed in Sect. 2. The main difference that can be observed is that the maximal effectiveness decreases as c increases. The available room makes that this can not be explored further here.

Let us move then to Jackson networks.

6   Effectiveness in a Jackson Network

We start by considering a network composed of $M/M/1$ in parallel, so independent from each other.

Theorem 2. *Consider a Jackson network composed of n nodes in parallel, where node i is an $M/M/1$ queue with arrival rate α_i and service rate $\beta_i > \alpha_i$. Then, the effectiveness of this network has maximum value when the loads at all the nodes are $1/2$, and this maximal value is $1/4$. In this case, the mean number of customers at each node is 1 (the "keep the pipe empty" property).*

Proof. The effectiveness of the network is

$$E_n = \left(\frac{\alpha_1 + \cdots + \alpha_n}{\beta_1 + \cdots + \beta_n}\right)^2 \frac{1}{\dfrac{\alpha_1}{\beta_1 - \alpha_1} + \cdots + \dfrac{\alpha_n}{\beta_n - \alpha_n}}.$$

We want to evaluate the maximum value of E_n for all possible arrival and service rates such that the model is stable. Let us make the following changes of variable:

$$q_i = \frac{\beta_i}{\beta_1 + \cdots + \beta_n}, \qquad x_i = \frac{\alpha_i}{\beta_i - \alpha_i}.$$

After some algebra, we get that the effectiveness becomes

$$E_n = \frac{\left(q_1 \dfrac{x_1}{1 + x_1} + \cdots + q_n \dfrac{x_n}{1 + x_n}\right)^2}{x_1 + \cdots + x_n},$$

where all the variables are positive, and where $q_1 + \cdots + q_n = 1$.

Finding the maximum possible value for this function is a constrained optimization problem. Dropping the subscript n for a while, the Lagrangian is simply $L = E - \lambda(q_1 + \cdots + q_n - 1)$. If we write the effectiveness as $E = F^2/G$, with $F = \sum_{i=1}^{n} q_i x_i (1 + x_i)^{-1}$ and $G = x_1 + \cdots + x_n$, we have, for all $i, j \in \{1, \ldots, n\}$,

$$\frac{\partial L}{\partial q_i} = \frac{2F}{G} \frac{x_i}{1 + x_i} - \lambda, \qquad \frac{\partial L}{\partial x_j} = \frac{F}{G}\left(\frac{2q_j}{(1 + x_j)^2} - \frac{F}{G}\right).$$

Let us look for the critical points. The equality $\partial L/\partial q_i = 0$ leads to $x_i = w$, that is, a value independent of i. Using this in the equality $\partial L/\partial x_j = 0$, we conclude that q_j does not depend on j neither. Using $q_1 + \cdots + q_n = 1$, we have $q_j = 1/n$ for all node j. Using this in $\partial L/\partial x_j = 0$, and observing that $F = w/(n(1 + w))$ and $G = nw$, we obtain that $w^* = 1$. So, the only candidate for a relative extremum in the interior of the domain is $x_i = 1$ and $q_i = 1/n$, for all i.

Now, let us look at the borders. If any of the $x_i \to \infty$, then $E \to 0$. Similarly, if for all i, $x_i \to 0$, $E \to 0$ as well. The case we must explore is when some of the x_i go to 0 but not all of them. So, assume that $x_i \to 0$ for $i = 1, \ldots, k$, for some k verifying $1 \le k < n$. Then,

$$E \to \frac{\left(q_{k+1} \dfrac{x_{k+1}}{1 + x_{k+1}} + \cdots + q_n \dfrac{x_n}{1 + x_n}\right)^2}{x_{k+1} + \cdots + x_n}.$$

We need now to look at $\min f_m$ where

$$f_m = \frac{\left(q_1 \dfrac{x_1}{1 + x_1} + \cdots + q_m \dfrac{x_m}{1 + x_m}\right)^2}{x_1 + \cdots + x_m},$$

where the variables are positive, but the constraint is $q_1 + \cdots + q_m \leq 1$. This needs to use the Karush-Kuhn-Tucker (KKT) conditions now, on the Lagrangian $L = f_m - \lambda(q_1 + \cdots + q_m = 1)$. The KKT conditions say that, for all i,

$$\frac{\partial L}{\partial q_i} \leq 0, \quad \frac{\partial L}{\partial x_i} \leq 0, \quad \frac{\partial L}{\partial \lambda} \geq 0, \quad q_i \frac{\partial L}{\partial q_i} = 0, \quad x_i \frac{\partial L}{\partial x_i} = 0, \quad \lambda \frac{\partial L}{\partial \lambda} = 0.$$

Solving this system as before leads to the same critical point given by $q_i = 1/m$, and $x_i = 1$, for all node i. Here, $f_m(1, \ldots, 1) = 1/(4m)$. In turn, we must look at the borders here, which will lead to the same min f_m problem but with a smaller m. By induction, the absolute minimal value of f_m is then $1/(4m) \leq 1/4$, which means that $1/4$ is the absolute maximum of E_n. ───────────■

We can now state the last result of the paper.

Theorem 3. *The maximum effectiveness of an arbitrary stable Jackson network with n nodes is $1/4$. It is reached when (i) the network is composed of a set of parallel queues, and (ii) when the load at each of the queues is $1/2$. Moreover, in such a case, the "keep the pipe full" property holds. If the network is not in this class, its effectiveness is always $< 1/4$.*

Proof. Consider now an arbitrary stable Jackson network $\mathcal{N}$ with n nodes, as described in Sect. 3 (and using the same notation). From the point of view of standard metrics such as the mean backlog N_i at node i, or the mean response time R_i, $\mathcal{N}$ can be replaced by a network $\mathcal{N}'$ made of n parallel $M/M/1$ queues with the same service rates but where the arrival rate at i is now T_i, the throughput through i in $\mathcal{N}$ (and thus, in $\mathcal{N}'$).

In $\mathcal{N}'$, the global throughput is obviously not λ (think of a series $\mathcal{N}$ of m identical $./M/1$ nodes, with arrival rate η: its global throughput is η, but in $\mathcal{N}'$, the global throughput is $m\eta$). We first observe that the global throughput $T_1 + \cdots + T_n$ in $\mathcal{N}'$, which we denote T' here, verifies $T' \geq \lambda$. For this purpose, write the flow conservation equation for node i, in $\mathcal{N}$, $T_i = \lambda_i + \sum_{j=1}^{n} T_j r_{j,i}$, where $r_{j,i}$ is the routing probability for going to i after visiting node j, and sum up in i. We obtain $T' = \lambda + \sum_{j=1}^{n} T_j(1 - r_{j,0})$. At least one of the nodes is connected to outside, and all the throughputs $T_1, \ldots, T_n$ are strictly positive, so, $T' > \lambda$.

Now, denote by E the effectiveness of $\mathcal{N}$, and by E' that of $\mathcal{N}'$. We have

$$E = \left(\frac{\lambda}{\gamma}\right)^2 \frac{1}{N} < \left(\frac{T'}{\gamma}\right)^2 \frac{1}{N} = E'.$$

This shows that all Jackson networks which are not in the parallel family, have necessarily effectiveness strictly less than $1/4$. This, together with the result of Theorem 2, ends the proof. ───────────■

Back at the Scholar Example at the Beginning of Sect. 4. Denoting by E_i the effectiveness of σ_i, we obtain $E_1 = b(1 - b)/2$, $E_2 = b(1 - b^2)/2$ and $E_3 = b(1 - b)$. We now have $E_3 \geq E_2 \geq E_1$, as intuitively desired. Observe that the effectiveness definition is different than the power both in the case of the multi-server system σ_2 and in the Jackson case of σ_1.

Back at the Tandem Example at the End of Sect. 4. We have $T = \lambda$, $\gamma = n\mu$ and $N = nb/(1-b)$, giving $E = b(1-b)/n^3$. Here, the effectiveness decreases quickly as the tandem's size increases, as expected (the throughput keeps unchanged, but *both* the cost and the delay increase linearly with n).

7 Conclusion

This paper presents some additional properties of the power function defined by Kleinrock, a performance metric for queueing systems capturing the tradeoff between work done and needed delay. This naturally leads to look for the maximal value of the power, and to say that the system operates at its optimal point when power is maximal. A nice property of this metric is that in the general $M/GI/1$ family, the optimal operating point is when the mean backlog is 1. This is intuitively consistent, but the property is not always true in the case of other queueing systems. We analyzed some families of $GI/M/1$ models where this *keep the pipe full* property doesn't hold, and others where again it is true. We showed that the property doesn't hold in general for finite storage capacity systems. In the paper, we also discussed the extension of the power metric to product-form open queueing networks, and we showed that the *keep the pipe full* property still holds.

Then, after showing that the power metric has some drawbacks if we want to take into account the cost associated with the processing capacity of the systems, we proposed a new metric, which we called *effectiveness*, that doesn't have this drawback. Effectiveness coincide with power on single server queues, but it is different on queues with many servers, and in queueing networks. Moreover, we proved that the optimal operating point of a network occurs again when the loads are equal to $1/2$, as for power, but here, the network must have the parallel structure for the power to reach its maximum. In future work, we will explore other properties of effectiveness, on other queues and networks of queues.

References

1. Kleinrock, L.: Queueing Systems, vol. I. Wiley Interscience, New York (1975)
2. Giessler, A., Hanle, J., Konig, A., Pade, E.: Free buffer allocation - an investigation by simulation. Comp. Net. **1**(3), 191–204 (1978)
3. Kleinrock, L.: Power and deterministic rules of thumb for probabilistic problems in computer communications. In: International Conference on Communications, pp. 43.1.1–43.1.10, Boston, June 1979
4. Kleinrock, L.: On flow control in computer networks. In: International Conference on Communications, vol. II, pp. 27.2.1–27.2.5, Toronto, July 1979
5. Gail, R., Kleinrock, L.: An invariant property of computer network power. In: International Conference on Communications, pp. 63.1.1–63.1.5, Denver, June 1981
6. Varshney, P.K., Dey, S.: Fairness in computer networks: A survey. J. Inst. Electron. Telecommun. Eng., May-June (1990)
7. Kleinrock, L., Huang, J.-H.: On parallel processing systems: amdahl's law generalized and some results on optimal design. IEEE Tran. Soft. Eng. **18**(5), 434–447 (1992)

8. Rosti, E., Smirni, E., Dowdy, L.W., Serazzi, G., Carlson, B.M.: Robust partitioning policies of multiprocessor systems. Perform. Eval. **19**, 141–165 (1993)
9. Bournas, R.M.: Bounds for optimal flow control window size design with application to high-speed networks. J. Franklin Inst. **1**, 77–89 (1995)
10. Kolias, C., Kleinrock, L.: Throughput analysis of multiple input-queueing in ATM switches. In: Mason, L., Casaca, A. (eds.) IFIP-IEEE Broadband Communications, pp. 382–393. Chapman and Hall, Boston (1996)
11. Hyytiä, E., Lassila, P., Penttinen, A., Roszik, J.: Traffic load in a dense wireless multihop network. In: Proceedings of the 2nd ACM International Workshop on Performance Evaluation of Wireless Ad Hoc, Sensor, and Ubiquitous Networks (PE-WASUN2005), pp. 9–17, Montreal (2005)
12. Inoie, A., Kameda, H., Touati, C.: A paradox in optimal flow control of M/M/n queues. Comput. Op. Res. **33**, 356–368 (2006)
13. Safaei, F., Khonsari, A., Fathy, M., Ould-Khaoua, M.: Communication delay analysis of fault-tolerant pipelined circuit switching in torus. J. Comput. Syst. Sci. **73**(8), 1131–1144 (2007)
14. Medidi, S., Ding, J., Garudapuram, G., Wang, J., Medidi, M.: An analytical model and performance evaluation of transport protocols for wireless ad hoc networks. In: Proceedings 41st Annual Simulation Symposium (ANSS-41), Ottawa, 14–16 April 2008
15. Rubino, G.: On Kleinrock's power metric for queueing systems. In: Proceedings of the 5th International Workshop on Performance Modelling and Evaluation of Computer and Telecommunication Networks (PMECT 2011), Maui (2011)
16. Chung, P.-T., Van Slyke, R.: Equilibrium analysis for power based flow control algorithm. Appl. Math. Sci. **6**(9), 443–454 (2012)
17. Yemini, Y.: A balance of power principle for decentralized resource sharing. Comput. Netw. **66**, 46–51 (2014)

A Solving Procedure for Stochastic Satisfiability Modulo Theories with Continuous Domain

Yang Gao[(✉)] and Martin Fränzle[(✉)]

Department of Computing Science, Carl von Ossietzky Universität Oldenburg,
Oldenburg, Germany
{yang.gao,fraenzle}@informatik.uni-oldenburg.de

Abstract. Stochastic Satisfiability Modulo Theories (SSMT) [1] is a quantitative extension of classical Satisfiability Modulo Theories (SMT) inspired by stochastic logics. It extends SMT by the usual as well as randomized quantifiers, facilitating capture of stochastic game properties in the logic, like reachability analysis of hybrid-state Markov decision processes. Solving for SSMT formulae with quantification over finite and thus discrete domain has been addressed by Tino Teige et al. [2]. In this paper, we extend their work to SSMT over continuous quantifier domains (CSSMT) in order to enable capture of continuous disturbances and uncertainty in hybrid systems. We extend the semantics of SSMT and introduce a corresponding solving procedure. A simple case study is pursued to demonstrate applicability of our framework to reachability problems in hybrid systems.

Keywords: Stochastic satisfiability modulo theory · Constraint solving · Hybrid system · Reachability analysis

1 Introduction

The idea of modelling uncertainty using randomized quantification was first proposed within the framework of propositional satisfiability (SAT) by Papadimitriou, yielding Stochastic SAT (SSAT) featuring both classical quantifiers and randomized quantifiers [3]. This work has been lifted to Satisfiability Modulo Theories (SMT) by Fränzle, Teige et al. [1,2] in order to symbolically reason about reachability problems of probabilistic hybrid automata (PHA). Instead of reporting true or false, an SSAT/SSMT formula Φ has a probability as semantics, which denotes the probability of satisfaction of Φ under optimal resolution of the non-random quantifiers. SSAT and SSMT permit concise description of diverse problems combining reasoning under uncertainty with data dependencies. Applications range from AI planning [4–6] to analysis of PHA [1].

This research is funded by the German Research Foundation through the Research Training Group DFG-GRK 1765: "System Correctness under Adverse Conditions" (SCARE, scare.uni-oldenburg.de) and the Transregional Collaborative Research Center SFB-TR 14 "Automatic Verification and Analysis of Complex Systems" (AVACS, www.avacs.org).

© Springer International Publishing Switzerland 2015
J. Campos and B.R. Haverkort (Eds.): QEST 2015, LNCS 9259, pp. 295–311, 2015.
DOI: 10.1007/978-3-319-22264-6_19

A serious limitation of the SSMT-solving approach pioneered by Teige [7] is that all quantifiers (except for implicit innermost existential quantification of all otherwise unbound variables) are confined to range over finite domains. As this implies that the carriers of probability distributions have to be finite, a large number of phenomena cannot be expressed within the current SSMT framework, such as continuous noise or measurement error in hybrid systems. To overcome this limitation, we relax the constraints on the domains of randomized variables, now also admitting continuous probability distributions in SSMT solving.

Our approach is based on a combination of the DPLL($\mathcal{T}$) [8] and ICP (Interval Constraint Propagation, [9,10]) algorithms, as first implemented in the iSAT solver for rich arithmetic SMT problems over the $\mathbb{R}^n$ [11,12], and on branch-and-prune rules for the quantifiers generalizing those suggested in [7,11]. We extend these methods so that they can deal with the SSMT formula with continuous quantifier domains. Our solving procedure therefore is divided into three layers: an SMT layer manipulating the Boolean structure of the "matrix"[1] of the formula, an interval constraint solving layer reasoning over the conjunctive constraint systems in the theory part of the formula, and a stochastic SMT layer reasoning about the quantifier prefix. Each layer is defined by a set of rules to generate, split, and combine so-called *computation cells*, where a computation cell is a box-shaped part of the $\mathbb{R}^n$, i.e. the problem domain of the constraints. The solver thereby approximates the exact satisfaction probability of the formula under investigation and terminates with a conclusive result whenever the approximation gets tight enough to conclusively answer the question whether the satisfaction probability is above or below a certain specified target.

Related Work. SSMT and its original solving procedure were proposed by Fränzle, Hermanns, and Teige [1] based on SSAT [3], SMT solving by DPLL($\mathcal{T}$) [8], and the iSAT algorithm [11] for solving non-linear arithmetic constraint systems by using interval constraint propagation. The original formulation of SSMT is confined to finite quantifier domains. The first paper extending SSMT to continuous domains is by Ellen et al. [13], proposing a statistical solving technique adopted from statistical AI planning algorithms and thus only being able to offer stochastic guarantees.

Structure of the Paper. The rest of this paper is organized as follows: in Sect. 2 we briefly recap the basic knowledge of interval arithmetic and probability theory, afterwards the syntax and semantics of stochastic SSMT with continuous domains are formalized in Sect. 3. The solving procedure along with a running example are detailed in Sect. 4. A case study is pursued in Sect. 5 to show the potential application of our framework, followed by a conclusion.

2   Preliminaries

2.1   Interval Arithmetic and Constraint Solving

Interval arithmetic is a mathematical technique systematized by Rosalind Cecily Young in the 1930s [14] and developed by mathematicians since the 1950s [15–17]

---

[1] In SSAT parlance, this is the body of the formula after rewriting it to prenex form and stripping all the quantifiers.

as an approach to putting safe bounds on errors in numerical computation. It has later been extended to handle arithmetic constraints over continuous domains [18], where it enables efficient search for approximate solutions to large-scale non-linear constraint systems. Our main framework is essentially based on interval arithmetic and interval constraint propagation, so we will recap the main concepts in this section.

Let $\mathbb{R}$ denote the reals. A real interval X is defined as the set of real numbers between (and including) a given upper and lower bound, i.e. $X = [\underline{X}, \overline{X}] = \{x \in \mathbb{R} \mid \underline{X} \leq x \leq \overline{X}\}$, and the set of intervals is denoted by $\mathbb{IR}$. Interval arithmetic is a lifting of real arithmetic. For an arithmetic operation $\circ$ like $+$ or $\times$, the corresponding interval operation on intervals X and Y is defined by $X \circ Y = \text{hull}(\{x \circ y \mid x \in X, y \in Y\})$, where $\text{hull}(X)$ denotes the smallest computer-representable interval covering X. This definition lifts immediately to expressions and function definitions over $\mathbb{R}^n$ in terms of the elementary arithmetic operations.

Interval arithmetic has been developed into interval constraint propagation (ICP, [9,10,19]) as a means of solving systems of real equalities and inequalities. Given a set of real constraints C and initial interval bounds $\rho : Vars \rightarrow \mathbb{IR}$ on their variables, ICP successively narrows the initial intervals to small intervals still covering all real solutions to the constraint system.

During the narrowing process, ICP will frequently reach a fix-point given by some consistency conditions, e.g. hull consistency (see Definition 1). In such a case, interval splitting will be performed to pursue further narrowing and recursively contract the sub-intervals. This framework is called *branch-and-prune* [10] and forms a core mechanism for ICP based constraint solving. A toy example is given below.

Definition 1. *A simple constraint* $c \equiv (p = q \circ r)$ *is hull consistent w.r.t.* $\rho : Vars \rightarrow \mathbb{IR}$ *iff* $\rho(p) = \rho(q) \circ \rho(r)$. *Likewise, a simple bound* $c \equiv (p \leq c)$, *where c is a constant, is said to be hull consistent w.r.t.* $\rho : Vars \rightarrow \mathbb{IR}$ *iff* $\rho(p) \subseteq (-\infty, \tilde{c}]$, *where $\tilde{c}$ is the smallest computer-representable number $\geq c$.*

A set C of constraints is hull consistent w.r.t. ρ iff all its constraints are. The notation $\rho \models_{hc} C$ is used.

Example 1. Consider the constraints $c_1 : p = q^2$ and $c_2 : p - q = 0$, and let the initial configuration be $\rho(q) = [-2, 2]$ and $\rho(p) = [-2, 2]$. We know that this problem has two isolated solutions $(q = 0, p = 0)$ and $(q = 1, p = 1)$. Starting by considering constraint c_1, the ICP process will give the following sequence:

$$(c_1, \rho(p) = [0, 2]), \quad (c_1, \rho(q) = [0, 1.414\ldots]),$$
$$(c_2, \rho(p) = [0, 1.414\ldots]), \quad (c_1, \rho(q) = [0, 1.189\ldots]),$$
$$\ldots$$

Iterative narrowing by ICP stops with a hull consistency set $\rho(q) = [0, 1.00\ldots]$ and $\rho(p) = [0, 1.00\ldots]$ (Converge to 1 but will not reach 1). In order to get close to the true results, we then choose a variable and split its domain into two parts, e.g. we could choose q and split into $\rho(q) = [0, 0.5]$ and $\rho(q) = (0.5, 1.00 \ldots]$. Then ICP will be used on each sub-interval, rapidly narrowing them to small boxes enclosing the actual solutions.

2.2 Probability Basis

Before we introduce the formal definition of Stochastic SMT over continuous domains, we recap necessary items from probability theory.

Let Ω be a *sample set*, the set of all possible outcomes of an experiment. A pair $(\Omega, \mathcal{F})$ is said to be a *sample space* if $\mathcal{F}$ is a σ-field of subsets of Ω. A triple $(\Omega, \mathcal{F}, \mu)$ is a *probability space* if μ is a probability measure over $\mathcal{F}$: 1) $0 \leq \mu(A) \leq 1$ for all $A \in \mathcal{F}$; 2) $\mu(\emptyset) = 0$ and $\mu(\Omega) = 1$; 3) $\mu(\bigcup_{k=1}^{\infty} A_k) = \Sigma_{k=1}^{\infty} \mu(A_k)$ for disjoint $A_k \in \mathcal{F}$. Let $(\Omega, \mathcal{F}, \mu)$ be a probability space. A function $X : \Omega \rightarrow \mathbb{R}$ is said to be a random variable iff $X^{-1}(B) \in \mathcal{F}$ for all $B \in \mathcal{B}$, where $\mathcal{B}$ is the Borel σ-algebra generated by all the open sets in $\mathbb{R}$. The *distribution function* of X is the function $F_X : \mathbb{R} \rightarrow [0, 1]$ defined by $F_X(x) := \mu(X \leq x)$ for all $x \in \mathbb{R}$. If there exists a nonnegative, integrable function $\pi : \mathbb{R} \rightarrow \mathbb{R}$ such that $F_X(x) := \int_{-\infty}^{x} \pi(y) dy$, then π is called the *density function* for X. It follows then that $\mu(X \in A) = \int_A \pi(x) dx$ for all $A \in \mathcal{F}$.

3 Stochastic Satisfiability Modulo Theories with Continuous Domain

The syntax of stochastic SMT formulas over continuous quantifier domains agrees with the discrete version from [1], except that continuous quantifier ranges are permitted.

Definition 2. *An SSMT formula with continuous domain (CSSMT) is of the form:* $\Phi = \mathcal{Q} : \varphi$, *where:*

- $\mathcal{Q} = Q_1 x_1 \in dom(x_1) \ldots Q_n x_n \in dom(x_n)$ *is a sequence of quantified variables,* $dom(x_i)$ *denotes the domain of variable* x_i, *which are intervals over the reals,* Q_i *is either an existential quantifier* $\exists$ *or a randomized quantifier* H_{π_i} *with integrable probability density function over the reals* π_i *satisfying* $\int_{dom(x_i)} \pi_i(x_i) dx_i = 1$.
- φ *is an SMT formula over quantifier-free non-linear arithmetic theory* $\mathcal{T}$. *Without loss of generality, we assume that* φ *is in conjunctive normal form (CNF), i.e.,* φ *is a conjunction of clauses, and a clause is a disjuction of (atomic) arithmetic predicates.* φ *is also called the* matrix *of the formula.*

Definition 3. *The semantics of a CSSMT formula* $\Phi = \mathcal{Q} : \varphi$ *is defined by the maximum probability of satisfaction* $Pr(\Phi)$ *as follows, where* ε *denotes the empty quantifier prefix:*

- $Pr(\varepsilon : \varphi) = 0$ *if* φ *is unsatisfiable.*
- $Pr(\varepsilon : \varphi) = 1$ *if* φ *is satisfiable.*

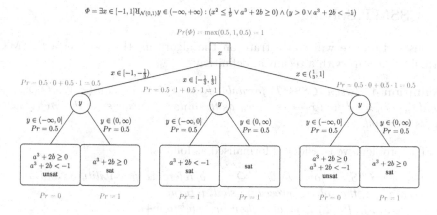

Fig. 1. Semantics of a CSSMT formula depicted as a tree.

- $Pr(\exists x_i \in dom(x_i) \ldots Q_n x_n \in dom(x_n) : \varphi)$
  $=\sup_{v \in dom(x_i)} Pr(Q_{i+1} x_{i+1} \in dom(x_{i+1}) \ldots Q_n x_n \in dom(x_n) : \varphi[v/x_i])$.
- $Pr(\text{Я}_{\pi_i} x_i \in dom(x_i) \ldots Q_n x_n \in dom(x_n) : \varphi)$
  $=\int_{v \in dom(x_i)} Pr(Q_{i+1} x_{i+1} \in dom(x_{i+1}) \ldots Q_n x_n \in dom(x_n) : \varphi[v/x_i]) \pi_i(v) dv$.

According to Definition 3, the maximum probability of satisfaction $Pr(\Phi)$ is computed by resolving the quantifiers from left to right, whereby existential quantifiers are resolved to yield highest (or actually approach the supremum) probability and randomized quantifiers yield the expectation of the remaining formulas. When all quantifiers are resolved, the satisfaction probability of the matrix ϕ is associated with its satisfiability.

Example 2. Figure 1 constructs a tree according to the semantics of CSSMT formula $\Phi = \exists x \in [-1, 1] \text{Я}_{\mathcal{N}(0,1)} y \in (-\infty, +\infty) : (x^2 \leq \frac{1}{9} \vee a^3 + 2b \geq 0) \wedge (y > 0 \vee a^3 + 2b < -1)$, where $\mathcal{N}(0, 1)$ refers to the standard normal distribution with mean value 0 and variance 1. Semantically, Φ determines the maximum probability s.t. there are values for x which are between $[-1, 1]$ s.t. for normal distributed values of y the matrix is satisfiable. As a start, we branch the domain of x into three parts, i.e., $x \in [-1, -\frac{1}{3})$, $x \in [-\frac{1}{3}, \frac{1}{3}]$ and $x \in (\frac{1}{3}, 1]$, again for each part, we branch the domain of y into two parts. Until now, all the quantified variables are resolved and we check satisfiability. Take the leftmost branch as an example, when x is in $[-1, -\frac{1}{3}]$ and y is in $(-\infty, 0]$, the matrix cannot be satisfied and we mark the probability of satisfaction as 0. When all the branches have been fully checked, we propagate the probability according to the corresponding quantifiers, for example, y is normal distributed, so the probability that y takes value from $(-\infty, 0]$ is 0.5. If we combine the probability from bottom to top and choose maximum value among the three branches for x (since x is bounded by existential quantifier), then we get that the probability of satisfaction of Φ is 1. Notice that if we start from a finer interval partition, i.e., split x or y into more branches, we will get a different structure, but the maximum probability keeps unchanged.

4   CSSMT Solving

In this section, we will concentrate on the algorithm that solving a CSSMT formula. We start with a technical definition:

Definition 4. *Given CSSMT formula* $\Phi = Q_1 x_1 \in dom(x_1) \ldots Q_n x_n \in dom(x_n) : \phi$, *we define* $x_1 \prec \cdots \prec x_n$ *as the order of variables in the prefix, with* x_1 *its minimal.*

The problem we consider is formalised as follows:

> *Given a CSSMT formula* $\Phi = Q : \phi$, *a reference probability* δ, *and an accuracy* ε, *the solving procedure shall return*
> - *"GE", if* $Pr(\Phi)$ *is greater than or equal to* $\delta + \varepsilon$;
> - *"LE", if* $Pr(\Phi)$ *is less than or equal to* $\delta - \varepsilon$;
> - *"GE" or "Inconclusive", if* $Pr(\Phi) \in [\delta, \delta + \varepsilon]$;
> - *"LE" or "Inconclusive", if* $Pr(\Phi) \in [\delta - \varepsilon, \delta]$.

The algorithm we will present is equipped with the following structures:

- C: a set collecting the constraints which must be satisfied in the current phase.
- ρ: an ordered list (corresponding to the order of variables in Q) which records the interval valuation for each variable.
- H: a set of computation cells.

In the following, we will use $(\Phi, \rho, C)^{(p,q)_i}$ to represent a computation cell. Intuitively, a computation cell can be understood as a subset of the whole searching space, i.e., subset of $\mathbb{R}^n$, $(p, q)_i$ is a probability estimation for this cell, where p and q are lower and upper bounds separately, and $(\cdot)_i$ means that the probability estimation is for the formula by chopping off the quantifier prefix before variable x_i, in the tree-like semantics, this can be regarded as the probability estimation for the subtree from the view of x_i.

The initial configurations w.r.t. the CSSMT formula $\Phi = Q_1 x_1 \in dom(x_1) \ldots Q_n x_n \in dom(x_n) : \phi$ are shown as below:

$$
\begin{cases}
C : = \emptyset \\
H : = \emptyset \\
\rho : = (\rho_1, ..., \rho_n) = (dom(x_1), \ldots, dom(x_n))
\end{cases}
$$

From this configuration, the algorithm will start its deduction sequence, which is given by the DPLL rules at the outermost level, which in turn builds on the rules at the constraint solving and the SSMT layer. All the rules share a common structure: they manipulate a set H which contains the relevant box-shaped computation cells within the search space, which itself is a subset of the $\mathbb{R}^n$. The rules select appropriate cells by mean of their premises, they update, split, or combine them according to their conclusions.

Remark 1. The rules presented in the following define a transition system in the tradition of structural operational semantics rather than a tree of inferences. The transitions transform the proof state from the left of the conclusion into the proof state to the right, while the premises denote the side conditions enabling that transition. A procedure thus consists of a finite sequence of state manipulating transitions, with each individual transition matching the conclusion of some rule, under the side condition that its source (and may be some elements of the target) satisfies the premise of the transition rule.

Example 3. We take advantage of a running example to demonstrate our algorithm.. We therefore consider the CSSMT formula (As the matrix is a CNF formula, we can regard it as a set of clauses):

$$\Phi = \exists x \in [-10, 10] \, \mathrm{U}y \in \mathcal{U}[5, 25] \, \mathrm{U}z \in \mathcal{U}[-10, 10] :$$
$$x > 3 \vee y < 1, z > x^2 + 2 \vee y \le 20, x^2 > 49 \vee y > 7x, x < 6 \vee y \ge z$$

where $\mathcal{U}[a, b]$ refers to uniform distribution with range $[a, b]$. The initial configurations are $C = \emptyset$, $H = \emptyset$ and $\rho = ([-10, 10], [5, 25], [-10, 10])$. Furthermore, we set $\delta = 0.45$ to be the reference probability.

SMT Level. Rule (INI) adds the first computation cell to H, which contains: (1) the formula $\mathcal{Q} : \phi$ to be decided; (2) ρ is an initial evaluation for each variable; (3) the constraints C which must be satisfied, initially an empty set; (4) a superscript $(p, q)_i = (0, 1)_1$ over-approximating the satisfaction probability of the formula when chopping off the quantifier prefix before variable x_i, here it means that no quantifiers have been resolved at the moment and the lower and upper estimations are 0 and 1 respectively.

$$\overline{H \to H \cup \{(\mathcal{Q} : \phi, \rho, C)^{(0,1)_1}\}} \tag{INI}$$

If all the disjuncts except one in some clause can not be satisfied w.r.t. the current evaluation ρ, then this remaining "unit" must hold. Rule (UP) corresponds to the unit propagation in the DPLL framework.

$$\frac{(L \vee l') \in \phi, \rho \nvDash L}{H' \cup \{(\mathcal{Q} : \phi, \rho, C)^{(p,q)_i}\} \to H' \cup \{(\mathcal{Q} : \phi, \rho, C \cdot \langle l' \rangle)^{(p,q)_i}\}} \tag{UP}$$

If the range of a variable x_j can be narrowed according to the constraints C and the current evaluation ρ by means of ICP, and if ρ is not yet hull consistent w.r.t. to the new bound (represented by $\nvDash_{hc}$, in this case interval narrowing can be performed by using interval constraint propagation), we update the evaluation set and the probability estimation according to the narrowing $\rho \overset{C}{\rightsquigarrow} (x_j \sim b)$ of x_j computed by ICP:

$$\frac{\rho \overset{C}{\rightsquigarrow} (x_j \sim b), \rho \nvDash_{hc} (x_j \sim b)}{H' \cup \{(\mathcal{Q} : \Phi, \rho, C)^{(p,q)_i}\} \to H' \cup \{(\mathcal{Q} : \Phi, update_\rho(x_j \sim b), C)^{renewal_{\rho_j}(p,q)_i}\}} \tag{ICP}$$

where

$$update_\rho(x_j \sim b)(x_i) = \begin{cases} \rho(x_j) \cap \{z|z \sim b\}, & if \ x_i = x_j \\ \rho(x_j), & otherwise \end{cases}$$

Intuitively, the *update* operator narrows the bound of variable x_j and leaves other variables unchanged. The corresponding change in the probability estimate induced by narrowing a —potentially randomized— variable x_j is reflected by

$$renewal_{\rho_j}(p,q)_i = \begin{cases} (p,q)_i, & if \ x_j \prec x_i \\ \mathbb{P}(\rho(x_i) \times \cdots \times \rho(x_j) \cap \{z|z \sim b\} \times \cdots \rho(x_n))_i, & otherwise \end{cases}$$

where $\mathbb{P}(I_i \times \cdots \times I_n)$ is a safe, interval-arithmetic based probability estimation which returns an interval over-approximating the measure of $I_i \times \cdots \times I_n$ under the distributions attached to the quantifiers.

When both rule (ICP) and rule (UP) do not yield further deductions, we say ϕ is inconclusive on ρ. We may then perform the splitting rule (SPL) to split the current computation cell into two cells (Any splitting strategies can be applied as long as the size of each interval is reduced, in practice, bisection is applied.) and update ρ as well as the probability estimation accordingly.

$$\frac{\rho_j \neq \emptyset, \rho_j^1 \cup \rho_j^2 = \rho_j}{H' \cup \{(Q:\phi,\rho,C)^{(p,q)_i}\} \rightarrow} \quad \text{(SPL)}$$
$$H' \cup \{(Q:\phi,\rho'\cdot\langle\rho_j^1\rangle\cdot\rho'',C)^{renewal_{\rho_j^1}(p,q)_j}, (Q:\phi,\rho'\cdot\langle\rho_j^2\rangle\cdot\rho'',C)^{renewal_{\rho_j^2}(p,q)_j}\}$$

Example 4. Let us reconsider the formula in Example 3. According to Rule (INI), we add the first computation cell $(\Phi, ([-10,10],[5,25],[-10,10]),\emptyset)^{(0,1)_1}$ to the set H. Considering the clause $x > 3 \vee y < 1$, we observe that $y < 1$ violates the current evaluation, so $x > 3$ must be satisifed, which we add to the set C (UP), this yields proof state $(\Phi, ([-10,10],[5,25],[-10,10]),\{x > 3\})^{(0,1)_1}$. By interval constraint propagation we can conclude that . According to the Rule (ICP), we update ρ and recalculate the probability interval accordingly. Since x is bound by $\exists$, the probability stays unchanged, yielding $(\Phi, ((3,10],[5,25],[-10,10]),\{x > 3\})^{(0,1)_1}$. The current evaluation makes $z > x^2+1$ unsatisfiable, so $y \leq 20$ will be added to C (UP), the domain of y is then narrowed to $[5,20]$ (ICP), since y is bounded by $\mho$, we need update the probability estimation, this yields $(\Phi, ((3,10],[5,20],[-10,10]),\{x > 3, y \leq 20\})^{(0,0.75)_1}$, we cannot guarantee that there are solutions in $[5,20]$, so the lower bound is 0, for the upper bound we can conclude that it will not exceed 0.75 since y is uniformly distributed and only the values in $[5,20]$ will be considered. The next step is to apply the rule (SPL). We choose x and split its interval into two parts, then take each part and update the corresponding evaluation and upgrade the probability estimation. As x is bound by $\exists$, the probability interval thereby remains unchanged, giving $H = \{(\Phi, ((3,7],[5,20],[-10,10]),\{x > 3, y \leq 20\})^{(0,0.75)_1}, (\Phi, ([7,10],[5,20],[-10,10]),\{x > 3, y \leq 20\})^{(0,0.75)_1}\}$.

Constraint Solving Level. When a conflict is obtained, i.e. if ICP under the current evaluation ρ and constraints C narrows some variables to empty sets, or if ρ violates every part in one clause, the current computation cell can be safely marked with probability 0. This is reflected by rule (CFL):

$$\frac{\rho \overset{C}{\rightsquigarrow} (x_i = \emptyset) \text{ or } L \in \phi \wedge \rho \not\models L}{H' \cup \{(Q : \phi, \rho, C)^{(p,q)_i}\} \rightarrow H' \cup \{(Q : \phi, \rho, C)^{(0,0)_n}\}} \quad \text{(CFL)}$$

If the current evaluation ρ is hull consistent w.r.t. the actual constraint set C, a paving procedure [19] can be invoked to generate an inner approximation and an outer approximation of the actual solution by sets of boxes (i.e., $\{(\cdot)\}^*$ means a number of cells). By computing safe upper (lower, resp.) approximations on the probability measures of the outer (inner, resp.) approximations of the solution sets, we obtain a safe interval estimate on the satisfaction probability. Rule (CNSIS) assigns these.

$$\frac{\rho \models_{hc} C}{H' \cup \{(Q : \phi, \rho, C)^{(p,q)_i}\} \rightarrow H' \cup \{(Q : \phi, \rho', C)^{(p',q')_n}\}^*} \quad \text{(CNSIS)}$$

Remark 2. The Rule (CNSIS) tells us that when the current evaluation ρ is hull consistent w.r.t. the constraints C, the boxes will be generated. In fact here we can benefit from a lot of state-of-art constraint solving techniques which can obtain both inner and outer approximation, i.e., interval arithmetic based techniques [20,21], affine arithmetic based techniques [22] etc. Our future work for implementation will be partially based on RealPaver [19], which is an interval solver using constraint satisfaction techniques.

Example 5. The computation cells in H are $(\varPhi, ((3,7), [5,20], [-10,10]), \{x > 3, y \leq 20\})^{(0,0.75)_1}$ and $(\varPhi, ([7,10], [5,20], [-10,10]), \{x > 3, y \leq 20\})^{(0,0.75)_1}$. We take the first computation cell into consideration and conclude that the evaluation violate the clause $x^2 > 49 \vee y > 7x$. According to rule (CFL) we mark this computation cell with probability 0. This gives $(\varPhi, ((3,7), [5,20], [-10,10]), \{x > 3, y \leq 20\})^{(0,0)_3}$. Now we turn to consider the second computation cell. By performing rule (UP) we get $(\varPhi, ([7,10], [5,20], [-10,10]), \{x > 3, y \leq 20, x^2 > 49, y \geq z\})^{(0,0.75)_1}$. We observe that the constraints in C are hull consistent w.r.t. the current evaluation ρ. In order to explain the decision procedure, here we generate one inner box and one outer box, according to Rule (CNSIS), we get two computation cells, $(\varPhi, ([7,10], [5,10], [-10,10]), \{x > 3, y \leq 20, x^2 > 49, y \geq z\})^{(0,0.33*0.75)_3}$ and $(\varPhi, ([7,10], (10,20], [-10,10]), \{x > 3, y \leq 20, x^2 > 49, y \geq z\})^{(0.66*0.75,0.67*0.75)_3}$, which over and under approximates the solutions for C w.r.t. ρ respectively. As has been depicted in Fig. 2, a light gray area is shown, where the formula $\varPhi$ is satisfiable. The red box is the corresponding outer box and blue is an inner.

Stochastic SMT Level. Two computation cells are *combinative* in that they estimate satisfaction probability w.r.t. adjacent intervals for the same variable x_i. In case that x_i is bound by $\exists$, combining the two cells yields the maximum probability (Rule ($\exists$-COM)); otherwise if bound by $\mho$, the two cells can be combined by adding their probabilities (Rule $\mho$-COM).

$$\frac{\rho_i^1 \uplus \rho_i^2 \text{ is the interval hull of } \rho_i^1 \text{ and } \rho_i^2}{\begin{array}{c} H' \cup \{(Q'\exists x_i Q'' {:} \phi, \rho' \cdot \langle \rho_i^1 \rangle \cdot \rho'', C)^{(p_1,q_1)_i}, (Q'\exists x_i Q'' {:} \phi, \rho' \cdot \langle \rho_i^2 \rangle \cdot \rho'', C)^{(p_2,q_2)_i}\} \rightarrow \\ H' \cup \{(Q {:} \phi, \rho' \cdot \langle \rho_i^1 \uplus \rho_i^2 \rangle \cdot \rho'', C)^{max((p_1,q_1)_i,(p_2,q_2)_i)}\} \end{array}} \quad \text{(}\exists\text{-COM)}$$

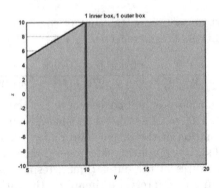

Fig. 2. Inner and outer approximations for constraint solving problem: $\{x > 3, y \leq 20, x^2 > 49, y \geq z\}$ where $x \in [7, 10]$, $y \in [5, 20]$ and $z \in [-10, 10]$.

$$\frac{\rho_i^1 \uplus \rho_i^2 \text{ is the interval hull of } \rho_i^1 \text{ and } \rho_i^2}{H' \cup \{(Q' \amalg x_i Q'' : \phi, \rho' \cdot \langle \rho_i^1 \rangle \cdot \rho'', C)^{(p_1,q_1)_i}, (Q' \amalg x_i Q'' : \phi, \rho' \cdot \langle \rho_i^2 \rangle \cdot \rho'', C)^{(p_2,q_2)_i}\} \rightarrow}{H' \cup \{(Q : \phi, \rho' \cdot \langle \rho_i^1 \uplus \rho_i^2 \rangle \cdot \rho'', C)^{(p_1,q_1)_i + (p_2,q_2)_i}\}} \quad \text{(Ꙋ-COM)}$$

where the *interval hull* of two sets I_1 and I_2 here is the smallest interval which contains I_1 and I_2.

For the rules (∃-COM) and (Ꙋ-COM), the order of combination is irrelevant, since the cells are parts of an overall search space and probability for each cell is safely bounded. So parallel computation is possible from this point of view, which will be considered in the tool implementation.

If all the computation cells w.r.t. the same variable have been tackled, the probability should be propagated to the preceding variable in the variable order. Rule (LFT) checks all the computation cells in H, and will propagate if all its siblings have been combined.

$$\frac{\forall (Q : \phi, \rho', C')^{(\cdot,\cdot)_j} \in H' : j \neq i}{H' \cup \{(Q : \phi, \rho, C)^{(p,q)_i}\} \rightarrow H' \cup \{(Q : \phi, \rho, C)^{(p,q)_{i-1}}\}} \quad \text{(LFT)}$$

Example 6. In running example we have previously reached a state where there are three computation cells in H. By successively applying rules (Ꙋ-COM), (∃-COM), and (LFT), we propagate the probability, as depicted in Fig. 3.

Termination. Whenever the estimated probability interval at the level of the first variable x_1 becomes less than or equal to the reference probability δ, the original formula is concluded to satisfy $P(\Phi) \leq \delta$. Rule (LE) then reports "LE"; rule (GE) does the equivalent for the converse case.

$$\frac{q \leq \delta}{H' \cup \{(Q : \phi, \rho, C)^{(p,q)_1}\} \rightarrow \text{LE}} \quad \text{(LE)}$$

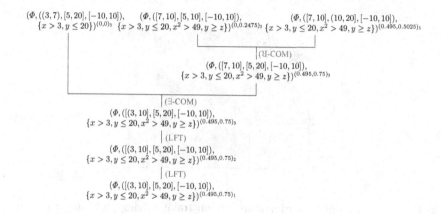

Fig. 3. Propagate probability by using combination rules.

$$\frac{p \geq \delta}{H' \cup \{(Q : \phi, \rho, C)^{(p,q)_1}\} \rightarrow \text{GE}} \qquad \text{(GE)}$$

If the above two cases cannot be be judged under the accuracy ε, the evaluation of the formula remains inconclusive w.r.t. δ:

$$\frac{q > \delta \wedge p < \delta \wedge |p - q| < \varepsilon}{H' \cup \{(Q : \phi, \rho, C)^{(p,q)_1}\} \rightarrow \text{INCON}} \qquad \text{(INCON)}$$

Whenever none of the above three termination rules applies, we have to go back to the SMT level and generate more cells by (SPL).

Remark 3. The whole procedure is performed on a number of computation cells maintained in a set H, which changes according to the rules, we name the configuration of set H at each decision step *snapshot*. We can cache the snapshots in main decision points: such as when the cells are split, or when boxes are generated by other inference mechanisms. If the termination test fails, the procedure can backjump to some snapshot. Backjumping to some restart points is heuristic. A straightforward, yet inefficient way is to simply backtrack to the latest snapshot. However, a more ingenious way can be performed by using Conflict Driven Clause Learning (CDCL) mechanism, allowing to backjump to the snapshot which leads to the conflict.

Example 7. The given δ for this running example is 0.45, according to the Rule (GE), we know that $Pr(\Phi) > 0.45$. The decision procedure terminates here.

In our running example, generating two boxes can reach δ. Now let us consider $\delta = 0.70$, a tighter approximation can be achieved by generating more boxes. As shown in Fig. 4, we use RealPaver [19], which is a modeling language implementing interval-based algorithms to process systems of nonlinear constraints over the real numbers, to generate the inner boxes and outer boxes so that a

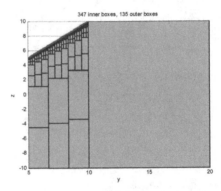

Fig. 4. Inner and outer approximations for constraint solving problem: $\{x > 3, y \le 20, x^2 > 49, y \ge z\}$ where $x \in [7,10]$, $y \in [5,20]$ and $z \in [-10,10]$ by RealPaver.

better result can be obtained. By doing so, we get a tighter approximation, which is $[0.7181, 0.7191]$. For this running example, the real probability can be computed easily, that is $23/32 \approx 0.71875\cdots$, which has been well approximated by the interval we have obtained.

Remark 4. The overall solving procedure starts from rule (INI), and then performs (UP), (ICP), (SPL) recurrently, thereby modifying the proof state. When a conflict is obtained or the evaluation is hull consistent regarding to the constraints, (CFL) and (CNSIS) are applied to refine the probability estimation. Combining rules are then applied to build the tree-like structure, if the under-approximation is smaller than the given reference number δ or the over-approximation is greater than δ, corresponding results will be returned according to (LE) or (GE), otherwise we should backtrack to the previous snapshot and perform this procedure again.

5   Case Study

We consider the problem of regulating the temperature of a room during some time horizon $[0, N]$ by a thermostat that can switch a heater on or off. This example is taken from [23], where the probabilistic reachability problem for controlled discrete time stochastic hybrid systems (DTSHS) was investigated by using dynamic programming (DP). In this work, it will be investigated using CSSMT.

Here we consider a room in which there is an thermostat with two modes $Q = \{ON, OFF\}$, and the thermostat issues switching commands $\mathcal{U} = \{0, 1\}$ to the heater: "0" means no switching command is issued and "1" oppositely. The average room temperature changes according to the laws $m_{OFF}(x) = x - \frac{a}{C}(x - x_a)\Delta t + \mathcal{N}(m, \sigma^2)$ and $m_{ON}(x) = m_{OFF}(x) + \frac{r}{C}\Delta t$, where a is the average heat loss rate, C is the average thermal capacity, x_a is the ambient temperature,

r is the rate of heat gain, and Δt is the discretization time interval. In order to capture the disturbance, we add a noise $\mathcal{N}(m, \sigma^2)$ term, which denote the probability measure over $(\mathbb{R}, \mathcal{B}(\mathbb{R}))$ associated with a Gaussian density function with mean m and variance σ^2.

We assume that a switching command will impact the heater state with some random time delay, which can be modelled by a probability α: each time when a switch action will be performed, it succeeds with probability α.

The regulation problem we consider here is the same with [23]: *determine a control strategy that maximizes the probability that the average room temperature x is driven close to a given temperature with an admissible tolerance.*

5.1 SSMT Formalization

To formalize the problem, we translate the initial condition, transition relations and desired sets into a CSSMT formula $\mathcal{Q} : \Phi$. Because of the continuity of some variables in the model, e.g., the noise described by Gaussian distribution, an SSMT formula with discrete domain [1] is not sufficient for our case.

Initial Condition. The initial room temperature starts from any value in the set $[T_l^0, T_u^0]$ and the heater is switched to OFF. The condition can be formulated as follow:

$$\mathcal{I} : (T_l^0 \leq x_0 \leq T_u^0) \wedge (q_0 = OFF) \tag{1}$$

Since the system is considered to be well behaved for any suitable initial value, we assume that it is uniformly distributed in the range $[T_l^0, T_u^0]$, i.e., x_0 is bound by $\maltese x_0 \in \mathcal{U}(T_l^0, T_u^0)$, q_0 is a discrete variable bounded by quantifier $\exists q_0 \in \{OFF, ON\}$.

Transition Relations. During each transition step the system will do two operations: (1) Choose a control action u_i by performing switching command $u_i = 0$ or $u_i = 1$, thereby obeying random delay for when the switching command takes effect; (2) Update the temperature x_i according to the thermostat state.

$$\mathcal{T}_i : ((u_i = 0 \wedge q_i = q_{i-1}) \vee (u_i = 1 \wedge delay_i \wedge q_i = q_{i-1}) \vee (u_i = 1 \wedge \neg delay_i \wedge q_i = \neg q_{i-1}))$$
$$\wedge \qquad ((q_i = ON \wedge x_i = x_{i-1} - \frac{a}{c}(x_{i-1} - x_a)\Delta t + \omega_{i-1} + \frac{r}{c}\Delta t) \vee$$
$$(q_i = OFF \wedge x_i = x_{i-1} - \frac{a}{c}(x_{i-1} - x_a)\Delta t + \omega_{i-1}) \tag{2}$$

u_i is a control action which can be either 1 (switching) or 0 (no switching), i.e., $\exists u_i \in \{0, 1\}$, *delay* is a discrete random variable to mimic the random time delay of the switch action, so a random quantifier $\maltese delay_i \in \{\alpha \rightarrow 1, (1 - \alpha) \rightarrow 0\}$ will be used. x_{i-1} and x_i are the previous and updated room temperature, they take value from a suitable temperature range: $\exists x_{i-1} \in [T_l^{i-1}, T_u^{i-1}]$ and $\exists x_i \in [T_l^i, T_u^i]$. ω_i is a continuous random variable with Gaussian distribution: $\maltese \omega_i \in \mathcal{N}(m, \sigma^2)$.

Target Sets. Target sets can be understood as the sets which we want to keep the room temperature inside. For each transition step, we attach a constraint to the SMT formula:

$$\mathcal{S}_i : C_l^i \leq x_i \leq C_u^i \tag{3}$$

which indicates that the room temperature x_i is expected to be in the set $[C_l^i, C_u^i]$.

Altogether the dynamic of the system can be formalized as an SSMT formula

$$\Game x_0 \in \mathcal{U}(T_l^0, T_u^0) \exists q_0 \in \{OFF, ON\} \exists u_i \in \{0,1\} \Game delay_i \in \{\alpha \to 1, (1-\alpha) \to 0\}$$

$$\vdots$$

$$\Game \omega_i \in \mathcal{N}(m, \sigma^2) \exists x_{i-1} \in [T_l^{i-1}, T_u^{i-1}] \exists x_i \in [T_l^i, T_u^i] \cdots : \mathcal{I} \wedge \bigwedge_{i=1}^{N} (\mathcal{T}_i \wedge \mathcal{S}_i), \qquad (4)$$

where N is the number of computation steps analyzed. Note that the length of the quantifier prefix is linear in N, necessitating sophisticated constraint solving to discharge the exponentially growing search spaces.

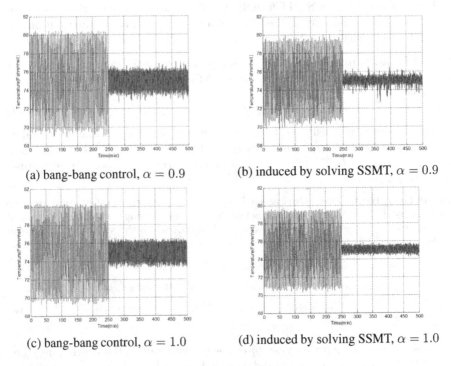

(a) bang-bang control, $\alpha = 0.9$ (b) induced by solving SSMT, $\alpha = 0.9$

(c) bang-bang control, $\alpha = 1.0$ (d) induced by solving SSMT, $\alpha = 1.0$

Fig. 5. Experiment results.

5.2 Experiment Results

The parameters we used are the same as in [23]: $x_a = 10.5°F$, $a/C = 0.1$ min^{-1}, $r/C = 10°/min$, all the $\mathcal{N}_i$ are independent and identically distributed with mean $m = 0$ and variance $\sigma = 0.33°F$; Initially, the room temperature is in $[70, 80]°F$. In the following we consider the system dynamics during the time interval $[0, 500]$, we specify that the room temperature will be kept in $[70, 80]°$F when time $t \leq 250$min and after that it is driven close to $75°$F, i.e., into $[74, 76]°$F.

In order to obtain an optimal control sequence, we first perform the CSSMT solving steps (Sect. 4) to obtain the maximum probability of satisfaction w.r.t. the formula (4). Then we extract the values of control action u_i from the branches which lead to the maximum probability. The prototype implementation is done in MATLAB with a discretization step $\Delta t = 0.1$min.

The results are depicted in Fig. 5, where the left column (a, c) shows the behaviour with classical threshold-driven bang-bang control, i.e., when the temperature reaches to the upper threshold we try to turn off the heater and vice versa. The results depicted in (b) and (d) are given by solving the corresponding CSSMT formula and extracting the optimal strategy with best possibility to stay in the desired temperature ranges. The effect is that the behavior is much closer to what we want, i.e., the temperature tends to stay with high probability within the given range - $[70, 80]°$F and $[74, 76]°$F, respectively.

Table 1 shows the probability approximations computed as they vary with the number of computation cells for each step. The more computation cells we generate, the tighter the bound we can get at the price of increased computation time.

Table 1. Variation of computed maximum probability of satisfaction with number of computation cells (for $\alpha = 0.9$).

♯ Computation cells	Probability	Time (s)
10^1	$[0.923550452972591, 1]$	0.007187
10^2	$[0.937388161337144, 1]$	0.016109
10^3	$[0.938914528816904, 1]$	0.168868
10^4	$[0.939042761643025, 1]$	11.032918
10^5	$[0.939055584603395, 1]$	1231.510472

6   Conclusion and Future Work

In this paper, we have extended the logic Stochastic Satisfiability Modulo Theory (SSMT) and its solving procedure [1, 7] to cover continuous quantifier domains and continuous distributions also. For the extended logic CSSMT (Continuous Stochastic Satisfiability Modulo Theory) we proposed an ICP-based procedure to compute safe bounds on the maximum probability of satisfaction of an arbitrary CSSMT formulas. A prototype implementation in Matlab has been pursued. We have demonstrated the expressive power of CSSMT on a simple case study of a stochastic hybrid system and have shown that both formalizing its dynamics in CSSMT and automatically analyzing its quantitative bounded reachability properties by our prototype solver are in scope of the method.

In this paper, only prototype MATLAB implementation is considered, however, it can not be generated and used for other case studies. As future work, we are aiming at more efficient and scalable implementation of the CCSMT solver by using C/C++, following the lines of and partially building on the iSAT/SiSAT tool sets [12,24]. More case studies and application scenarios will be considered, applying CSSMT in particular to problems of highly automated driving maneuvers under uncertain perception, as pertinent to automotive assistance systems relying on complex sensors like computer vision.

References

1. Fränzle, M., Hermanns, H., Teige, T.: Stochastic satisfiability modulo theory: a novel technique for the analysis of probabilistic hybrid systems. In: Egerstedt, M., Mishra, B. (eds.) HSCC 2008. LNCS, vol. 4981, pp. 172–186. Springer, Heidelberg (2008)
2. Teige, T., Fränzle, M.: Stochastic satisfiability modulo theories for non-linear arithmetic. In: Trick, M.A. (ed.) CPAIOR 2008. LNCS, vol. 5015, pp. 248–262. Springer, Heidelberg (2008)
3. Papadimitriou, C.H.: Games against nature. J. Comput. Syst. Sci. **31**(2), 288–301 (1985)
4. Majercik, S.M., Littman, M.L.: Maxplan: A new approach to probabilistic planning. AIPS **98**, 86–93 (1998)
5. Littman, M.L., Majercik, S.M., Pitassi, T.: Stochastic boolean satisfiability. J. Autom. Reasoning **27**(3), 251–296 (2001)
6. Majercik, S.M., Littman, M.L.: Contingent planning under uncertainty via stochastic satisfiability. In: AAAI/IAAI, pp. 549–556 (1999)
7. Teige, T.: Stochastic satisfiability modulo theories: a symbolic technique for the analysis of probabilistic hybrid systems. Ph.D thesis, Universität Oldenburg (2012)
8. Nieuwenhuis, R., Oliveras, A., Tinelli, C.: Solving sat and sat modulo theories: from an abstract davis-putnam-logemann-loveland procedure to DPLL($\mathcal{T}$). J. ACM (JACM) **53**(6), 937–977 (2006)
9. Rossi, F., Van Beek, P., Walsh, T.: Handbook of Constraint Programming. Elsevier, Amsterdam (2006)
10. Van Hentenryck, P., McAllester, D., Kapur, D.: Solving polynomial systems using a branch and prune approach. SIAM J. Numer. Anal. **34**(2), 797–827 (1997)
11. Fränzle, M., Herde, C., Teige, T., Ratschan, S., Schubert, T.: Efficient solving of large non-linear arithmetic constraint systems with complex boolean structure. JSAT **1**(3–4), 209–236 (2007)
12. iSAT Homepage. https://projects.avacs.org/projects/isat/. Accessed February 2015
13. Ellen, C., Gerwinn, S., Fränzle, M.: Statistical model checking for stochastic hybrid systems involving nondeterminism over continuous domains. Int. J. Softw. Tools Technol. Transfer **17**(4), 485–504 (2015)
14. Young, R.C.: The algebra of many-valued quantities. Mathematische Annalen **104**(1), 260–290 (1931)
15. Sunaga, T., et al.: Theory of an interval algebra and its application to numerical analysis. Jpn. J. Ind. Appl. Math. **26**(2–3), 125–143 (2009). [reprint of res. assoc. appl. geom. mem. 2 (1958), 29–46]

16. Moore, R.E., Moore, R.: Methods and Applications of Interval Analysis, vol. 2. SIAM, Philadelphia (1979)
17. Alefeld, G., Herzberger, J.: Introduction to Interval Computation. Academic press, New York (1984)
18. Benhamou, F., Granvilliers, L.: Continuous and interval constraints. Handb. Constraint Prog. **2**, 571–603 (2006)
19. Granvilliers, L., Benhamou, F.: Realpaver: an interval solver using constraint satisfaction techniques. ACM Trans. Math. Softw. (TOMS) **32**(1), 138–156 (2006)
20. Benhamou, F., Languénou, F.G.É., Christie, M.: An algorithm to compute inner approximations of relations for interval constraints. In: Bjorner, D., Broy, M., Zamulin, A.V. (eds.) PSI 1999. LNCS, vol. 1755, pp. 416–423. Springer, Heidelberg (2000)
21. Vu, X.-H., Sam-Haroud, D., Silaghi, M.-C.: Approximation techniques for nonlinear problems with continuum of solutions. In: Koenig, S., Holte, R. (eds.) SARA 2002. LNCS (LNAI), vol. 2371, pp. 224–241. Springer, Heidelberg (2002)
22. Goubault, E., Mullier, O., Putot, S., Kieffer, M.: Inner approximated reachability analysis. In: Proceedings of the 17th international conference on Hybrid systems: computation and control, pp. 163–172. ACM (2014)
23. Abate, A., Prandini, M., Lygeros, J., Sastry, S.: Probabilistic reachability and safety for controlled discrete time stochastic hybrid systems. Automatica **44**(11), 2724–2734 (2008)
24. SiSAT Homepage. https://projects.avacs.org/projects/sisat/. Accessed March 2015

Bayesian Statistical Analysis for Performance Evaluation in Real-Time Control Systems

Pontus Boström[1]([⊠]), Mikko Heikkilä[2], Mikko Huova[2],
Marina Waldén[1], and Matti Linjama[2]

[1] Åbo Akademi University, Turku, Finland
{pontus.bostrom,marina.walden}@abo.fi
[2] Tampere University of Technology, Tampere, Finland
{mikko.heikkila,mikko.huova,matti.linjama}@tut.fi

Abstract. This paper presents a method for statistical analysis of hybrid systems affected by stochastic disturbances, such as random computation and communication delays. The method is applied to the analysis of a computer controlled digital hydraulic power management system, where such effects are present. Bayesian inference is used to perform parameter estimation and we use hypothesis testing based on Bayes factors to compare properties of different variants of the system to assess the impact of different random disturbances. The key idea is to use sequential sampling to generate only as many samples from the models as needed to achieve desired confidence in the result.

1 Introduction

Model-based design in e.g. Simulink is a popular way to design control software, where the discrete controller controls a continuous-time system. In this approach, the control algorithms are designed together with a simulation model of the system to be controlled. This allows use of system simulation for controller validation even before the system is built. Typically the models are synchronous, i.e. computation is assumed to take no time. Delays are often modeled deterministically with the worst case or average case. This simplifies modelling and simulation is also typically faster. However, random delays and computation times might have a large impact on the system behaviour in practise. Including these delays makes analysis of the models harder.

Even if safety is important, sufficient performance of the system is crucial. One important question to answer is how different modifications to a model impact performance. Here we are mainly interested in the impact of stochastic disturbances on the system compared to the ideal synchronous system. To compare models we rely on hypothesis testing. In [17] they use Bayesian hypothesis testing based on Bayes factors for Statistical Model Checking (SMC). They have developed a procedure to generate samples that for a BLTL property ϕ

The work has been partially funded by the EDiHy project (no. 139540 and no. 140003) funded by the Academy of Finland.

J. Campos and B.R. Haverkort (Eds.): QEST 2015, LNCS 9259, pp. 312–328, 2015.
DOI: 10.1007/978-3-319-22264-6_20

that holds with a probability p choose to either accept the hypothesis $H_0 : p \geq \theta$ or $H_1 : p < \theta$, where θ is a user defined bound. Compared to numerical model checking techniques, the advantage of using SMC is that it is fast and easy to implement for various modelling frameworks, as one only needs to sample traces. We extend the approach to hypothesis testing in [17] to compare parameters for statistical models derived from the different system models. Sequential sampling is used to draw samples until a desired confidence in the results has been achieved. As simulations can take a very long time this is very useful. Small sample sizes are also important when applying the methodology to the actual final system. We check hypotheses such as, e.g., is the mean of the mean square error over a time interval greater in one model or the other, or, is the rate of events greater in one model or the other. By using several different hypotheses we can build a comprehensive suite of checks to *automatically* compare how different versions of models behave in the presence of stochastic disturbances with desired confidence in the result.

We are also interested in estimation of parameters for different random variables, in particular for validating the statistical models used in hypothesis testing. Bayesian statistical model checking [17] provides methods to estimate probabilities for desirable properties expressed in some logic to hold in stochastic models. They consider the satisfaction of a Bounded Linear Time Logic (BLTL) formula as a Bernoulli random variable and estimate the probability that the property holds. The procedure ensures that the real probability is within given bounds with given probability chosen by the user. The estimation can be carried out to an arbitrary level of precision. In this paper we use the same approach as [17] to do more general Bayesian inference. We use the same idea of sequential sampling to sample from random variables to estimate parameters of statistical models, in particular random variables that can be seen as approximately Normal and Poisson distributed. This is used to estimate properties such as the average mean square error of a signal over a time interval or the rate that events occur. The approach can be extended to other distributions.

In this paper we compare two different variants with different disturbances of a model of digital hydraulic power management system (DHPMS) [15]. This is done to get a better understanding of how this system will perform in real life under non-ideal conditions. Our contribution is:

- A significant extension of the approach in [17] to compare properties of different models and to analyse more general properties than the probability of a BLTL formula to hold. Both parameter estimation and hypothesis testing is discussed.
- Application of the methodology to a case study with a discussion on possible extensions and limitations. The case study demonstrates issues often encountered when running controllers under non-ideal conditions.

In Sect. 2, we present the case study. In Sect. 3 we briefly discuss Simulink models and SMC and in Sect. 4 we describe the performance metrics used in the paper. Section 5 presents the statistical techniques used and Sect. 6 shows

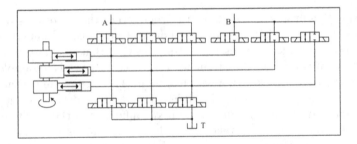

Fig. 1. The hydraulic diagram of a DHPMS with three pistons. The figure shows three pistons connected to a rotating shaft on the left. Each piston is connected via valves to high pressure lines A and B, as well as a tank line T [15].

the application to the case study. Sections 7 and 8 present related work and conclusions.

2   Case Study

The example used in this paper is a model of a six-piston digital hydraulic power management system (DHPMS) with two independent outlets. The hydraulic diagram of the machine is shown in Fig. 1 but, for simplicity, only three pumping pistons are presented. Each piston can be connected to either one of the outlets A or B, or to the tank line T via on/off control valves. Hence, the DHPMS can operate as a pump, motor, and transformer. Furthermore, it can provide independent supply line pressures for the actuators. The pressure levels are kept as close as possible to user-defined reference values by utilising a model-based control approach to select optimal pumping and motoring modes for each piston [12]. Furthermore, the valve timing must be accurate to avoid too high and too low pressures (cavitation) in the piston chambers which can damage the system.

2.1   Test Systems

The goal is to investigate the impact of random time delays on system performance. This is used to assess how control algorithms constructed in an ideal model with deterministic timing will behave in practise. We compare two models:

System 1. In this model, the controller is assumed to be synchronous and all delays deterministic. Also the rotation speed of the motor is constant and known to the controller. This makes the model simpler and simulation is fast. The sampling period T_s is $50\,\mu s$, the delay d of opening and closing the valves are $1ms$. The rotation frequency of the motor is a fixed $25\,Hz$.

System 2. In this model computation time is taken into account. The delay before reading inputs is assumed to be a uniformly distributed variable in the interval $[0.01T_s..0.1T_s]\,\mu s$. The computation time between the input is read until outputs are produced is assumed to be a uniformly distributed random variable in the interval $[0.1T_s..0.55T_s]\,\mu s$. The rotation speed of the motor is fixed at 25 Hz, but the controller estimates the speed with an incremental rotary encoder [2]. Additionally, the valves open and close with a delay in the interval $[0.9d..1.1d]ms$.

3 Modelling and Specification

We have used Simulink as our Modelling tool. A Simulink model can be assumed to be a probabilistic *discrete-time hybrid automaton* (DTHA) [17]. Hence, there is a well-defined probability measure over the trace space produced by simulating a model. Simulink/Stateflow is a complex language and a full formal semantics as a DTHA that accurately considers all features is difficult to define. However, we do not need a formal semantics, since we only sample traces. The sampling uses the built-in simulation capabilities of Simulink. In [17] they use properties that can be checked on finite prefixes of simulations of a model, i.e. their truth-value depend only on a finite prefix. All statistics we compute apply only for finite prefixes of predefined length of simulation traces. However, the considered simulation traces can be arbitrarily long. The models considered in this paper are open models, where the input signals are random signals with some properties, representing possible workloads of the system. The models are closed by also modelling the random input signals. Generating random input signals that are representative of real workloads in both the time and frequency domain can be challenging. However, a thorough investigation of this topic is outside the scope of this paper.

4 Definition of Performance

We analyse performance using several different metrics. The goal of the metrics is to capture key performance properties of the system, where the metric for a model can be seen as a random variable. These metrics obtained from different versions of the system can then be used to compare them.

Mean Square Error. The DHPMS should provide pressure to the A- and B- lines as close as possible to the reference pressures. Hence, we need a characterisation of the deviation. One common way to characterise the size of the deviation is the mean square of the error signal v:

$$J = \lim_{T \to \infty} \int_0^T v(t)^2 dt \tag{1}$$

We cannot simulate the system for an infinite time and we therefore compute the mean square error for a finite time interval $[t_0, t_1]$, $J_n = \int_{t_0}^{t_1} v(t)^2 dt$ for each run n of the system. Then J_n is a random variable. We do not know the shape of $v(t)^2$, but if the system is time invariant then according to the *Central Limit Theorem* J_n is approximately normally distributed. This follows from the fact that J_n can be seen as the average of mean square errors of smaller intervals that then, if sufficiently long, are approximately independent and identically distributed (iid). Then if time invariance is again assumed, J is the average of all J_i with $E[J] = \lim_{N \to \infty} \frac{1}{N} \Sigma_{i=0}^{N} E[J_i] = E[J_i]$. Note that in special cases we can actually compute $E[J]$ exactly [14].

Safety Properties. Safety properties can be formulated as BLTL properties and the probability that they hold can be directly estimated by the approach in [17]. One safety relevant performance property is to ensure that pressure peaks occur sufficiently rarely. We can analyse this property by computing the probability of a pressure peak during a certain time interval.

Rate of Events. We are also interested in analysing how often good or bad events happen in the system. E.g. for the pressure in the pumping cylinders, we know that pressure peaks can occur. Furthermore, we are more interested in the event that the pressure becomes too high than the time the pressure stays too high. We can analyse this by analysing the rates of pressure peaks or low pressure (cavitation) events. If we assume the events are independent from each other, the number of events per time unit follows a Poisson distribution.

5   Statistical Analysis

We cannot compute the properties of the random variables described earlier exactly due to system complexity and we instead use Bayesian statistics to analyse them. Bayesian inference is based on using the Bayes rule (2) [8] to fit a probability model to a set of data possibly using some prior information. The result is probability distribution on the parameters of the model.

$$p(\theta|y) = \frac{p(y|\theta)p(\theta)}{p(y)} \tag{2}$$

Bayes rule gives the posteriori probability $p(\theta|y)$ where θ are the parameters of the model and y is the data. The notation $p(\theta|y)$ denotes the conditional probability for θ given y. The distribution $p(y|\theta)$ is called the sampling distribution and gives the probability distribution for observing y. The prior distribution $p(\theta)$ describes prior knowledge of θ, while $p(y)$ is a normalisation factor $p(y) = \int p(y|\theta)p(\theta)d\theta$ for continuous θ.

When data y has been observed we can use this to make predictions about an unknown observable $\hat{y}$ from the same process [8].

$$p(\hat{y}|y) = \int p(\hat{y}|\theta)p(\theta|y)d\theta \tag{3}$$

Hence, the prediction uses the posteriori probability density estimated for θ to estimate the probability for $\hat{y}$. Predictive distributions are used here to validate the models against the data.

The goal of parameter estimation is to estimate the probability of a certain property for any random simulation of a system. Each property can be evaluated on a prefix with fixed length of a simulation. When evaluating a BLTL formula we like to estimate the probability θ that the property holds in a random simulation. That the property ϕ holds for a simulation σ can be associated with a Bernoulli distributed random variable X. The conditional probability that the property holds is $p(x|\theta) = \theta^x(1-\theta)^{1-x}$ where $x = 1$ if the property holds ($\sigma \models \phi$) and $x = 0$ if it does not ($\sigma \not\models \phi$). The (unknown) probability for $x = 1$ is given by θ. For inference, we need a prior probability density for θ. If no information is available a non-informative prior can be used.

To simplify computation we can use a so called conjugate prior [8]. Using a conjugate prior means intuitively that the posteriori distribution has the same form as the prior distribution. The conjugate prior for the Bernoulli distribution is the Beta-distribution $Beta(\theta|\alpha,\beta) = \frac{1}{B(\alpha,\beta)}\theta^{\alpha-1}(1-\theta)^{\beta-1}$ where $B(\alpha,\beta)$ is the Beta function. If we have iid random variables then $p(x_1,\ldots,x_n|\theta) = \Pi_{i=1}^n p(x_i|\theta)$ where $x_1,\ldots,x_n$ is the results of n simulations. The posteriori distribution $p(\theta|x_1,\ldots,x_n)$ is then $p(\theta|x_1,\ldots,x_n) = Beta(\theta|\alpha+x,\beta+n-x)$ where x is the number of 1:s in the samples and n is the number of samples. α and β are given by the prior Beta-distribution. If $\alpha = \beta = 1$ the Beta-distribution equals the uniform distribution, which is considered a non-informative prior.

We are not limited to only Bernoulli distributed random variables. However, to avoid computational problems and thereby automate the approach, it is extremely useful if the sampling distribution has a conjugate prior. In this paper we use Poisson distribution as a sampling distribution where the Gamma distribution is a conjugate prior, as well as the normal distribution with unknown mean and variance. Given the prior distribution of the rate of events r, $Gamma(r|\alpha,\beta)$ where $\alpha,\beta > 0$ are user defined parameters and samples $k_1,\ldots,k_n$ drawn from a Poisson distribution with unknown rate r, $Poisson(k_i|r)$, the posteriori probability distribution for the rate r is [8]:

$$p(r|k_1,\ldots,k_n) = Gamma(r|\alpha + n\bar{k}, \beta + n) \tag{4}$$

Here $\bar{k}$ is the average of $k_1,\ldots,k_n$. In the case of normally distributed data $N(y_i|\mu,\sigma^2)$, where μ is the mean and σ^2 is the variance, the marginal posteriori distribution for the mean μ is given by the Student-t distribution:

$$p(\mu|y_1,\ldots,y_n) = t_{\nu_n}(\mu|\mu_n, \sigma_n^2/\kappa_n) \tag{5}$$

with ν_n degrees of freedom [8] where the parameters are:

$$\begin{aligned}
\mu_n &= \frac{\kappa_0}{\kappa_0+n}\mu_0 + \frac{n}{\kappa_0+n}\bar{y} \\
\kappa_n &= \kappa_0 + n \\
\nu_n &= \nu_0 + n \\
\nu_n\sigma_n^2 &= \nu_0\sigma_0^2 + (n-1)s^2 + \frac{\kappa_0 n}{\kappa_0+n}(\bar{x}-\mu_0)^2
\end{aligned} \tag{6}$$

The parameters $\kappa_0,\nu_0,\mu_0,\sigma_0$ are chosen by the user for the prior distribution, $\bar{y}$ is the average and s^2 is the computed variance of the samples.

Table 1. The interval $[t_0, t_1]$ for the three posteriori distributions used in the paper.

Distribution	lower bound t_0	upper bound t_1
Beta	$\hat{p} + \delta \leq 1? \max(0, \hat{p} - \delta) : 1 - 2\delta$	$\hat{p} - \delta \geq 0? \min(1, \hat{p} + \delta) : 2\delta$
Student-t	$\hat{\mu} - \delta\hat{\sigma}$	$\hat{\mu} + \delta\hat{\sigma}$
Gamma	$\hat{r} - \max(\delta, \delta\sqrt{\hat{r}}) \geq 0?$	$\hat{r} - \max(\delta, \delta\sqrt{\hat{r}}) \geq 0?$
	$\hat{r} - \max(\delta, \delta\sqrt{\hat{r}}) : 0$	$\hat{r} + \max(\delta, \delta\sqrt{\hat{r}}) : 2\max(\delta, \delta\sqrt{\hat{r}})$

5.1 Bayesian Estimation Algorithm

The goal is to estimate parameters θ of a chosen statistical model. The idea here is to draw samples until the desired confidence in the result has been achieved. The confidence in the estimate is defined as an interval for $[t_0, t_1]$ in which the estimated parameter θ should be with the probability of at least c. The probability γ of θ being in the interval is given by

$$\gamma = \int_{t_0}^{t_1} p(\theta|y)d\theta \tag{7}$$

where $p(\theta|y)$ is the posteriori probability of θ given the data y. This integral can be computed efficiently numerically in the case of Beta, Gamma and Student-t distribution using e.g. MATLAB. However, for more complex distribution this is typically not the case [8]. The algorithm to estimate θ, which is a straightforward extension of the algorithm in [17], is shown in Fig. 2. The first step in the loop is to draw a sample by simulating the model, then the function f is used to calculate the data needed in the parameter estimation from the generated simulation trace. Finally, the probability that θ is between t_0 and t_1 is computed. If the probability is high enough the loop terminates.

The interval $[t_0, t_1]$ used for the different distributions are given in Table 1, where δ is a user defined parameter determining the width of the interval. We use the mean of the posteriori distribution as the center of the interval. For the Poisson distributed variables and for the normally distributed random variables, the width of the interval is defined as a fraction δ of the estimated standard deviation of the sampling distributions. Note that for a Poisson process, the mean equals the variance. Note also the handling of boundary cases. One could also consider a (user-defined) static lower bound for the width $t_1 - t_0$ in the case of normally distributed data, since width now approaches zero when the variance approaches zero. Other choices of interval bounds are possible, as long as the width of the interval $t_1 - t_0$ converges to a strictly positive value.

5.2 Bayesian Hypothesis Testing

We compare models by using Bayesian hypothesis testing with Bayes factors [13]. Here we decide between two mutually exclusive hypotesis H_0 and H_1.

Input :
  $f(\sigma, y)$ – A function that computes a statistic on a trace and adds it to existing statistics
  $p(\theta|y, \theta_0)$ – A posteriori probability density function
  $t_0(y, \theta_0)$ – A function that computes lower bound for the interval
  $t_1(y, \theta_0)$ – A function that computes upper bound for the interval
  $c \in (0.5, 1)$ – The interval coverage coefficent
Output :
  y – The statistic y needed for the posteriori distribution
repeat
  $\sigma :=$ Draw a sample trace from the system model
  $y := f(\sigma, y)$
  $t_0 := t_0(y, \theta_0)$
  $t_1 := t_1(y, \theta_0)$
  $\gamma := \int_{t_0}^{t_1} p(\theta|y, \theta_0)d\theta$
until $\gamma \geq c$
return y

Fig. 2. The algorithm for parameter estimation to a desired precision

For a parameter θ defined on two models M_1 and M_2 we will use the hypotheses:

$$H_0 : \theta_1 \geq \theta_2 \qquad\qquad H_1 : \theta_1 < \theta_2 \qquad\qquad (8)$$

$$H_0 : |\theta_1 - \theta_2| \leq \epsilon \qquad\qquad H_1 : |\theta_1 - \theta_2| > \epsilon \qquad\qquad (9)$$

The hypotheses in (8) are used to test if θ in model M_1 is greater or less than θ in M_2. The second set of hypotheses in (9) are used to check if θ in both models differ from each other with less than a user defined value ϵ.

For normally distributed data, θ will typically be the mean μ and for Poisson distributed data θ is the rate r. Based on the data d Bayes theorem then gives the posteriori probability hypothesis H_i

$$p(H_i|d) = \frac{p(d|H_i)p(H_i)}{p(d|H_0)p(H_0) + p(d|H_1)p(H_1)} \quad i = 0, 1 \qquad\qquad (10)$$

Here the prior probabilities must be strictly positive and $p(H_1) = 1 - p(H_0)$. The posteriori odds for hypothesis H_0 is

$$\frac{p(H_0|d)}{p(H_1|d)} = \frac{p(d|H_0)p(H_0)}{p(d|H_1)p(H_1)} \qquad\qquad (11)$$

The Bayes factor B is then defined as $B = p(d|H_0)/p(d|H_1)$. When the priors are fixed, the Bayes factor is used to measure the confidence in the hypothesis H_0 against H_1. Guidelines for interpreting the Bayes factor are given in [13]. A Bayes factor $B \geq 100$ can be seen as strong evidence in favour for H_0 and a value of $B \leq 0.01$ as strong evidence in favour for H_1. We adapt the algorithm from [11,17] to dynamically chose the number of samples so that either H_0 is accepted with a fixed threshold T or H_1 is accepted with a threshold $1/T$.

We use the posteriori probabilities $p(H_0|d)$ and $p(H_1|d)$ to compute the Bayes factor. Below we show the definition of the Bayes factor for normally distributed data and the hypotheses $H_0 : \mu_1 \geq \mu_2$ and $H_1 : \mu_1 < \mu_2$. In this case, the posteriori distribution of μ_1 and μ_2 are Student-t distributions. The probability for $\mu_1 \geq \mu_2$ is obtained by integrating joint probability distribution $p(\mu_1, \mu_2)$

over the area satisfying this condition. The same applies to $\mu_1 < \mu_2$. Assuming μ_1 and μ_2 are *independent*, the posteriori odds then becomes:

$$\frac{p(H_0|d)}{p(H_1|d)} = \frac{\int_{-\infty}^{\infty} \int_{-\infty}^{\mu_1} t_{\nu_{n1}}(\mu_1|\mu_{n_1}, \sigma_{n_1}^2/\kappa_{n_1}) t_{\nu_{n2}}(\mu_2|\mu_{n_2}, \sigma_{n_2}^2/\kappa_{n_2}) d\mu_2 d\mu_1}{\int_{-\infty}^{\infty} \int_{\mu_1}^{\infty} t_{\nu_{n1}}(\mu_1|\mu_{n_1}, \sigma_{n_1}^2/\kappa_{n_1}) t_{\nu_{n2}}(\mu_2|\mu_{n_2}, \sigma_{n_2}^2/\kappa_{n_2}) d\mu_2 d\mu_1} \quad (12)$$

where μ_n is the mean estimated from the sample average, σ_n^2 the variance estimated from the sample variance, ν_n and κ_n are the degrees of freedom. For H_0 and H_1 in (9) the posteriori odds become:

$$\frac{p(H_0|d)}{p(H_1|d)} = \frac{\int_{-\infty}^{\infty} \int_{\mu_1-\epsilon}^{\mu_1+\epsilon} t_{\nu_{n1}}(\mu_1|\mu_{n_1}, \sigma_{n_1}^2/\kappa_{n_1}) t_{\nu_{n2}}(\mu_2|\mu_{n_2}, \sigma_{n_2}^2/\kappa_{n_2}) d\mu_2 d\mu_1}{\substack{\int_{-\infty}^{\infty} \int_{-\infty}^{\mu_1-\epsilon} t_{\nu_{n1}}(\mu_1|\mu_{n_1}, \sigma_{n_1}^2/\kappa_{n_1}) t_{\nu_{n2}}(\mu_2|\mu_{n_2}, \sigma_{n_2}^2/\kappa_{n_2}) d\mu_2 d\mu_1 + \\ \int_{-\infty}^{\infty} \int_{\mu_1+\epsilon}^{\infty} t_{\nu_{n1}}(\mu_1|\mu_{n_1}, \sigma_{n_1}^2/\kappa_{n_1}) t_{\nu_{n2}}(\mu_2, \mu_{n_2}, \sigma_{n_2}^2/\kappa_{n_2}) d\mu_2 d\mu_1}}$$

$$(13)$$

The Bayes factor then becomes $B = \frac{p(H_1)p(H_0|d)}{p(H_0)p(H_1|d)}$. For other forms of the posteriori distributions (Gamma distribution and Beta distribution in this paper) the Bayes factor for the hypothesis can be formulated in a similar manner. Note that the Bayes factor is notorious for being difficult to compute for arbitrary distributions [13]. In our cases numerical integration techniques in MATLAB work on the tests we have done when the number of samples are sufficiently large. However, more easily computed approximations, such as the Schwarz criterion, exists [13]. Markov Chain Monte-Carlo (MCMC) techniques [8] to compute approximate the integrals can also be used, but due to the large number of samples needed to accurately approximate the integrals, they can be very slow.

The Bayesian hypothesis testing algorithm in Fig. 3 modifies the one in [17] to test hypotheses concerning two models. The algorithm draws iid sample traces from the models on which the desired hypothesis is tested. Samples are drawn until either the Bayes factor is greater than a predefined threshold indicating that H_0 should be accepted or the Bayes factor is smaller than $1/T$ indicating that H_1 should be accepted.

Note that it is useful to combine the estimation algorithm with the hypothesis testing algorithm. As the statistical models are approximations, the estimated θ is needed to validate that the statistical model approximates the data sufficiently well. Additionally, when deciding between hypotheses H_0 and H_1 in (8) and $\theta_0 \approx \theta_1$ then a huge number of samples might be needed to accept one of the hypothesis even if the values of the parameters are almost the same. Hence, one can also stop the iteration when the parameters θ_0 and θ_1 has been estimated to desired precision, while the hypothesis testing is still inconclusive.

5.3 Algorithm Analysis

We essentially use the same estimation and hypothesis testing algorithms as in [17], but use different probability distributions. They have proved termination (with high probability) of both parameter estimation and hypothesis testing. Termination of parameter estimation is straightforward to prove and the proof

Input :
  $f(\sigma, y)$ − A function that computes a statistic on a trace and adds it to existing statistics
  $p(\theta|y, \theta_0)$ − A posteriori probability density function
  $T \geq 1$ − The threshold to accept H_0
Output :
  $H_0 : \theta_1 \geq \theta_2$ accepted or $H_1 : \theta_1 < \theta_2$ accepted
do
  $\sigma_1 :=$ Draw a sample trace from the system model 1
  $\sigma_2 :=$ Draw a sample trace from the system model 2
  $y_1 := f(\sigma_1, y_1)$
  $y_2 := f(\sigma_2, y_2)$
  $B :=$ BayesFactor(p, y_1, y_2, θ_0)
  if $(B > T)$ **then return** H_0 accepted
  if $(B < 1/T)$ **then return** H_1 accepted
end do

Fig. 3. The Bayesian hypothesis testing algorithm

in [17] is straightforward to adapt here. Termination of hypothesis testing is more difficult, see [10] for a proof in the case of Bernoulli sampling distribution with a Beta prior. In the case of hypothesis testing, termination is in our case perhaps more of a theoretical interest, due to numerical problems when deciding between hypotheses where the evidence in favour of one or the other is weak.

The upper bound $1/T$ on the probability of making type I and type II errors in hypothesis testing has also been proved [17]. By type I error we mean that we reject the H_0 hypothesis even if it is true. A type II error is the error of accepting H_0 even if it is false. The bound on the probability on the type I/II error in hypothesis testing is straightforward to derive for the Poisson distribution (same proof as in [17]) and it works with minor modifications for the normal distribution. The probability of estimation errors [17] can be analysed in terms of Type I/II error by using the hypothesis $H_0 : \theta \in [t_0, t_1]$ and $H_1 : \theta \notin [t_0, t_1]$. The hypothesis H_0 then represents the case that θ is within the desired interval. The probability of a type I or II error is bounded above by $\frac{(1-c)\pi_0}{c(1-\pi)}$ where c is the coverage coefficient and π_0 is the prior probability of H_0. Note that this applies when we sample from random variable with the assumed distribution. However, here we typically approximate the probability distribution of a random variable with an easy to use distribution. Hence, the error bounds provide only an idealised bound and the real bounds are unknown.

Figure 4 shows the number of iterations needed when estimating rate r for Poisson distributed data (left) and when estimating the mean (here the mean $\mu = 0$) of normally distributed data (right) using the interval bounds in Table 1. As the interval $[t_0, t_1]$ becomes smaller the number of iterations required increases rapidly. The increase in c does not have as strong impact, which is also noted in [17]. The number of iterations required for Poisson distributed data is almost independent of the rate r except for small rates, due to the fixed minimum size of the interval. In the case of normally distributed data, the number of samples is the same regardless of the variance, except when the variance becomes small when the number of samples required approaches infinity as the interval width approaches zero. Hence, proper scaling of the problem becomes essential.

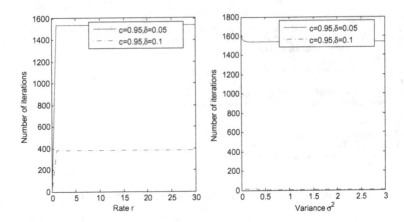

Fig. 4. The number of iterations required to achieve desired precision using the parameter estimation algorithm. The number of iterations for Poisson distributed data is shown to the left and for normally distributed data to the right.

Table 2. Results from parameter estimation in the five tests for system 2

Test	Post. dist	c	δ	γ	Iterations
Test 1	Student-t	0.9	0.1	0.9	274
Test 2	Beta	0.9	0.02	0.9	539
Test 3	Gamma	0.9	0.1	0.9	271
Test 4	Gamma	0.9	0.1	0.99	73
Test 5	Student-t	0.9	0.1	0.9	273

6 Application

The testing methodology has been applied to the case study to evaluate 5 different properties in the tests *Test 1, ..., Test 5* described below (see the technical report [3] for more tests). The metrics in Sect. 4 are used. The results are summarised in Tables 2 and 3. Recall the bounds on the probability of Type I and II errors and the bound on the probability of estimation errors discussed in Sect. 5.3.

Mean Square Error of Pressure Tracking (Test 1). When analysing pressure tracking performance, we focus on the pressure in the A-line of the system. We are interested in the difference $p_A - p_{A,ref}$, where p_A is the (continuous) pressure signal and $p_{A,ref}$ is the (discrete) reference pressure.

This test consider the mean square error (1) of the difference $p_A - p_{A,ref}$ over a time window of 5–15 s. This is a performance property. The square error is scaled by a factor of 10^{-6} to avoid numerical problems. The time windows have been chosen to avoid transients at the start of the system in order to focus on steady-state behaviour.

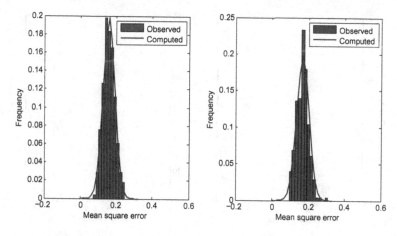

Fig. 5. The estimated and observed distribution of the mean square error in system 1 (left) and system 2 (right).

Figure 5 shows the observed and estimated distributions of mean square errors for system 1 and system 2, respectively. As can be seen from the figures, the normal distribution is a fairly good approximation of the observed distribution. A surprising result is that the control quality does not decrease even with the delays. The hypothesis H_0 in (9) stating that the mean square errors differ at most by ϵ, where ϵ is the estimated standard deviation of system 1, is accepted with the threshold $T = 100$ in 21 iterations.

Probability of Pessure Paks (Test 2). Absence of peaks is described as the BLTL property, $\mathbf{G}^{15s}(p_i(1) \leq 21MPa)$. The probability of a property violation is assumed to be a Bernoulli distributed random variable. The estimated probability for the property to hold for system 1 and system 2 is shown in Fig. 6. The hypothesis that the property holds with greater probability in system 1 than in system 2 is accepted after 76 iterations.

The Rate of Cylinder Pressure Peaks (Test 3) and Low Pressure (Test 4). A pressure peak event is defined as a pressure rising above 21 MPa and a low

Table 3. Results from hypothesis testing comparing system 1 and system 2

Test	Post. dist	T	Result (iterations)
Test 1, hypotheses (8)	Student-t	100	H1 accepted (233)
Test 1, hypotheses (9), $\epsilon = 0.033$	Student-t	100	H0 accepted (21)
Test 2, hypotheses (8)	Beta	100	H0 accepted (76)
Test 4, hypotheses (8)	Gamma	100	H1 accepted (50)
Test 5, hypotheses (8)	Student-t	100	H1 accepted (9)
Test 5, hypotheses (9), $\epsilon = 82$	Student-t	100	H0 accepted (46)

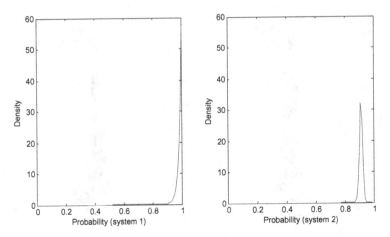

Fig. 6. The estimated the probability that the that property $\mathbf{G}^{15s}(p_i(1) \leq 21MPa)$ holds in system 1 (left) and system 2 (right).

pressure event is defined as the pressure falling below 0.1 MPa. All cylinders are analysed separately and events are counted for each type of violation (low pressure, pressure peak) separately. A Poisson distribution is assumed for the rate of events.

Figure 7 shows estimated and observed distribution of low pressure events for the first pumping cylinder in system 1 and system 2, respectively. The rates for all 6 cylinders are very similar in both cases. In system 1 the events occur rarely, while in system 2 they occur with an average rate of above 20 events per 10 s. The hypothesis that the rate in system 2 is greater than in system 1 is accepted after 50 iterations with threshold $T = 100$. This is actually the main difference between system 1 and system 2. However, the high pressure peaks do not follow a Poisson distribution very well, which indicates that the events are not independent (see [3] for the observed and estimated distribution of events). This is actually an interesting discovery. Hence, that a probabilistic model does not fit the data can be an interesting result in itself. However, the hypothesis testing approach cannot be used to compare models in this case. A more accurate approximation, recommended in [8] when events are clustered, is a negative binomial distribution.

Energy Losses (Test 5). We can estimate the energy losses in the pumping process. Test 5 compares the energy losses in the two systems. As the total energy losses are the sum of smaller random losses, the central limit theorem motivates the assumption that the energy losses over a fixed time interval is a normally distributed random variable if the system is time invariant. The energy loss during the time interval 0–15 s is similar in both systems (see Fig. 8), the estimated mean in system 1 is $\hat{\mu}_1 = 5370J$ and $\hat{\mu}_2 = 5380J$ in system 2. The hypothesis that the means are within one standard deviation $\hat{\sigma} = 87$ from each other is accepted in 46 iterations.

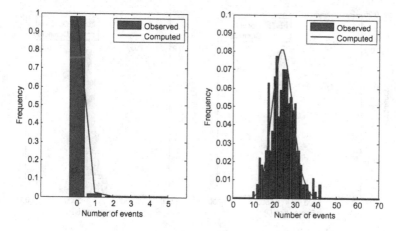

Fig. 7. The rate at which low pressure events occur in system 1 (left) and in system 2 (right).

Conclusions. The control performance given by the mean square error and the energy consumption are not significantly negatively affected by the additional disturbances in system 2. The only significant effect is the increase of low pressure events. However, when testing several hypotheses, the probability to have at least one type I/II error is $p(n_{error} \geq 1) = 1 - p(n_{error} = 0)$. If hypothesis i has threshold T_i and the errors are independent, then this probability is bounded above, $p(n_{error} \geq 1) \leq 1 - \Pi(1 - \frac{1}{T_i})$. Hence, the probability of type I and II errors can be significant, if we have many hypotheses and a relatively small T.

7   Related Work

Statistical model checking has been an active area of research. The focus has been on checking various kinds of (bounded) temporal logic formula on stochastic models. Here we directly extend the methodology in [17] to more general properties than can be expressed by temporal logic formulas using well-known Bayesian techniques. Younes and Simmons [16] have used hypothesis testing to determine if the probability that a temporal logic property holds is above a desired limit. They use the *sequential probability ratio test* (SPRT) to adaptively sequentially sample only as many times as needed. Additionally, they can also bound the probability of Type I and II errors. However, they perform no parameter estimation, which is here important for model validation. For parameter estimation, techniques based on the Chernoff-Hoeffding bound have been used by Hérault et.al. [9]. According to experiments by Zuliani et. al. [17], this estimation approach can be significantly slower than the Bayesian approach used here. Both hypothesis testing using SPRT and parameter estimation using the Chernoff-Hoeffding bound has been implemented in UPPAAL-SMC [4, 7]. Additionally they implement comparisons of probabilities from different models by

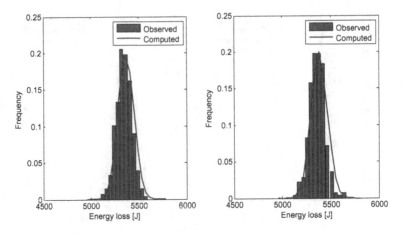

Fig. 8. The estimated and observed distribution of the energy loss in system 1 (left) and system 2 (right).

hypothesis testing based on SPRT. David et.al. [6] have also used ANOVA for model comparison. The goal is similar to how we compare parameters in different models using Bayesian hypothesis testing. However, an in-depth comparison is further work.

Jitterbug [5] can be used to compute the mean square error (1) numerically for a special kind of systems. However, in our case the systems do not fulfil the requirements (the system dynamics is not linear). We have also analysed worst case timing for a version of the case study system [2] by applying model-checking of timed automata in the TIMES tool [1]. That paper focuses on the improvements obtained by using interrupt driven tasks instead of periodic tasks. Here the focus is on the impact of random delays on system performance.

8   Conclusions

This paper presents an approach to statistically analyse and compare stochastic models, based on Bayesian parameter estimation and hypothesis testing. We demonstrated the approach to compare two versions of a model of a digital hydraulics power management system. The differences in the models concern stochastic delays and disturbances. The results from the hypothesis testing show that we can gain confidence in the result with relatively few samples. The parameter estimation was used to validate that data fits the used probability model. However, comprehensive model validation [8] is outside the scope of the paper.

In [17], they give bounds on probability of Type I and Type II errors in hypothesis testing and bounds on the error of estimated probabilities. Although the same results apply here, the Poisson and Normal distributions are approximations of the real processes and, hence, model validation becomes an essential step in order to draw valid conclusions. Therefore, parameter estimation is an important complement to hypothesis testing. To automate the analysis there need to be efficient ways to accurately compute the needed integrals. The need for certain types of statistical models then also limits the approach to certain properties, where the models are expected to provide a good fit or where a poor fit provides some insight into the behaviour of the process.

The approach in the paper is not limited to analysing impact of random delays, but can be used for other purposes as well. The key advantage is to generate only as many samples as needed to gain a desired confidence. Additionally, prior information can be included by using the prior distributions. Future work include using more complex statistical models. Incremental model validation would also be useful to ensure that the checking process returns an accurate result.

References

1. Amnell, T., Fersman, E., Mokrushin, L., Pettersson, P., Yi, W.: TIMES: a tool for schedulability analysis and code generation of real-time systems. In: Larsen, K.G., Niebert, P. (eds.) FORMATS 2003. LNCS, vol. 2791, pp. 60–72. Springer, Heidelberg (2004)
2. Boström, P., Alexeev, P., Heikkilä, M., Huova, M., Waldén, M., Linjama, M.: Analysis of real-time properties of a digital hydraulic power management system. In: Lang, F., Flammini, F. (eds.) FMICS 2014. LNCS, vol. 8718, pp. 33–47. Springer, Heidelberg (2014)
3. Boström, P., Heikkilä, M., Huova, M., Waldén, M., Linjama, M.: Bayesian statistical analysis for performance evaluation in real-time control systems. Technical Report, 1136, TUCS (2015)
4. David, A., Larsen, K.G., Legay, A., Mikučionis, M., Poulsen, D.B., van Vliet, J., Wang, Z.: Statistical model checking for networks of priced timed automata. In: Fahrenberg, U., Tripakis, S. (eds.) FORMATS 2011. LNCS, vol. 6919, pp. 80–96. Springer, Heidelberg (2011)
5. Cervin, A., Henriksson, D., Lincoln, B., Eker, J., Årzén, K.E.: How does control timing affect performance? - analysis and simulation of timing using Jitterbug and truetime. IEEE Control Syst. Mag. **23**(3), 16–30 (2003)
6. David, A., Du, D., Guldstrand Larsen, K., Legay, A., Mikučionis, M.: Optimizing control strategy using statistical model checking. In: Brat, G., Rungta, N., Venet, A. (eds.) NFM 2013. LNCS, vol. 7871, pp. 352–367. Springer, Heidelberg (2013)
7. David, A., Larsen, K.G., Legay, A., Mikučionis, M., Wang, Z.: Time for statistical model checking of real-time systems. In: Gopalakrishnan, G., Qadeer, S. (eds.) CAV 2011. LNCS, vol. 6806, pp. 349–355. Springer, Heidelberg (2011)
8. Gelman, A., Carlin, J.B., Stern, H.S., Rubin, D.B.: Bayesian Data Analysis, 2nd edn. Chapman & Hall/CRC, New York (2004)
9. Hérault, T., Lassaigne, R., Magniette, F., Peyronnet, S.: Approximate probabilistic model checking. In: Steffen, B., Levi, G. (eds.) VMCAI 2004. LNCS, vol. 2937, pp. 73–84. Springer, Heidelberg (2004)
10. Jha, S.K., Clarke, E.M., Langmead, C.J., Legay, A., Platzer, A.: Statistical model checking for complex stochastic models in systems biology. Technical Report, CMU-CS-09-110, School of Computer Science, Carnegie Mellon University (2009)

11. Jha, S.K., Clarke, E.M., Langmead, C.J., Legay, A., Platzer, A., Zuliani, P.: A bayesian approach to model checking biological systems. In: Degano, P., Gorrieri, R. (eds.) CMSB 2009. LNCS, vol. 5688, pp. 218–234. Springer, Heidelberg (2009)
12. Karvonen, M., Heikkilä, M., Huova, M., Linjama, M.: Analysis by simulation of different control algorithms of a digital hydraulic two-actuator system. Int. J. Fluid Power 15(1), 33–44 (2014). Mar
13. Kass, R.E., Raftery, A.E.: Bayes factors. J. Am. Stat. Assoc. 90(430), 773–795 (1995)
14. Lincoln, B., Cervin, A.: Jitterbug: a tool for analysis of real-time control performance. In: 41st IEEE Conference on Decision and Control, IEEE (2002)
15. Linjama, M., Huhtala, K.: Digital pump-motor with independent outlets. In: The 11th Scandinavian International Conference on Fluid Power, SICFP 2009, Linköping, Sweden (2009)
16. Younes, H.L.S., Simmons, R.G.: Statistical probabilistic model checking with focus on time-bounded properties. Inf. Comput. 204, 1368–1409 (2006)
17. Zuliani, P., Platzer, A., Clarke, E.M.: Bayesian statistical model checking with application to stateflow/simulink verification. Formal Methods Syst. Des. 43, 338–367 (2013)

Author Index

Printed in the United States
By Bookmasters